The Dynamic Earth:
Textbook in Geosciences

Peter J. Wyllie

John Wiley & Sons, Inc.

New York · London · Sydney · Toronto

The Dynamic Earth:

Textbook in Geosciences

Preface

This book developed from a course called "The Solid Earth" that I organized at the University of Chicago in 1967. The aim of the course was to give first-year graduate students a birds-eye review of geology, and a global geophysical framework within which they could locate their specialized research topics in mineralogy, petrology, or geochemistry. There are many elementary texts in "Physical Geology" giving fine over-views of the Earth, but after a course using one of these, students usually take progressively more specialized subjects. By the time they reach the B.S. level they have learned enough geology to begin to appreciate the significance of the material that they had read with little comprehension a few years earlier. With this in mind, I prepared a more advanced course parallel in some respects to the elementary "Physical Geology" texts, dealing specifically with the major problems in earth sciences that the students might not otherwise meet in their subject-oriented Ph.D. courses.

The revolution in earth sciences was proclaimed in 1967, as I was preparing the course for the first time. The many new developments in global tectonic theory since then have convinced me that J. Tuzo Wilson is right when he states that geological curricula must be revised. I offer this text as one "revolutionary" design for geology courses, a design which emphasizes major units of the Earth, and major geological processes, without regard for the conventional subjects of curricula. The text is designed for graduate students in geology, geochemistry, and geophysics, but it could be used by well-prepared senior undergraduates either as the basis for a course, or as a supplementary text.

A single text incorporating all of the material that is appropriate for this volume would be enormous. I have selected and organized topics and material in what appears to me to be a useful sequence, knowing that no textbook is complete for a good teacher, and hoping that teachers will be able to use my selection of material as a basis for their courses, with the balance being supplied from their own experience, their favorite research topics, and the current literature. Specialists in most areas will probably find their subjects slighted by inadequate attention. For example, paleontologists, stratigraphers, and isotope geochemists will certainly feel this way, and with justification. Geophysicists may be disconcerted by the rather shallow treatment of physical topics but, on the other hand, geologists and geochemists may be relieved that the presentation is more comprehensible to them than that in the available texts on geophysics. The lack of reference to literature in languages other than English will be noted and deplored by some. For these and other shortcomings I can only apologize, plead that the life of an author is short, and claim that selectivity is essential.

I have adopted a didactic approach throughout. It is impossible for students to learn everything: the important thing is for them to learn how to examine evidence and ideas critically. For each major topic introduced I have tried to show how the associated ideas developed, and to present both sides of controversies where they exist. I made no attempt to review all pertinent literature. On the contrary, I made a special effort to keep literature citations down to a minimum, selecting specific papers for examination. Yesterday's theory is of historical interest only, and today's theory is tomorrow's history. I have therefore attempted to examine today's theories as part of the spectrum of history rather than as the definitive solutions so often claimed by enthusiastic researchers.

It is currently fashionable to denigrate advanced textbooks and monographs with the phrase: "It was out of date before it was published." I started the final draft of this volume in October 1969, and certainly by the time I submitted it to the publishers in August 1970, many of the facts recorded were already superceded by research papers in press. Minor revisions in November did not permit appreciable up-dating, despite my urge to include an additional chapter on the recent JOIDES results. This volume contains a record of the revolution in earth sciences during the 1960's, surely the most remarkable decade in the history of geology. I trust that this record of the accumulation of data and the development of ideas must have lasting value to students and teachers despite the

flood of new data appearing annually. I anticipate also that many professional scientists, busily engaged in their own work during the 1960's, will find this volume a useful substitute for hours of library research. The historical account of the revolution will provide them with a sound background against which to view the new developments of the 1970's.

Among the many people who have helped me during the preparation of this text, I wish to thank in particular those who read parts of the manuscript at various stages. This does not make them responsible for the validity of my treatment and interpretations. Thanks to: A. L. Boettcher, F. Chayes, D. P. Gold, J. R. Goldsmith, D. H. Green, K. Ito, G. C. Kennedy, E. D. Jackson, I. B. Lambert, R. B. Merrill, A. A. Meyerhoff, R. C. Newton, M. J. O'Hara, J. K. Robertson, T.J. M. Schopf, A. M. Ziegler, and A. N. Onymous who reviewed the entire manuscript for the publisher. For their speedy, cheerful typing I must also thank Mrs. Irene Baltuska and Mrs. Glenda York.

University of Chicago, Illinois
February, 1971

PETER J. WYLLIE

Contents

The Dynamic Earth:
Textbook in Geosciences

1. *Introduction: A Global Approach*

The Earth is a near-spherical body in space. It moves. It rotates about its own axis and it follows an elliptical path through the solar system about the sun, which is one of at least 100 billion stars rotating around the center of the Milky Way galaxy.

The structure of the Earth may be considered as a series of concentric shells with an inner core, an outer core, a mantle with several shells, and a crust. The hydrosphere, atmosphere, and magnetosphere form an envelope around the solid Earth and shield it from much of the radiation and many of the meteoritic particles that bombard the Earth from space. Most of the hydrosphere fills shallow depressions on the Earth's surface forming the oceans, but this shell overlaps into the atmosphere and into the crust. Movements occur within each of the shells.

Motions within the atmosphere such as wind are familiar; the circulation of the atmosphere on a global scale and its more localized rotary motions are essential ingredients of weather and climate at the surface of the Earth. Wind-driven surface currents in the oceans contribute toward the circulation of the oceans. The ocean currents carry heat energy from equatorial regions toward the poles, which influences climate and the energy balance in the hydrologic cycle. The hydrologic cycle involves the circulation of water through the hydrosphere, with evaporation of water from the oceans being followed ultimately by its precipitation from the atmosphere and by gravity-controlled flow over and through the crust back to the oceans. Weathering and downward erosion of the Earth's solid surface is accomplished largely by the moving atmosphere and by moving water within the hydrologic cycle.

Solar radiation provides the energy for the circulation of the atmosphere and the hydrosphere and thus for the erosion of the Earth's surface. Erosion would reduce the exposed solid surface to sea level within a very short period, in terms of the geological time scale, if there were no forces within the Earth causing uplift and repeated exposure of rocks to the action of solar radiation, the atmosphere, and the hydrosphere.

Geological study of rocks confirms that they have been folded and uplifted; local motions within the crust are thus established. Records of the Earth's former magnetic field from paleomagnetic studies and magnetic anomalies provide evidence that continental masses and oceanic crust have moved with respect to the poles. The existence of the Earth's magnetic field is explained by the dynamo theory which requires motions in the liquid outer core. These motions are dominated by the Coriolis force which arises from rotation of the Earth. Attempts to explain the origin and present distribution of the ocean basins, continents, and mountain ranges suggest that motions also occur in the upper mantle, possibly of a circulatory nature. The energy source for these motions is heat, probably derived from the decay of radioactive materials within the Earth.

It appears that the major features of the surface of the Earth are produced by the interactions of processes driven by two energy sources. The internal heat within the Earth causes motions in the mantle which influence the distribution and elevation of the conti-

1

nental masses. The external solar radiation drives the circulation of the atmosphere and hydrosphere, which produces the detailed sculpture of the solid Earth surface. The Earth's crust is a thin, brittle shell sandwiched between the Earth's mantle and the envelope comprising the atmosphere and hydrosphere.

Geology involves study of the physiographic features and rocks at the Earth's surface and deduction of the processes and history of their formation. The direct observations of this accessible part of the Earth lead to the conclusion that the present shapes and positions of the continents, distribution of features such as mountain ranges, as well as distribution of various rock types are second-order results of major processes occurring within the mantle. Thus, when we consider the Earth's surface, we find ourselves concerned with the Earth's interior, and hence with the origin of the Earth, which leads us to the origin of the solar system and the universe. These are complex problems and we cannot devote many pages to cosmology. Our attention will be concentrated on the crust of the Earth and the upper mantle which appears to dominate its behavior.

This book was conceived as a text for a course that provides an overview of the whole Earth, and simultaneously reviews the various subjects that are usually taught in geology and Earth sciences. Figure 1-1 is a "Tetrahedron of physical sciences," which includes all of the subjects applied to the study of the Earth. Geology remained largely descriptive and historical in its approach until recently when the application of physical and chemical methods and techniques led to the development of geophysics and geochemistry. We will be concerned for the most part with the front face of this tetrahedron and less with paleontology and the life sciences.

The tremendous advances made in Earth sciences have led in many universities to a proliferation of courses. Many specialized topics are now recognized as "subjects," each occupying a very small space in Figure 1-1,

but each with appropriate courses being taught and textbooks written. Students are burdened with more and more facts and more and more instruments for gathering data.

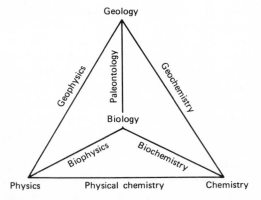

Figure 1-1. Tetrahedron of physical sciences. From "Introduction to Geophysics" by B. F. Howell. Copyright 1959. Used with permission of McGraw-Hill Book Company.

A. D. Stewart drew attention to this problem in a letter to *Nature* on March 9, 1968, entitled "Geology in British Universities." The fact that the article was reprinted in *Geotimes* eight months later indicates that the problem is not limited to Britain. Stewart pointed out that most university geology departments offer a range of semi-independent, specialized courses within the field from which the student is expected to construct his own view of the Earth. Departments are usually staffed with one teacher for each subject segment, with each segment completely separated from its neighbors. Standard segments include mineralogy, petrology, geochemistry, structural geology, stratigraphy, geophysics, sedimentology, and paleontology. The system tends to be self-perpetuating because as data accumulate original segments become split with the appointment of a new teacher. Paleontology, for example, may be split into macropaleontology and micropaleontology with an additional segment of vertebrate paleontology. What is needed, according to Stewart, is for teaching and research to be directed toward synthesis of geological information at levels above that of these segments. He proposed as an alternative a three-tier hierarchical scheme in which scale plays a key part, and teaching is process-oriented rather than descriptive. Level III is concerned with the standard subject segments and with techniques; level II is concerned with broader topics such as the Earth's core, mantle, and crust, the hydrosphere and atmosphere, and the evolution of life; level I is concerned with the Earth as a planet and with planetary studies.

Recent developments in the Earth sciences, especially those related to marine geophysics and paleomagnetism, have led to a great deal of synthetic research effort at level II, and this has certainly filtered through to undergraduate courses in universities. Several excellent introductory textbooks are available which present material more closely tied to the framework of level II than to the traditional segments. Unfortunately, however, after this there usually comes a series of successively more specialized courses. A presentation of material at level II should mean more to students when they have already learned something about geological processes and rock types than when they are first being introduced to the Earth sciences and that is what this volume attempts to provide. I have tried to present material appropriate for a student who is about to graduate with a bachelor's degree or who is beginning his graduate work in the Earth sciences.

We read a great deal today about the scientific revolution in the Earth sciences, and J. Tuzo Wilson considers that this revolution is shaking the very foundations of classical historical geology. In a revolutionary era the traditional subject-oriented instruction programs certainly require modification. Even if we are not living through a major revolution I consider it essential for students to be aware of a global, geophysical framework in which the various specialized research topics can be located, and to be familiar with the hypotheses

of world geology which embrace all of the subjects.

Geology is the science of the Earth, and the Earth is a chemical system subject to physical processes. I have therefore tried to incorporate both geophysics and geochemistry into consideration of geology on a global scale. The study can be pursued on three different scales.

1. We have the standard approach of regional geology, with rocks being examined in the range from local outcrops to mountains or other major tectonic units.

2. The history of such units is dependent on global geology, which is concerned with the size, distribution, and possibly the movements of continents and ocean basins.

3. The third scale relates to the Earth as a sphere, and is concerned with movements in the mantle and core which control the movements of the crust.

For complete understanding of geology we have to decipher the processes occurring within the Earth.

This volume is concerned mainly with the second scale, the development and distribution of the continents and ocean basins. The next six chapters review the chemistry, physics, and geology of the Earth's mantle and crust. The materials of the Earth are discussed in Chapter 4 together with a general review of geological processes. Most of these processes are involved in the mountain building cycle. The classical concepts of geosynclines and orogeny are reviewed in Chapter 9. Theories of the behavior of the Earth's interior must be capable of explaining the complexity of geosynclinal history.

Changes in temperature at depth in the mantle and crust cause changes of phase which may have significant effects at the surface and in Chapters 8 and 10 we consider the products and effects of partial melting at depth and the tectonic significance of solid state phase transitions. Vertical movements induced by deep-seated phase transitions may contribute to mountain building, but during the 1960's evidence supporting continental drift, in a modified guise of sea-floor spreading and plate tectonics, forced geologists to consider seriously the implications of global movements for the interpretation of mountain ranges and more localized problems. This is the stage at which revolution was proclaimed.

Chapters 11, 12, 13, and 14 include a history of the development of ideas related to continental drift and wandering poles, with special attention being paid to the concept of sea-floor spreading, the interpretation of linear magnetic anomalies of the ocean basins, and the formulation of the theory of plate tectonics. The new global tectonics involves the origin and evolution of the continents and of mountain ranges and the use of all of the traditional subject segments in geology. In Chapters 14 and 15 we examine the effects of revolution on reexamination and reinterpretation of classical geological concepts, and note the beginning of a counterrevolution by geologists unconvinced by the new evidence and ideas.

2. *Surface Features of the Earth*

INTRODUCTION

It is inevitable that a book written at the end of the geological revolution of the 1960's, which saw development of the global theories of sea-floor spreading and plate tectonics, must be greatly concerned with whether or not the oceans have grown and spread from a central rift, and whether or not the ancient continental masses have drifted with a young, spreading ocean floor. Nevertheless, we must begin with the state of the Earth as it is today.

Theories for the origin and history of the Earth must explain first the present distribution of the continents and ocean basins and second the major physiographic and geologic features of the interface between the Earth's solid surface and its envelope of atmosphere and hydrosphere. Interpretation of geophysical evidence, especially from paleomagnetic studies, linear magnetic anomalies of the ocean basins, and seismic studies, indicates that continents and ocean basins have undergone significant changes in relative position.

Evidence that relative motion may be in progress at the present time is provided by the observation that stable areas of the Earth's surface are traversed by elongated belts of instability. The distribution of volcanoes and earthquakes defines the active belts, and geological studies of folded orogenic zones define the location of earlier active belts on the continental masses. The existence of activity does not necessarily imply mobility and some geological evidence can be adduced to show that the relative positions of the continents and ocean basins have not changed for several hundred million years at least.

In this chapter we introduce the major physical features of the Earth's surface and the active belts that traverse them. These features are merely the superficial expressions of deep-seated processes, and therefore we will review the physical features of the Earth's interior in the following chapter. Together these two chapters provide the global framework for the topics discussed in subsequent chapters.

DISTRIBUTION OF CONTINENTS AND OCEANS

The world maps reproduced in figures through this book give a rather misleading picture of the relative areas and the distribution of the continents and oceans, and examination of a world globe is necessary for appreciation of the significance of the statistics reviewed. Figure 2-3 shows the distribution of the continents and oceans and Table 2-1 lists their areas.

Asia and Europe are often considered separately, but in a physiographic sense we must treat them as combined in the continent of Eurasia, which is almost twice as large as Africa, the next largest continent. Various authors subdivide the oceans in different ways; Table 2-1 lists areas for the Pacific, Atlantic, Indian, and Arctic Oceans. Separate areas are given in the table for each ocean and for each ocean plus marginal seas, gulfs, and straits. Other systems subdivide the Pacific and Atlantic into North and South Oceans at the equator, and include a Southern (Antarctic) Ocean arbitrarily bounded by the parallel 55°S.

More than 70% of the Earth's surface is covered by the oceans, and each of the three major oceans is larger than Eurasia. The Pacific Ocean constitutes just over half of all ocean surface. It is larger than all of the continents combined, and together with adjacent seas it covers 35.4% of the Earth's surface.

The continents are not distributed evenly over the Earth's surface. More than 65% of the land is in the northern hemisphere, with a concentration of continental material just south of the Arctic Circle. The continents of North and South America, Africa, and Asia with its Indian appendage, are roughly triangular in shape with their apices pointing south. About 81% of all of the land surface is situated on a land hemisphere with pole near Spain at 0° E-W, 38°N; this hemisphere contains 47% land and 53% sea. The opposite hemisphere with pole in New Zealand, the water hemisphere, contains 11% land and 89% sea. On the Earth as a whole about 45% of the surface has sea opposite sea, and only about 1.5% has land opposite land. Of the total land surface 95% is antipodal to sea. A striking example is provided by the opposite characters of the Arctic and Antarctic. The Antarctic continent is centered over the south pole and is surrounded by ocean, whereas the north pole lies in the Arctic Ocean which is almost completely surrounded by land.

TABLE 2-1   Areas of Continents and Oceans and Mean Ocean Depths[a]

	Area (10^6 km²)	Percent of land or ocean	Percent of world surface	Mean depth (km)
World surface	510	—	100	—
All continents	148	100	29.2	—
All oceans	362.0	100	70.8	3.729
Eurasia	54.8	36.8	10.8	—
Asia	44.8	29.8	8.7	—
Europe	10.4	7.0	2.1	—
Africa	30.6	20.5	6.0	—
North America	22.0	14.8	4.3	—
South America	17.9	12.0	3.5	—
Antarctica	15.6	10.5	3.1	—
Australia	7.8	5.2	1.5	—
Pacific Ocean	166.2	—	—	4.188
—with adjacent seas	181.3	50.1	35.4	3.940
Atlantic Ocean	86.6	—	—	3.736
—with adjacent seas	94.3	26.0	18.4	3.575
Indian Ocean	73.4	—	—	3.872
—with adjacent seas	74.1	20.5	14.5	3.840
Arctic Ocean	9.5	—	—	1.330
—with adjacent seas	12.3	3.4	2.4	1.117

[a] Data for oceans from Menard and Smith (1966).

SURFACE RELIEF OF THE SOLID EARTH

If the ocean water were removed from the surface of the Earth, we would see the continental masses rising abruptly from the ocean floor. The reason for the existence of the continents is one of the problems we seek to answer. Removal of the ocean water would also reveal the relief of the solid surface beneath the oceans and expose the system of ridges and rises encircling the globe.

The distribution of the elevations of the surface of the solid Earth has been the subject of many studies since measurements of land heights and ocean depths became available. The distribution of elevations has been well

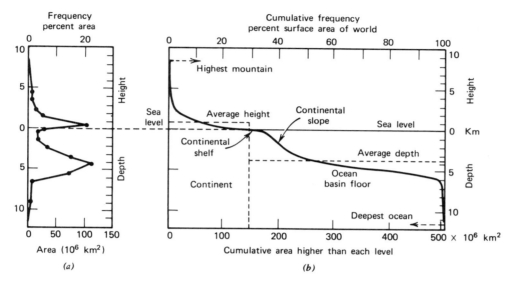

Figure 2-1. Distribution of areas of the solid Earth between successive levels. (*a*) Frequency distribution. (*b*) Cumulative hypsographic curve.

established for more than 50 years, but the number of depth soundings in the ocean basins has increased by many orders of magnitude during the last decade or two. Menard and Smith (1966) therefore re-examined the distribution of ocean depths using the modern bathymetric charts and considering specific physiographic and structural provinces.

Data for construction of the frequency distribution curve and the cumulative hypsographic curve of Figure 2-1 are listed in Table 2-2. The areas of the Earth's solid surface within 1 km intervals above and below sea level are given, as well as the percentage of the total surface area that is within each class interval. In Figure 2-1*a* these percentages are plotted at the midpoints of the level intervals, and the lines joining them show that the surface of the solid Earth is strongly concentrated at two levels: one corresponds to the continental platform and the other to the ocean basin floor.

The cumulative areas listed in Table 2-2 for total surface area above the lower limit of each height or depth interval are obtained by

TABLE 2-2 Hypsometry of the Earth's Surface[a]

Height or depth interval (km)	Area × 10^6 km²	Percent of total	Cumulative area (higher than lower limit of interval) × 10^6 km²	Percent of total
Above sea level: greatest height, Mt. Everest, 8.848 km; average height, 0.875 km				
>5	0.5	0.1	0.5	0.1
4–5	2.2	0.4	2.7	0.5
3–4	5.8	1.1	8.5	1.6
2–3	11.2	2.2	19.7	3.8
1–2	22.6	4.5	42.3	8.3
0–1	105.8	20.8	148.1	29.1
Below sea level: greatest depth, Marianas trench, > 11 km; average depth, 3.729 km				
0–0.2	27.1	5.3	175.2	34.4
0.2–1	16.0	3.1	191.2	37.5
1–2	15.8	3.1	207.0	40.6
2–3	30.8	6.1	237.8	46.5
3–4	75.8	14.8	313.6	61.5
4–5	114.7	22.6	428.3	84.0
5–6	76.8	15.0	505.1	99.0
6–7	4.5	0.9	509.6	99.9
7–11	0.5	0.1	510.1	100.0

[a] Data for oceans from Menard and Smith (1966). For continents from Scheidegger (1963) after E. Kossina.

adding the separate areas for all higher intervals; for example, the total area in millions of square kilometers with height greater than 3 km above sea level is given by 8.5, the sum of 0.5, 2.2, and 5.8. The cumulative areas are plotted against height or depth in Figure 2-1*b* and the line drawn through the points gives the hypsographic curve. The curve gives directly the area of surface above any selected level, expressed either in square kilometers, or as percentage of total surface. This figure also shows that the surface of the Earth is dominated by two levels. The mean elevation of land above sea level is 0.875 km, and the mean depth of the ocean is 3.729 km. Thus the average level of the continents is 4.604 km above the average level of the ocean floor. If the relief at the surface of the solid Earth were smoothed out, then the depth of the world-wide ocean so formed would be 2.44 km.

The total vertical relief on the Earth's solid surface is about 20 km, the difference between the peak of Mt. Everest, at 8.85 km, and the greatest known ocean depth of more than 11 km in the Marianas trench. This vertical relief is more than half the thickness of the average continental crust.

MAJOR STRUCTURAL UNITS AND PHYSIOGRAPHIC FEATURES

The hypsographic curve in Figure 2-1*b* illustrates in a general way the major physiographic provinces. The two predominant structural units are (a) the continental platforms and (b) the ocean basin floors. These are linked through (c) the continental margins, which comprise the continental shelf and slope, and other provinces to be reviewed in more detail below. Figure 2-1*b* does not show the existence of (d) the elevated oceanic

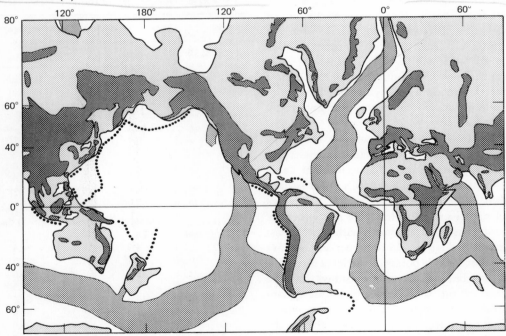

Figure 2-2. Major physiographic features of the world. Key: white, abyssal ocean floor; medium shading, oceanic ridge system; heavy dotted lines, oceanic trenches; light shading, continental platform and continental shelf; dark grey, mountains, intermontane basins, associated hills, and some elevated plateaus (various sources).

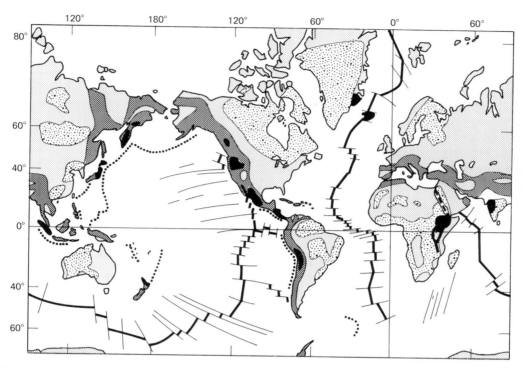

Figure 2-3. Major tectonic features of the world. Key: heavy lines, active rift systems of oceanic ridges; light lines, oceanic faults; dotted lines, oceanic trenches; light shading, continental platforms; ornamented, continental shields; dark grey, Tertiary folded mountain chains; black, Cenozoic volcanic regions (various sources).

ridges and rises, because their depths are included in the region labelled "continental slope." Small percentages of the oceanic and continental platforms are occupied by (e) ocean deeps and (f) high mountains respectively. The surface distribution of these six units is shown in figures 2-2 and 2-3, and surface areas are listed in Table 2-3. Note that although the maximum depths and maximum heights are at opposite ends of the diagram in Figure 2-1b, the ocean trenches are always located close to and almost parallel with land of high elevation such as an island arc or mountain range.

Continents

The position of the shore line between oceans and continents is not the boundary between the structural units. The continents are partly covered by water, and the flooded continental shelf and slope constitute 10.9% of the area of the whole Earth (Table 2-3). This is about one-quarter of the total area that is structurally continental. The sides of the continental slabs are represented by the continental slopes, which drop off abruptly from the edges of the continental shelves at angles of 2 to 3.5°, and continue down to the ocean basin floors lying in the depth range 3 to 6 km.

There is good evidence for worldwide changes in sea level, as well as areally restricted changes caused by local phenomena. Figure 2-1b shows that a small rise in sea level would produce a significant shift in the shore line position, an increase in area of continental shelf, and a decrease in area of sub-

TABLE 2-3   Areas of Structural Units and Physiographic Provinces[a]

	10^6 km²	Percent of land or ocean	Percent of world surface
Continent	149.0	100.0	29.1
(1) Precambrian shields	29.4	19.7	5.8
(2) Platforms	66.9	44.9	13.1
(3) Orogenic folded zones			
Riphean-Paleozoic	24.4	16.4	4.8
Mesozoic-Cenozoic	28.3	19.0	5.5
Submerged continent			
Continental shelf and slope	55.4	15.3	10.9
Ocean	306.5	84.6	60.0
(1) Abyssal ocean basin floor	151.5	41.8	29.7
(2) Ridge and rise	118.6	32.7	23.2
(3) Continental rise	19.2	5.3	3.8
(4) Island arc and trench	6.1	1.7	1.2
(5) Volcanic ridges, and volcanoes	5.7	1.6	1.1
(6) Other ridges and elevations	5.4	1.5	1.1

[a] Data for continents from Ronov and Yaroshevsky (1969). Data for oceans from Menard and Smith (1966).

aerial continent; for example, if the 25×10^6 km³ of ice stored in ice sheets and glaciers were to melt rapidly, sea level would rise by 50 to 70 m, before settling down at about 35 m after isostatic adjustment of land levels.

Figure 2-3 shows the main geological divisions of the continents.

1. The Precambrian shields.
2. The platform areas which are shields covered by a thin veneer of flat-lying younger sediments and the continental borderlands which are contiguous with the continental shelf.
3. The mountain ranges extending along orogenic belts of folded rocks.

The geologically young mountain chains follow two orogenic belts, each of which follows approximately a great circle. The circum-Pacific belt extends through the Philippines, Japan, Alaska, the Rocky Mountains, the Andes, and Antarctica. The Mediterranean-Asian belt follows the Alps, the Himalayas, Indonesia, New Guinea, and New Zealand. The continental areas with abundantly developed volcanic rocks of Cenozoic age are also indicated in Figure 2-3.

Ocean Basins

Table 2-3 lists six physiographic provinces for the oceans excluding the continental shelves and slopes because these, although covered by ocean, are not oceanic. Only three of these provinces occupy more than 2% of the ocean basin. (1) The abyssal ocean basin floor, and (2) the ridges and rises together cover more than 74%, and (3) the continental rise occupies 5.3%. The oceanic ridges and rises cover 32.7% of the ocean basin and 23.2% of the total Earth surface. This is much more than the folded orogenic belts of the continents. They form a worldwide system with many branches (Figures 2-2 and 2-3) and with mean depth of 3.970 km compared with a mean depth of 4.753 km for the ocean basin floor.

The continental rise, including partially filled sedimentary basins, consists of coalescing alluvial fans and piedmont plains deposited on the ocean floor beneath the continental slope, by turbidity currents and slumps. The continental rise, where recognized, is gently sloping or almost flat, and its characteristic features are similar to those of thick accumulations of sediments in partially enclosed basins such as the Gulf of Mexico.

Figure 2-4 shows the hypsometry of the world ocean basins and compares this with the depth distribution of the three major

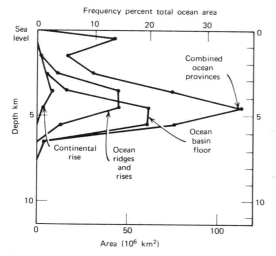

Figure 2-4. Hypsometry of all ocean basins for all physiographic provinces combined, and for individual major provinces. (From H. W. Menard and S. M. Smith, Jour. Geophys. Res., **71**, 4305, 1966, with permission).

physiographic provinces of the oceans; the ocean basin floor, the ridges and rises, and the continental rise. The areas of each province within the 1-km depth intervals can be read from the diagram, as well as the percentage of the total ocean area (compare Figure 2-1a). The combined curve for the oceans shows a double peak with maximum values at depths of 4.5 and 0.5 km, corresponding to the level of the deep ocean floor, and to the continental shelf and slope. Notice that the individual hypsometric curves for the ocean basin floor and for the oceanic

rises and ridges have similar shapes, which could be interpreted as indicating that an oceanic rise represents normal oceanic basin which has been elevated by an average of 1 km. Elevation or depression of the rises could affect sea level quite markedly.

Table 2-3 gives the areas of two more oceanic provinces and a third group of poorly defined features: (4) island arc and ocean trench, (5) composite volcanic ridges and individual oceanic volcanoes, and (6) long, narrow, steep-sided ridges not known to be volcanic, and poorly defined elevations.

Island arcs and trenches, including the whole system of low swells subparallel to the trenches, are considered as a single province covering only 1.2% of the Earth's surface, if continental equivalents or extensions of island arcs, such as Japan, are excluded. The median depth of 3.97 km is less than that for all ocean basins; the process which forms trenches and related features apparently elevates the oceanic crust on a regional scale.

The individual volcanoes and volcanic ridges rising from the ocean basin floor constitute a physiographic province distinct from the volcanoes associated with the oceanic rises and ridges. The volcanic ridges are formed from overlapping volcanoes. This province covers about the same surface area as the island arc and trench province, but it may be larger, because probably some of the remaining 1.5% of the ocean floor consisting of ridges and elevated regions not known to be volcanic will eventually prove to be of volcanic origin.

Continental Margins

The continental margin is the region between continent and ocean basin. The edge of the continental shelf is usually sharply defined by a marked increase in slope. The width of the continental shelf ranges from zero to 1500 km, with an average of 78 km, and the ocean depth at the shelf edge ranges from 20 to 550 m, with an average of 133 m. The continental margins are deeply incised by submarine canyons, which serve as routes for sediment transportation away from the continental shelves. The sediments are deposited in ocean trenches, or on the abyssal ocean floor, where they build up continental rises.

Three types of continental margin can be distinguished in Figure 2-3. The *Atlantic type* comprises the continental shelf, continental slope, and continental rise. The *Andean type*

has a narrow continental shelf, with an oceanic trench below the continental slope. The *Island-arc type* consists of a volcanic island arc with oceanic trench, separated from the main continental mass by a small ocean basin. Some island arc systems are so far from continental masses, as much as 2000 km, that they can not be considered as features of continental margins.

TECTONICALLY ACTIVE ZONES: OROGENIC BELTS AND THE WORLD RIFT SYSTEM

The platforms of the continents and ocean basins (Figures 2-1*b* and 2-2) are stable, but the abrupt change in elevation at the continental margins makes them potential or actual sites of instability. The Earth is in a state of activity, as demonstrated by the occurrence of earthquakes and the eruption of volcanoes. The existence of high mountain ranges composed of intensely folded rocks confirms that the Earth has been active in the past. The unstable or active parts of the Earth occur along elongated belts bordering or traversing the stable platform areas. The distribution of orogenic belts has already been reviewed in Figures 2-2 and 2-3. The post-Precambrian orogenic belts can be subdivided into four regions with different periods of folding. Orogenic activity apparently is not continuous in any given location, but intermittent. Another major zone of activity follows the crest of the midoceanic ridge-rise system.

The World Rift System and Major Faults

During the last 20 years evidence has accumulated that a worldwide rift system follows the crest of oceanic ridges and rises and extends across parts of the continents. There is a striking rift valley in the center of the mid-Atlantic ridge, but there remains. some uncertainty about the physiography of other rises. There appears to be nothing comparable to the Atlantic rift valley and

associated ridges on the East Pacific Rise. Figure 2-3 shows the distribution of the rift system according to various authorities, although many more depth soundings are required before the continuity of the rift system on a worldwide scale can be considered absolutely established. Note that the rift system is not restricted to the oceans. It extends across the African continent, for example.

The map shows that the rift is discontinuous on a local scale because of displacement along fracture zones which were at first believed to be transcurrent faults, but which are now interpreted as transform faults (Chapter 14). In the East Pacific there is a series of major fracture zones about 100 to 200 km wide, approximately following great circles, which extend for a few thousand kilometers, with a vertical relief of a few kilometers. Individual ridges and troughs within the zones are several hundreds of kms long and a few tens of kilometers wide. Within the troughs occur the greatest ocean depths of the central Pacific, 6 to 7 km. Figure 2-3 shows that these enormous fracture zones are transverse to the crest of the East Pacific Rise, and it appears that they lie entirely on the broad flanks of the Rise.

Another group of fractures is the series of large transcurrent fault systems that parallel the margins of the Pacific. These are not shown in Figure 2-3 because it would complicate the diagram too much. Movements on these faults, such as the San Andreas fault in California, are of the order of 1 cm/year, which would provide long-term displacements large enough to be consistent with the hypothesis of continental drift. In the early 1960's many geologists considered trans-current faulting to be the dominant mode of tectonics in the circum-Pacific belt, with the evidence suggesting that the entire Pacific basin was rotating anticlockwise or that the whole basin was moving northwest. It now appears that the San Andreas fault is a transform fault (Chapter 14), and attention is directed toward sea-floor spreading (Chapter 12) rather than rotation of the sea floor.

Figures 2-2 and 2-3 show the distribution of major physiographic and geological features of the Earth's surface. The sites of current activity are defined by volcanoes in eruption and earthquakes.

Distribution of Volcanoes

Nearly 800 volcanoes are active today or known to have been active in historical times. More than 75% are situated in the circum-Pacific belt, which is known as the Ring of Fire. This belt coincides with the young mountain ranges of western America, and the volcanic island arcs fringing the north and western sides of the Pacific. The Mediterranean-Asian orogenic belt has volcanoes distributed sparsely, except for Indonesia and the Mediterranean where they are more abundant. The oceanic volcanoes have already been described as a physiographic province, and the remaining active volcanoes are strung along the oceanic ridges and associated with the African rift valleys, which both belong to the world rift system.

J. H. Tatsch presented a three-dimensional least-squares analysis of the distribution of volcanoes in 1964. He found that approximately 93% of the active volcanoes appear to lie along belts defined by the traces of three mutually orthogonal planes passing through the center of the Earth. The circle traces of these planes intersect each other at the

points: lat. 5°N, long. 95°E; 30S-175W; 55N-165W; 5S-85W; 30N-5E; and 55S-15E.

Distribution of Earthquakes

Earthquakes are of three kinds, shallow, intermediate, and deep. Shallow earthquakes are those occurring at depths above 70 km. If the focus is between 70 and 300 km the earthquake is intermediate. Deep-focus earthquakes are those originating at depths between 300 and 700 km. The earthquake belts of the Earth are defined in Figure 2-5a by the epicenters of all earthquakes recorded during the period 1961–1967. The intermediate- and deep-focus earthquakes are restricted in their distribution, as shown by Figures 2-5b, and 14-3.

The earthquake zones follow closely the distribution of volcanoes, although not necessarily in detail. More than 80% of the world's shallow earthquakes occur in the circum-Pacific zone. The same zone experiences about 90% of intermediate shocks and nearly all of the deep shocks. Other earthquakes occur along the Mediterranean-Asian mountain system and on the world rift system. The relation of earthquakes to continental margins and ocean trenches is shown in Figure 14-3.

Examination of Figures 2-5 and 14-3 shows that the deeper earthquakes of the circum-Pacific belt are displaced toward the continents compared to the shallow-focus earthquakes. According to H. Benioff, in 1955, these earthquake epicenters lie within fracture zones about 250 km wide, dipping beneath the continent or island arcs, and extending to depths of 650 to 700 km. He illustrated the two types of deep-seated seismic zones in Figures 2-6a and 2-6b. The oceanic fault zones extend downward at 61° from the ocean floor beneath the island arcs, some distance from the continental borders. The marginal or continental fault zones are more complex. Figure 2-6 shows an example extending downward from an ocean trench bordering a continental mountain range. A shallow component

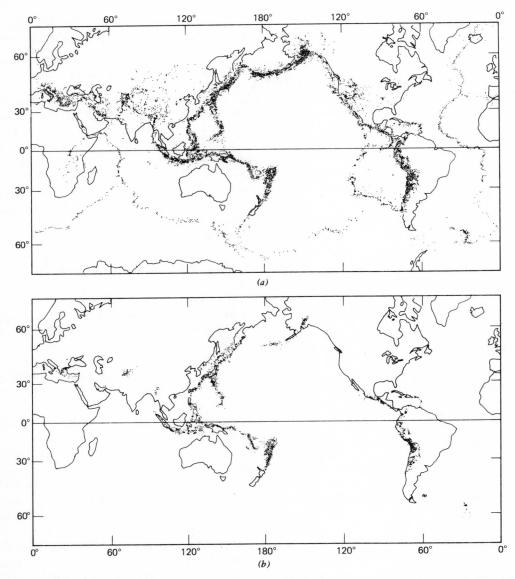

Figure 2-5. Seismicity of the Earth, 1961–1967. Plot of all earthquake epicenters recorded by U.S. Coast and geodetic Survey. (*a*) All epicenters. (*b*) Epicenters with depths between 100 and 700 km. (From M. Barazangi and J. Dorman, Bull. Seismol. Soc. Amer., **59**, 369, 1969, with permission.)

extends to a depth of 70 km, an intermediate component extends to about 300 km with an average dip of 32°, and a deep component extends from 300 to about 700 km with a dip of 60°. These seismic zones are now known as Benioff zones.

The use of digital computers in recent years to redetermine the hypocenters of earthquakes in Benioff zones indicates that the earthquake foci in some island arc regions are confined to a zone only 50 to 100 km thick. It also appears that many deep earthquake zones extending under island arcs have dips of about 45°. Recent results show considerable

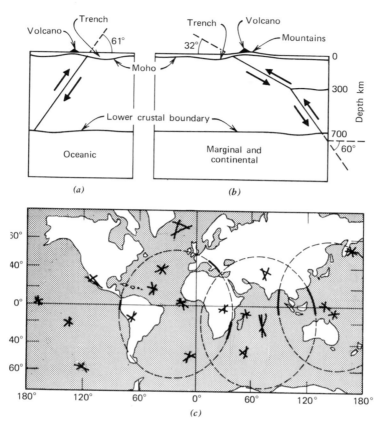

Figure 2-6. (*a*) and (*b*) Oceanic and marginal continental faults and crustal structure, showing deep fault zones marked by intermediate and deep-focus earthquakes (after Benioff, 1955, with permission of The Geological Society of America). (*c*) Three examples of the 21 belts that cover better than 90% of linear seismic zones longer than 700 km, according to Rouse and Bisque. These belts show the 19 multiple intersections plotted (after Rouse and Bisque, 1968, with permission of *The Mines Magazine*).

variation in the dips of Benioff zones from one location to another. L. R. Sykes concluded in 1966 that the shallow-, intermediate-, and deep-focus earthquakes associated with the Benioff zones dipping beneath island arcs are part of a continuous zone of tectonic activity. The earthquake foci under young folded regions are less deep than those under island arcs, and they are concentrated in relatively small areas. The Benioff zones constitute an important part of the New Global Tectonics reviewed in Chapter 14. They also form the basis of a global scheme presented by Rouse and Bisque in 1968.

Global Tectonics of G. E. Rouse and R. E. Bisque

There have been many attempts to explain the distribution of the active belts of the world within a comprehensive global tectonic system, and a recent example used Benioff zones as the basis for correlation. Rouse and Bisque noted that if a Benioff zone is extended through the Earth with an average dip of 60°, it turns out that it is in tangential contact with the surface of the core. This observation led them to examine the relationships of other tectonic features, including ocean ridges, island arcs, and major fault systems, to the circles described at the surface of the earth by planes tangential to the core-mantle interface.

They found that of the infinite number of planes possible 21 were sufficient to cover more than 90% of the known seismic belts. They defined the positions of 16 circle zones or belts by listing for each circle at least four major linear structural systems. A few circles are shown in Figure 2-6c. They reasoned that if the belts do have global structural significance then zones of multiple intersections might be of particular interest; they located 19 of them, as shown in Figure 2-6c. All but five of these do occur at zones of notable seismic or volcanic activity. Several of them occur on belts whose positions are defined by features in other quadrants of the globe, which suggests that the planes themselves do have global tectonic significance.

Rouse and Bisque proposed tentatively that the fluid motions in the core, which are invoked to explain the Earth's magnetic field, cause stress patterns in the mantle which are transmitted to the surface along planes tangential to the core-mantle interface. It was mentioned above, however, that the dip of many Benioff zones is now known to be 45° and not the 60° shown in Figure 2-6. Extrapolation of these zones through the Earth would provide no contact with the core-mantle interface. The significance of the multiple intersections plotted in Figure 2-6c therefore remains uncertain. What is certain, however, is that we must seek within the Earth for explanations of the active belts at the Earth's surface. Let us now turn to the physical features of the Earth's interior.

3. *Physical Properties of the Earth and its Interior*

INTRODUCTION

The Earth is a sphere in space, and physical properties of the sphere as a whole can be determined. The size, shape, and mass of the Earth have been measured with precision, and its magnetic and gravity fields and the geothermal flux from the sphere have also been measured. The Earth's interior is inaccessible, and we can only infer its chemical composition from indirect evidence (Chapter 6). However, the passage of earthquake waves through the Earth provides a kind of X-ray of its interior, and these waves bring to the surface information about the physical properties of material within the Earth. Interpretation of countless earthquake waves shows that the Earth has the concentric structure illustrated in Figure 3-1. The

Earth's crust, ranging in thickness from about 5 km beneath the ocean to 35 km or more beneath the continents, is separated from the mantle by the Mohorovicic discontinuity (the Moho). Inside the mantle and separated from it by the Gutenberg-Wiechert discontinuity is the Earth's spherical core, with a radius of 3473 km.

This topic merits a whole book in itself. In this chapter we can only review broad outlines, with sufficient detail to demonstrate that interpretations have changed considerably within a decade. We should therefore be prepared to assume that further changes in interpretation will arise from new data gathered during the next decade.

SEISMOLOGY AND THE EARTH'S INTERIOR

The internal structure of the Earth depicted in Figure 3-1 is based on seismology, the study of earthquake waves. The use of underground nuclear explosions in the study of seismological problems is one of the outstanding advances of recent years. They have the advantage that both site and time of the event can be selected for the convenience of the observers.

Seismic Waves

An earthquake involves the sudden release

of energy which is transmitted through the Earth in all directions, in the form of seismic waves. In order to track the paths followed by the waves and to determine their times of travel the precise location and time of the earthquake must be known. Because this information can only be derived from the seismic waves themselves, there is some uncertainty in the procedures. Underground nuclear explosions provide point sources of seismic energy with all ambiguities about the location of an earthquake focus removed.

Consider a nuclear explosion set off

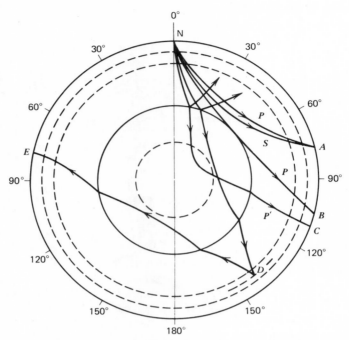

Figure 3-1. Cross section through the Earth, showing paths of seismic waves from earthquake epicenter, N.

precisely at the North Pole N in Figure 3-1. If the Earth were a homogeneous, elastic, isotropic sphere, elastic waves would spread in all directions through the Earth, following straight lines, until they reached the surface again. The velocity of the waves could easily be calculated from the time taken for a wave to reach any point on the surface.

We know, however, that the Earth is not homogeneous. Because the mean density of the Earth is 5.517 g/cm^3, compared with densities of 2.6 to 3.0 for crustal rocks, part of the Earth's interior must be composed of material with density greater than 5.517 g/cm^3. The moment of inertia of the Earth provides information about the mass distribution. For a sphere of uniform density of radius, r, and mass, m, rotating about an axis, the moment of inertia is 0.4 mr^2. The moment of inertia of the Earth, as determined from satellites in orbit, is 0.331 MR^2 (where M and R are the mass and the radius of the Earth), which indicates that mass is concentrated toward the

center of the Earth. Therefore the seismic waves are refracted as they pass deeper into the Earth, through material with different properties, and it has been established that they follow curved paths, or rays, like those shown in Figure 3-1. Where the property changes are sufficiently rapid, the waves may be reflected as well as refracted, and interpretation of the seismic records then reveals a surface of discontinuity at the depth where reflection occurred.

The release of energy at an earthquake focus or nuclear explosion produces several different seismic waves. There are body waves, which follow rays through the Earth like those shown in Figure 3-1, and surface waves, which travel around the outside of the Earth. The primary (P) wave is a compressional body wave, and it travels with about twice the speed of the secondary (S) wave, which is a transverse body wave. These waves are propagated because the Earth material is deformable, and their study thus depends upon elastic theory. For a perfectly elastic body, the velocities of P and S waves are given by:

$$V_P^2 \rho = K + 4\mu/3 \qquad (3\text{-}1)$$

$$V_S^2 \rho = \mu \qquad (3\text{-}2)$$

where ρ is density, K is bulk modulus or incompressibility, and μ is the modulus of rigidity. The consistency between these equations and seismic observations confirms that the theory of perfect elasticity is relevant to seismic wave propagation.

Surface waves (L waves) are strongly developed in the Earth. Rayleigh surface waves consist of motion in a plane perpendicular to the surface and parallel to the direction of transmission. Love surface waves are shear waves, with vibration perpendicular to direction of transmission and in a horizontal plane, which travel in a thin surface layer. The energy of surface waves is distributed through an appreciable depth into the Earth. Energy of long wavelength penetrates to greater depth than shorter wavelength energy, and the deeper energy travels faster. The variation of velocity with wavelength is called dispersion. Study of the dispersion of surface waves is yielding detailed information about the Earth's crust and upper mantle. Observations now extend out to the fundamental periods of free oscillations of the whole Earth.

Travel-time Curves

Consider again a nuclear explosion set off at the North Pole in Figure 3-1, at a precisely known time, and consider the transmission of seismic waves from this impulse to a series of points on the surface of the earth, whose positions lie on a great circle passing through the Pole. The position of each point is denoted by the angle subtended at the center of the Earth by the surface between the point and the energy source, as shown by the abscissa in Figure 3-2. In this example the angle is equivalent to degrees of latitude. At point A, denoted by the angle 75° subtended

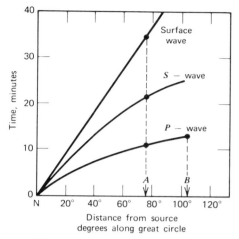

Figure 3-2. Travel-time curves for seismic waves emanating from epicenter N in Figure 3-1.

by the arc NA, energy is received first from the P wave, followed by the S wave; the waves follow the refracted rays shown passing through the mantle in Figure 3-1. At a later time surface waves which have travelled around the surface NA are received.

From the seismic records at the station A the time taken for each wave to reach A is determined and plotted on the travel-time diagram as shown in Figure 3-2. These points, along with similar sets of points for a series of recording stations between and south of NA, permit the construction of time-distance curves, or travel-time curves, for the different seismic waves. The graph obtained for the surface wave is a straight line, because the velocity does not change as the wave travels around the surface of the Earth. However, the graphs for the P and S waves are not linear; they have downward curvatures. This shows that with increasing distance of the recording station from the impulse source, and increasing length of the ray path within the Earth, the average velocities of the P and S waves increase. Figure 3-1 shows that this velocity increase is a function of depth of penetration of the rays into the Earth.

The ray NB in Figure 3-1 just penetrates to the core-mantle boundary and point B (103°) is thus a limit for seismic waves passing directly through the mantle. Figure 3-2 shows that for points more distant than *B* on the great circle *NB* the normal *P* waves disappear. Figure 3-2 also shows that the travel-time curve for the *S* wave terminates at about the same distance from *N*, but somewhat nearer to *N* because the *S* waves received at a given point such as *A* follow a somewhat deeper trajectory through the mantle than the *P* wave (Figure 3-1).

Many sets of travel-time tables giving the average times of travel along corresponding rays in terms of angular distances were prepared between 1920 and 1940, and by 1940 the errors for many distances were reduced from the order of a minute or more to only one or two seconds. In recent years major effort has been directed toward detecting differences in travel-time data for different environments.

Seismic Velocity Profiles and the Earth's Internal Structure

The velocities of seismic waves at depth can be calculated from the travel-time data, although the calculations are complex. The velocities of *P* and *S* waves have been determined with an accuracy of 2% at most depths. Figure 3-3 shows the velocities of *P* and *S* waves within the Earth, according to H. Jeffreys and B. Gutenberg. These curves agree closely, and they have been widely quoted for about 30 years. The principal uncertainties are in the velocity gradients at several depths in the Earth. Using Jeffreys' curves K. E. Bullen divided the Earth into a series of layers, with boundaries associated with levels where the velocities, or the velocity gradients, changed abruptly. These layers are designated *A*, *B*, *C*, etc., in Table 3-1 and Figure 3-3.

Figure 3-3 shows that at a depth of 2900 km, the *S*-wave velocity curve terminates, and there is an abrupt decrease in the velocity of the *P* wave. The maximum depth of penetration into the Earth for *P* waves following the ray *NB* in Figure 3-1 is 2900 km, and at

TABLE 3-1 First Approximation to the Earth's Internal Layering (Bullen, 1967)

Region	Range of depth (km)	Name	Characteristics of *P* and *S* velocities
A	0–33	Crust	Complicated
B	33–410	Upper mantle	Normal gradients
C	410–1000	Transition region	Greater than normal gradients
D'	1000–2700	Lower	Normal gradients
D''	2700–2900	Mantle	Gradients near zero
E	2900–4980	Outer core	Normal *P* gradient
F	4980–5120	Transition region	Negative *P* gradient
G	5120–6370	Inner core	Smaller than normal *P* gradient

distances between *N* and *B* a pulse is received at times corresponding to reflection from a surface at 2900 km depth. This surface is the boundary between the mantle and the core. At distances greater than *NB* (103°C), the *S* wave is not received at all, indicating that the core is composed of material that cannot maintain a shear stress. The core thus behaves like a liquid with respect to the transmission of elastic waves.

P waves are rarely observed at angles greater than 103°, and the area behind the core is thus called the shadow zone. *P* waves which reach the core are in part reflected and part refracted, as shown by the ray *NC* in Figure 3-1, and there is a focusing effect. The refracted *P* waves are slowed down. The material in the core has markedly different elastic properties than the material in the mantle. There are at least two layers in the core, the inner core probably being solid. Many more complex models for the core have been proposed since 1940.

Most of the discrepancies between the wave velocities of Jeffreys and Gutenberg are within the upper mantle and the transition zone between the upper and lower mantle (Figure 3-3), that is, in the depth interval of 40 to 1000 km. On the basis of Jeffreys' velocities,

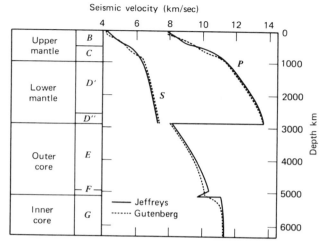

Figure 3-3. Velocities of *P*-waves and *S*-waves in the Earth according to Jeffreys and Gutenberg, showing Bullen's (1949) layers. See Bullen (1967) for review (with permission of Academic Press).

Bullen defined three subdivisions of the mantle, *B*, *C*, and *D*, with the latter subdivided into *D'* and *D''*. Neither of the boundaries *B-C* at 410 km and *C-D* at 1000 km appeared to be sharp, and several different depths have been assigned to them by various investigators. Gutenberg's velocity distributions in Figure 3-3 show a well defined low-velocity layer at 100–150 km depth, the existence of which he first proposed in 1926. Gutenberg has proposed that the *B-C* discontinuity is the outermost limit of the shadow zone caused by the low velocity layer.

The reality of a low-velocity zone is now established (Figures 3-4 and 3-5) although there is less certainty about its depth in relationship to tectonic environment. Surface wave studies have provided significant contributions. For the study of body waves computers have been a convenience but not a necessity. The study and interpretation of surface waves and free oscillations, on the other hand, could not have proceeded beyond the qualitative stage without the computer. Figure 3-4 compares layered shear velocity models depicting low-velocity zones in three different tectonic environments. These models were derived by analysis of Rayleigh wave dispersion.

The classic picture of the Earth's internal structure, shown in Figures 3-1 and 3-3 and Table 3-1, has been changed by the discovery of several discontinuities of first order in the mantle, and others which may be of first or second order. The depths to the discontinuities

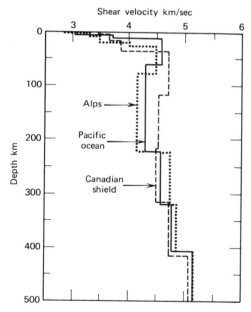

Figure 3-4. Models for *S*-wave velocities in three different tectonic environments. (From J. Dorman, Geophysical Monograph **13**, 257, 1969, with permission of Amer. Geophys. Union.)

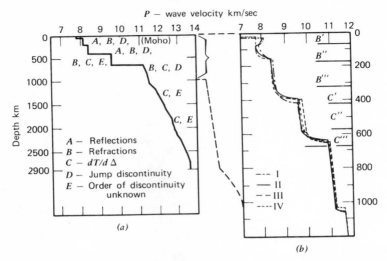

Figure 3-5. Models for *P*-wave velocity distributions in the mantle. (*a*) Schematic distribution after Knopoff (1967, with permission of Academic Press). The velocity gradients within the upper layers are not known and are shown as zero; there may be separate discontinuities at 640 and 800 km. (*b*) Models for four tectonic provinces of the United States. The most highly variable regions of the mantle are the *B* zones. Variations between the models in the *C* and *D* regions are not considered significant. The transition zones *C'* and *C'''* are well established, but the *D''* transition is questionable. I. East Basin and Range Province and North Rocky Mountains. II. Colorado Plateau and Rocky Mountains. III. Basin and Range Province. IV. Snake River plains and west Rocky Mountains. (From C. B. Archambeau, *et al.*, Jour. Geophys. Res., **74**, 5825, 1969, with permission.)

appear to vary in different parts of the world, and a schematic *P* wave velocity distribution is shown in Figure 3-5*a*. The discontinuities have been detected by various methods.

Figure 3-5*b* shows *P*-wave velocity profiles for the mantle within four different structural provinces of the western United States. In all provinces there are high-velocity gradients (sharp zones of transition) near 150, 400, 650, and possibly near 1000 km. The low-velocity zone is clearly defined beneath the Rocky Mountains and Colorado plateau. Beneath the Basin and Range province abnormally low velocities extend from near the base of a thin crust to depths of 150 km, but there is no lid with high velocities capping the zone. These detailed studies show that there is considerable variation in the properties of the upper mantle down to at least 150 or 200 km from one province to another and even within each province.

An example of the sensitivity of modern

seismic techniques is illustrated by the rays *NC* and *NDE* in Figure 3-1. The ray *NC* passes through the core, and we have already referred to the fact that part of the ray is reflected at the core-mantle boundary and recorded at a seismic station between *N* and *A*. The line at *D* represents the 650 km discontinuity shown in Figure 3-5*b*. After diffraction through the core, a portion of the ray *ND* is reflected from the underside of the discontinuity *D*, and this passes back through the core along the ray *DE*. Pulses received at *E* as a result of the seismic waves emanating from *N* can be detected and interpreted in terms of reflection from *D*.

Figure 3-6 shows the results of a detailed analysis of mantle structure by R. Z. Tarakanov and N. V. Leviy. They proposed that the upper mantle has a complicated layered structure like that of the crust, and they distinguished four layers of low velocity and low strength at depths of 60–90 km, 120–

160 km, 220–300 km, and 370–430 km. The model correlates the change with depth of magnitudes of earthquakes, velocities and amplitudes of P waves, and the derivative of an empirical travel-time curve.

Figure 3-6a shows the distribution of the maximum earthquake shock intensities plotted against the depth of the focus, for major earthquakes recorded throughout the world during 1896–1962. The maximum earthquake magnitudes fluctuate within the shaded band. There are significant minima in this profile, which are interpreted in terms of layers of reduced strength at these depths.

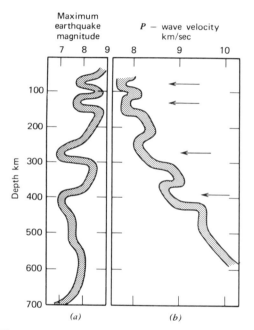

Figure 3-6. (*a*) Change of the maximum magnitude of earthquakes with depth. (*b*) The distribution of *P*-wave velocities with depth. Arrows show four low strength and low-velocity layers (after R. Z. Tarakanov and N. V. Leviy, Geophysical Monograph **12**, 43, 1968, with permission of Amer. Geophys. Union).

Empirical curves were constructed from data collected at more than 100 observation stations in the transitional zone from the Asiatic continent to the Pacific Ocean, across the Kuril and Japanese Islands. Using a method devised by Gutenberg, they determined the P-wave velocity at the focal depths of earthquakes in the depth range from 30 to 680 km. The velocity profile so obtained is shown in Figure 3-6b, with the uncertainty in velocity determinations being shown by the shaded band. The four zones of reduced velocity are well correlated with the depths of weak layers shown in Figure 3-6a by the minima in the earthquake magnitude distribution.

Lateral Variations in the Upper Mantle

Figure 3-5b shows that there are lateral variations in the upper mantle. Figures 3-7 and 3-8 illustrate the pattern of these variations on a continental and a global scale.

The velocity of the P wave just below the Moho discontinuity, P_n, was generally taken to be near 8.1 km/sec until about 1960. By then it was discovered that the crust and upper mantle of the western United States is different from that of the eastern United States, and values of P_n were measured between 7.7 and 8.4 km/sec. Low values for P_n are also found beneath island arcs and midoceanic ridges, with the higher velocities under deep ocean basins (see Chapter 5). Figure 3-7a shows the estimated P_n velocity for the United States, based on data from deep seismic soundings, underground nuclear explosions, and earthquakes; P_n velocities are lower beneath the mountainous west than beneath the plains of the east.

The correlation of topography with P_n on a regional scale is matched approximately by crustal thickness, and by mean crustal seismic velocity. Figure 3-7b shows the variations in crustal thickness, with the depth to the Moho contoured at 10 km intervals, compared with the regions having P_n greater than, or less than 8 km/sec. The mean crustal P-wave velocity is also illustrated. The Rocky Mountain system appears to divide the United

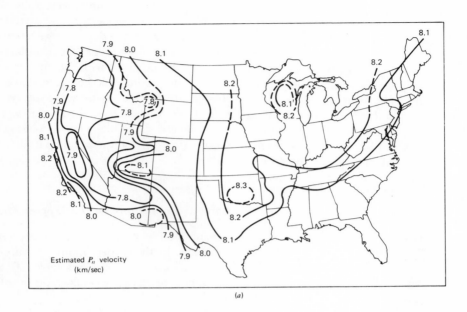

(a)

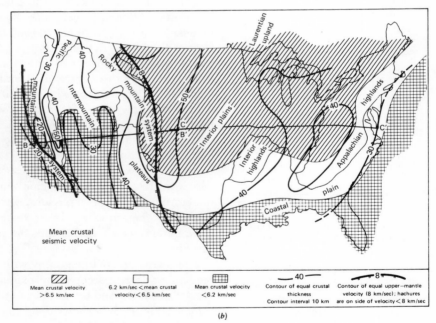

(b)

Figure 3-7. (a) Estimated P_n velocity for upper mantle in the United States, based on data from deep seismic soundings, underground nuclear explosions, and earthquakes (after E. Herrin, Geophysical Monograph **13**, 242, 1969, with permission of Amer. Geophys. Union). (b) Variations in crustal thickness, mean crustal velocity, and upper mantle velocity in the United States, showing locations of aeromagnetic profiles, *B-B'* and *C-C'* of Figure 7-2a. (From L. C. Pakiser and I. Zietz, Rev. Geophys., **3**, 505, 1965, with permission.)

States into two superprovinces involving both crust and upper mantle

1. In the eastern superprovince with lower elevation, P_n is greater than 8 km/sec, the mean crustal velocity is generally greater than 6.4 km/sec, and the crust is generally thicker than 40 km.

2. In the western mountainous superprovince P_n is less than 8 km/sec, the mean crustal velocity is generally less than 6.4 km/sec, and the crust is generally thinner than 40 km.

These correlations indicate that crust and upper mantle are closely linked in a tectonic sense, and that both have to be considered in isostasy and gravity calculations.

S_n is a short-period seismic shear wave that propagates in the uppermost mantle and does not penetrate the low-velocity zone. The efficiency of transmission of S_n on a world-wide scale is summarized in Figure 3-8. This shows that S_n propagates very efficiently across the stable continental shields and ocean basins. The areas for inefficient propagation include the oceanic ridges and rises, some orogenic belts, and the concave side of island arcs. Paths crossing the concave side of island arcs or the crests of oceanic ridges do not transmit S_n. The rigid outer shell of the upper mantle, part of the lithosphere, is assumed to correlate with low attenuation of S_n (or high Q), and the results then imply that the uppermost mantle is weaker beneath the island arcs and oceanic ridges than elsewhere. These locations can be interpreted as discontinuities in the lithosphere.

There is considerable current interest in the hypothesis of mantle convection (Chapters 12 and 14). If flow has occurred in the upper

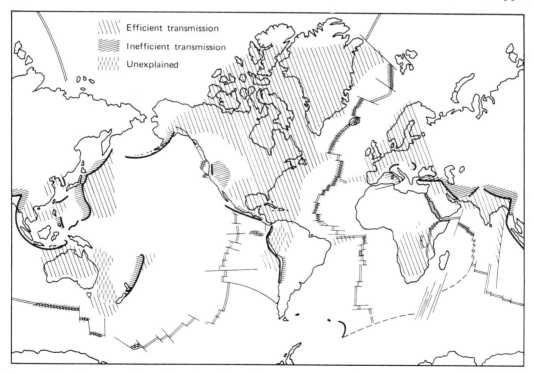

Figure 3-8. Summary of world regions where S_n waves propagate inefficiently in the upper mantle. Island-arc structures are represented by bold dark lines; crests of midocean ridges by double lines; and fracture zones by single lines. (From P. Molnar and J. Oliver, Jour. Geophys. Res., **74**, 2648, 1969, with permission.)

mantle this could have produced anisotropy in a horizontal plane. Recent seismic velocity measurements have therefore been directed toward identification of anisotropy in the uppermost mantle beneath the oceans. The seismic results were reviewed by H. G. Ave'Lallemant and N. L. Carter (1970) in their discussion of modes of flow in the upper mantle, based on experimental syntectonic recrystallization of olivine. The metamorphic texture and fabrics produced in the recrystallized grains are similar to those occurring in some peridotites and dunites believed to be mantle-derived tectonites; these include nodules in kimberlites and basalts, and orogenic peridotites (see Chapter 6).

PROPERTIES OF THE EARTH'S INTERIOR

The study of seismic waves yields the distribution of seismic wave velocities within the Earth, and several properties of the Earth's interior can be calculated from these profiles, the average density of the Earth, and its moment of inertia. The distributions of ρ, μ, k, g, and pressure, p, within the Earth are closely interlocked, and Equations 3-1 and 3-2 show that the velocities of body waves are expressed in terms of the ratios of density and elastic properties. Most investigators have concentrated on deducing a consistent density distribution first and deriving the other property distributions from this.

The temperature distribution within the Earth is not well known, and no unique solution is possible for the geothermal gradient at depth. Knowledge of the thermal state of the Earth and its variation with time, however, is of fundamental importance for geophysical and geological problems, and there have been many estimates for the positions of geotherms in various tectonic environments.

Other properties of the Earth's interior which have been estimated as a function of depth include viscosity and electrical conductivity.

Density Distribution

The general method for deducing the density distribution involves many assumptions. It requires that the Earth be treated as a series of concentric layers, each assumed to have homogeneous composition with no abrupt changes in physical properties. If m is the mass of material within a sphere of radius r, and G is the universal constant of gravitation, then the value of g at distance r from the Earth's center is

$$g = \frac{Gm}{r^2} \qquad (3\text{-}3)$$

and since the stress in the earth's interior is essentially equivalent to hydrostatic pressure, the variation of p with depth is given by

$$\frac{dp}{dr} = g\rho = -\frac{Gm\rho}{r^2} \qquad (3\text{-}4)$$

The incompressibility is linked with the density and the pressure by

$$k = \rho\left(\frac{dp}{d\rho}\right) \qquad (3\text{-}5)$$

These equations, together with 3-1 and 3-2, give the Williamson-Adams equation:

$$\frac{d\rho}{dr} = -\frac{Gm\rho}{r^2}\left(V_p{}^2 - \frac{4V_s{}^2}{3}\right) \qquad (3\text{-}6)$$

This cannot be integrated directly, but it can be applied to each of the assumed homogeneous layers, giving the rate of change of density at a particular depth. If a value for the density of the upper mantle is assigned to material beneath the crust, its density deeper in the layer can be calculated. A new value for $d\rho/dr$ is calculated for the new density, and the calculation is repeated, layer by layer, until a discontinuity in the wave velocity

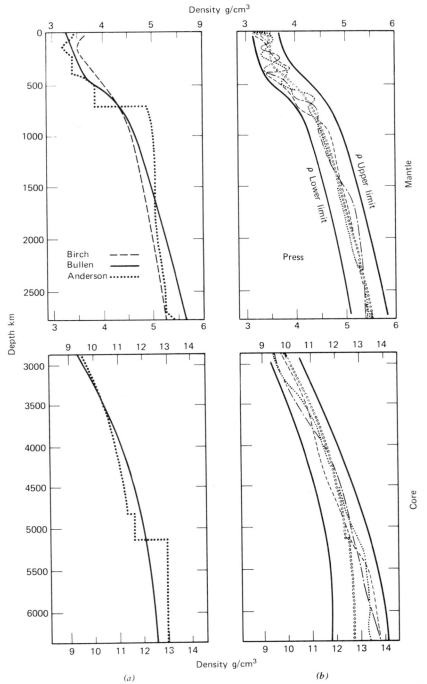

Figure 3-9. Density distributions in the mantle. (*a*) Recently proposed density models (1961, 1965) compared with the standard Bullen *A*(*i*) model, as compared by Anderson (1967, with permission of Academic Press). (*b*) Three plausible, random density distributions obtained by the Monte Carlo procedure, compared with the Bullen model (after Press, 1968, with permission from Science. Copyright 1968 by the American Association for the Advancement of Science). Bullen curve, open circles.

profile is reached, indicating a discontinuity in density. A higher density is assumed for the material just below the discontinuity, and the successive calculations are continued. Limits are placed on the density discontinuities by the requirement that the complete density distribution must give the correct moment of inertia and average density.

Figure 3-9a shows the density distribution derived in a series of calculations by K. E. Bullen since 1936, assuming a density of 3.32 g/cm³ for the material immediately below the Moho (Figure 3-10). This is representative of all models based on Equation 3-6. The results shown for the core represent only one possible hypothesis, and there is still no generally accepted density distribution for this part of the Earth. Perhaps the most striking feature of the diagram is the abrupt increase in density at the mantle-core boundary, between layers D and E (Table 3-1). The assumptions necessary for these calculations depend upon preconceived models of the composition and structure of the Earth. It would obviously be advantageous to have more direct methods of determining the density at depth.

Two other density models are shown in Figure 3-9a. That of F. Birch is based on an assumption of linear relationship between density and the P-wave velocity. D. L. Anderson has shown how long-period surface wave dispersion and free oscillations of the Earth can yield densities in a direct manner,

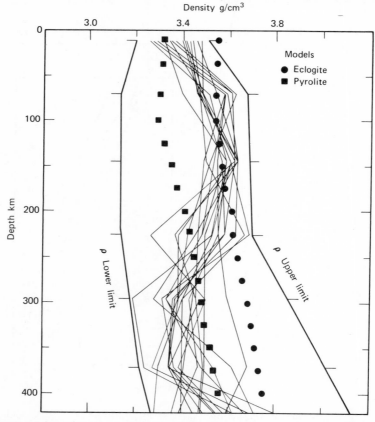

Figure 3-10. Successful density models for suboceanic upper mantle using Monte Carlo method; bounds define range permitted in selection. Points show density values according to Clark and Ringwood (1964) for "pyrolite" and eclogite models (after Press, 1969, with permission from Science. Copyright 1969 by the American Association for the Advancement of Science).

if the seismic wave velocities are completely and accurately known from body wave studies. One of his models is shown in Figure 3-9a. Note the low-density zones and the two regions of very rapid increase in density, compared with the seismic velocity profiles in Figure 3-5.

A completely different approach was adopted by F. Press, involving the Monte Carlo statistical procedure, which is quite independent of any of the usual assumptions. Approximately five million Earth models were randomly generated by feeding into the computer values for Earth parameters including P-wave velocity, S-wave velocity, density distribution in the mantle (which is limited between upper and lower bounds), core density, and core radius (limited within the range 3473 ± 25 km). These models were tested against available geophysical data including the mass of the Earth, its dimensionless moment of inertia, travel times of seismic waves, and Earth eigenperiods for free oscillation modes. Only six of the five million models satisfied the known geophysical data, and only three of these were plausible.

Every successful model required an increase in the radius of the Earth's core by 18 to 22 km. Figure 3-9b compares Bullen's standard density distribution with the three random distributions found to be plausible by Press's method. Inner core densities are significantly higher than in the standard model. In the mantle deeper than 1000 km the successful solutions have a surprisingly narrow spread of densities within the permitted range. Within the transition zone of the mantle, between about 400 and 1000 km depth, the density distributions exhibit higher density gradients than the standard model. Within the upper mantle the successful models show large variability, with surprisingly large density fluctuations, and with one or two density minima. The range of acceptable upper mantle densities, 3.34 to 3.54 g/cm^3, corresponds to the range covered by peridotite to eclogite. Applying these randomly generated density variations to existing concepts of

mantle composition leads to the hypothesis that the upper mantle is laterally and radially variable, with extensive zones, or layers, of eclogite within a peridotite mantle. The results also have implications for the composition of the core and mantle, which will be considered in Chapter 6.

A similar study gave the first independent determination of density in the suboceanic upper mantle, as shown in Figure 3-10. The successful models show that in the vicinity of 100 km the densities occupy a narrow band between 3.5 and 3.6 g/cm^3, in the upper part of the permissible range. In the depth range 250 to 400 km the models have densities reduced to the range 3.3 to 3.5 g/cm^3. The figure also shows densities computed by S. P. Clark and A. E. Ringwood for petrological models of mantle composed of either eclogite, or of a hypothetical peridotite, termed "pyrolite" (Chapter 6). The eclogite model alone is consistent with the results between 80 and 150 km, and the pyrolite model is favored in the region near 300 km. Press found a low velocity zone for shear waves in all successful models, with center between 150 and 250 km, beneath a lid about 100 km thick. He concluded that the mantle between 80 and 150 km, including the lower part of the lithosphere, consists of peridotite with about 50% eclogite.

Elastic Properties, Pressure, and Gravity

From the seismic wave velocities (Figures 3-3 and 3-5) and the density distribution (Figure 3-9), all of the elastic constants may be calculated. Their variation with depth is shown in Figure 3-11a. From the density distribution, the mass m of material within a sphere of radius r can be calculated, and this permits determination of the acceleration due to gravity at any level within the Earth from Equation 3-3, and it also gives the pressure distribution within the Earth, as in Equation 3-4. The pressure distribution is compared with the elastic constants in Figure 3-11a and with the gravity distribution in Figure 3-11*b*. Pressure increases continuously with depth. Gravity tends to increase below the surface of the Earth, but there is little change to a depth of 2400 km. It increases slowly from here to the mantle-core boundary, and within the core it decreases steadily to zero at the Earth's center. The figure also shows one estimate of the temperature distribution within the Earth, and therefore provides values for pressure and temperature at any selected depth.

The results shown in Figures 3-3, 3-5, 3-9, 3-10, and 3-11, depicting the variation with depth of seismic wave velocities, density, elastic properties, gravity, and pressure, include all that is known with any degree of certainty about the physical properties of the Earth's deep interior. No hypothesis concerning the mantle and core can be considered acceptable unless it is consistent with these facts.

Temperatures within the Earth

There is great uncertainty about the temperature distribution within the Earth, as shown by the selected estimates of geotherms for the outer mantle in Figure 3-12. The present thermal state of the Earth depends upon its thermal history. Hypotheses for the thermal history depend critically upon assumptions about the origin and early history of the Earth, the chemistry of material composing the Earth, and the physical properties of this material. Estimated geotherms therefore depend on these same assumptions.

Limits are placed on the temperature distribution by several factors. Surface heat flow measurements relate to the thermal gradients only within the upper few tens of kilometers. Inferences about temperature distributions at greater depths are based on observed seismic velocity profiles and variations in electrical conductivity, but these depend on assumptions about the physical properties of postulated mantle material at high pressures and temperatures. Seismic data show that the mantle is essentially crystalline and that the outer core is liquid. The temperature within the mantle is therefore below the solidus curve for mantle

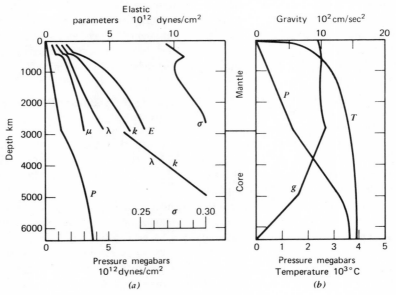

Figure 3-11. Properties of the Earth's interior. (*a*) Variation of pressure and elastic properties with depth. (*b*) Variation of pressure, gravity, and temperature with depth. (Various sources; summaries in Howell, 1969, and Jacobs *et al.*, 1959.)

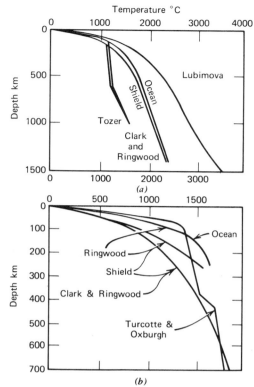

Figure 3-12. Estimates of temperature distribution with depth in the Earth.

material, but the fact that volcanoes exist and have erupted at frequent intervals and in most locations through the long span of geological time shows that the solidus temperature is locally exceeded. Therefore the geotherms probably remain close to the solidus for the upper mantle at least.

The present temperature distribution within the Earth depends upon its temperature when formed, the amount of heat generated as a function of depth and time, and the rate of outward flow of heat. It seems probable that significant cooling from the Earth's surface has not occurred much below 100 km. The Earth's initial temperature is not known, but current hypotheses favor accretion of a cold body (Chapter 6). Sources of internal heat include:

1. Gravitational energy converted to heat during formation of the Earth's core; heat from gravitational energy during accretion of the Earth was probably generated near the surface and radiated away into space.

2. Energy of friction arising from Earth tides, dissipated as thermal energy probably in zones of reduced strength, such as the low-velocity zone.

3. Decay of radioactive elements; this is the principal contribution to the Earth's thermal budget and the present heat flow.

Processes for the outward flow of heat from within the Earth include (a) conduction, (b) radiation, and (c) convection. The transfer of heat by thermal conduction is a very slow process in silicate rocks, but it is the main process in the outer layers of the Earth. At higher temperatures, possibly at depths of 150 km and more, radiation with exciton and photon transfer becomes a dominant mode of heat transfer. This could easily increase the effective conductivity by a factor of 10, but the magnitude of the effect is largely conjectural. Convection, if it occurs in the mantle, is an efficient mechanism for transferring deep-seated heat to the surface, and it provides a large increase in effective thermal conductivity. Heat is also transported directly to the surface by magmas and hydrothermal solutions, but this appears to represent a small fraction of the thermal budget.

Four different approaches have been used to estimate the temperature distribution in the Earth:

1. Methods based on deduction of the thermal history of the Earth.

2. Methods based on variations in physical quantities such as seismic wave velocities and electrical conductivity.

3. Methods based on geochemical and petrological data and inferences.

4. Methods based on convection models.

Figure 3-12 shows the range of temperature profiles deduced in recent years by some of these approaches.

The geotherm calculated by E. A. Lubimova in Figure 3-12a is based on conduction and the Earth's thermal history. Figure 3-11b shows the temperature distribution extended down to the Earth's center.

S. P. Clark and A. E. Ringwood developed a petrological model involving the formation of continents by vertical differentiation of the upper mantle. Because of the approximate equality of heat flow from oceans and continents, the temperature beneath the oceans is higher, at a given depth, than that beneath the continents, as shown in Figure 3-12a. The two geotherms must converge at depth. For consistency the model requires a very high thermal conductivity at high temperatures, and they concluded that this is produced by radiative transfer. Clark and Ringwood presented curves to 400 km and D. C. Tozer extended them to 1400 km. At 1000 km depth these curves give temperatures about 800°C lower than the conduction geotherm of Lubimova. Ringwood later modified these conduction-radiation geotherms by changing the contribution of radiative transfer. This had the effect of raising the geotherms by about 250°C at 250 km depth, as shown in Figure 3-12b. These are the geotherms used in the following chapters.

The effect of convection superimposed on the conduction-radiation model of Clark and Ringwood was examined by D. C. Tozer, with the results shown in Figure 3-12a. He first extended the geotherms from 400 to 1400 km, and concluded that for these temperature distributions there was a viscosity increase by a factor of 10^4 at a depth of 600 km, and that the mantle was convectively unstable above this depth. He therefore proceeded on the assumption that the lower boundary of a convecting layer occurred at 600 km; the upper boundary was taken as 50 km beneath the oceans and 150 km beneath the continental shields. The effect of convection is to lower the temperature of the bottom of the layer by about 600°C, according to Figure 3-12a. Below 600 km he calculated a new conduction equation solution producing the change in geothermal gradient shown in the figure. He found that the temperature distribution in the convecting region was very insensitive to the heat source distribution, which is why the geotherms for oceanic and shield regions are so similar.

The temperature distribution for another convection model, in the oceanic regions, is given in Figure 3-12b. D. L. Turcotte and R. Oxburgh considered a model for mantle convection using boundary-layer theory, with rates of flow estimated from magnetic anomaly data (Chapter 13). The temperature profile through the cold conduction boundary layer above the convection cell, at about 100 km from an oceanic ridge, was estimated by comparing measured values of heat flow as a function of distance from midoceanic ridges with theoretical models calculated using assumed mantle properties. This extends down to about 50 km. Below 150 km they concluded that the thermal gradient would be close to the adiabatic value, and they adopted an earlier proposal of B. Gutenberg that the geotherm in the upper mantle would follow a 1400°C adiabatic gradient. They matched the estimated boundary layer profile to a 1400°C adiabatic profile through the zone of main horizontal flow and into the slow moving core of the convection cell. The rather abrupt increase in temperature at about 400 km depth is due to the increase in adiabatic gradient through a phase transition zone (Chapter 6). At 350 km depth the estimated temperature is about 350°C higher than Tozer's convection model, and very close to the conduction-radiation model of Clark and Ringwood.

Viscosity

Estimates of the viscosity of the upper mantle have been based on the rates of uplift of large crustal regions, such as Fennoscandia after removal of its ice sheet. Values lie in the range 10^{21} to 10^{22} poises, increasing to about 2×10^{22} at a depth of about 1000 km. Two different interpretations for the cause of the nonequilibrium figure of the Earth give estimates for the average viscosity of the mantle as either 10^{26} poises, or 5×10^{22}

poises. Seismic anelasticity measurements of the mantle give estimates for the average viscosity of 2.4×10^{22} poises, with 8×10^{21} poises for the upper mantle and 10^{23} poises for the lower mantle.

There has been considerable discussion in the literature about whether the mantle behaves as a plastic body with a yield strength which would involve a power law or exponential dependence of strain rate on stress or whether it behaves as a viscous fluid with linear or nonlinear (Newtonian) relationships between stress and strain. Diffusion creep is likely to be a significant factor for deformation in the Earth's mantle at high temperatures, and assuming that diffusion creep alone occurs, Turcotte and Oxburgh used their temperature profile for the mantle (Figure 3-12*b*) to estimate the dependence of viscosity with depth. They found a distinct minimum in viscosity of 1.09×10^{21} poises near 100 km depth. A second minimum of 3.10×10^{21} poises at 425 km is due to the temperature increase associated with the phase transition. Other values tabulated include: 49 km, 3.9×10^{22} poises; 300 km, 7.75×10^{21} poises; 700 km, 1.18×10^{23} poises.

In 1970 N. L. Carter and H. G. Ave'Lallemant experimentally deformed dunite and peridotite at mantle pressures and temperatures and at constant strain rates. They found that the steady-state deformation is best fit by a power-creep equation, and extrapolation gives estimates of 10^{20} to 10^{21} poises for the viscosity over most of the upper mantle.

Uncertainties in estimates of mantle viscosity are confirmed by the range of estimates listed above. There is increasing evidence, however, for the existence of a low viscosity layer at depths corresponding to the zone of low seismic velocity in the upper mantle, and many estimates for this layer are about 10^{21} poises.

MAGNETISM, GRAVITY, AND HEAT FLOW

We have discussed the physical properties of different parts of the Earth's interior, and now we can review some physical properties of the Earth as a whole, which are measured at the surface. There is a magnetic field associated with the Earth which approximates to that of a dipole magnet with an axis slightly offset compared to the rotation axis. This is illustrated in Figure 3-13*a*, which shows also the distribution of the magnetic lines of force near the Earth's surface. The study of paleomagnetism and anomalies in the present magnetic field has contributed greatly to current concepts of global geology. The Earth has an external gravity field which arises from the distribution of mass within the Earth. Gravity variations or anomalies therefore provide information about the density in different parts of the Earth. Most major geophysical and geological processes are controlled by, or at least influenced by, the generation, distribution, and transfer of heat within the Earth. Measurement of surface heat flow thus provides a boundary condition of fundamental significance.

Measurements of the magnetic field, the gravity field, the heat flow, and similar compilations for the travel times of seismic waves, crustal thickness, and the surface topography of the Earth, can be expanded in terms of spherical harmonics. Then the correlation coefficients between the different sets of data can be compared with their harmonic coefficients. This provides the means for investigation of the structure and specifically the inhomogeneities in the Earth's mantle, as reviewed in 1969 by M. F. Toksöz, J. Arkani-Hamed, and C. A. Knight.

The Magnetic Field

At any point on the surface the Earth's magnetic field is defined by its strength and direction, which are usually expressed in

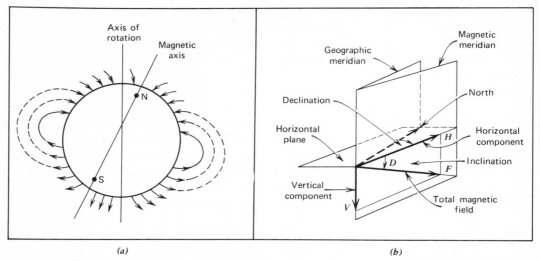

(a)   (b)

Figure 3-13. (a) Earth's magnetic field relative to axis of rotation and magnetic axis. (b) Horizontal and vertical components of the Earth's total magnetic field, resolved in the vertical plane through the magnetic meridian.

terms of the magnetic elements shown in Figure 3-13b. The total magnetic field is represented by a vector F making an angle of dip D (the inclination) with the horizontal. The vector is resolved into a horizontal component H and a vertical component V. The local magnetic meridian is the vertical plane containing F and H. The angle measured from the geographic meridian to the magnetic meridian is the declination or variation.

Isomagnetic maps show contours connecting points on the Earth's surface with equal values of the magnetic elements. Isogonic maps show declinations, isoclinic maps show inclinations, and isodynamic maps show magnetic intensity, either the total field or a resolved component. Figure 3-14a is an isoclinic map. The zero isocline, called the magnetic equator, is a curve quite close to the geographic equator. The angle of dip becomes vertical at two magnetic dip poles, presently located in northern Canada (73°N, 100°W) and Antarctica (68°S, 143°E); their positions change with time. Each dip pole is about 2300 km from the antipodal point of the other, and therefore a line joining the poles does not pass through the center of the Earth. The

magnetic field is strongest near the dip poles (0.7 oersted in the south) and weakest near the magnetic equator (about 0.3 oersted). The magnetic dip poles are not coincident with the geomagnetic poles which are defined below.

Magnetic anomalies. The irregularity of contours in isomagnetic maps shows that the concept of a single dipole (Figure 3-13a) is too simple. The observed magnetic field varies from place to place owing to the permeability and magnetization of rocks beneath the surface and to the ionosphere. The difference between the simple dipole field and the observed field is called the nondipole field or the magnetic anomaly. Anomalies are measured in gammas (γ), where $1\gamma = 10^{-5}$ oersted.

Small-scale anomalies are caused by near-surface features, and mapping local magnetic anomalies is a useful prospecting technique for ore bodies. In addition to these local anomalies there are large-scale anomalies extending over thousands of square kilometers. The anomaly of the vertical intensity V (Figure 3-13b) is shown in the isodynamic map of figure 3-14*b*. These anomalies must be caused by the properties of materials deep within the Earth.

The regular field which best approximates the Earth's field is obtained by spherical harmonic analysis of the observations for any year from all parts of the Earth, which resolves the field into components from internal (94%) and external sources. The potential of the total magnetic field is expanded in spherical harmonics. This leads to representation of the observed field by a number of magnetic dipoles, each with a different orientation, distributed at the center of the Earth as required by the coefficients of the harmonic analysis. The first and most powerful, which accounts for the regular dipole part of the Earth's field, is called the

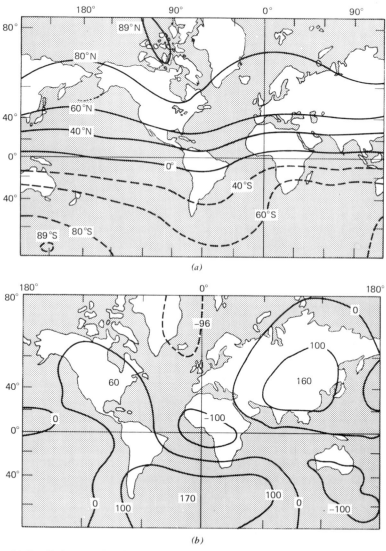

Figure 3-14. (*a*) Isoclinic map showing lines of equal magnetic inclination in 1945 (from Vestine *et al.*, 1947, courtesy of Carnegie Institution). (*b*) Map showing lines of equal anomalies of vertical magnetic intensities (*V* in Figure 3-13*b*) in 10⁻³ gauss (after Bullard *et al.*, 1950, with permission of author).

inclined geocentric dipole. The geomagnetic axis of this dipole makes an angle of 11.5° with the rotation axis and intersects the surface at the geomagnetic poles. The positions of these poles, unlike those of the dip poles, have changed little in the past century: 79°N, 70°W, and 79°S, 110°E. Latitude relative to this axis is called the geomagnetic latitude. The higher harmonics constitute the irregular or nondipole field, or the geomagnetic anomaly. Its strength is about 5% of the dipole field.

Variations with time. The intensity and direction of the magnetic field not only varies from place to place, but also from time to time. Direct records of the geomagnetic field have been kept at magnetic observatories during the last 400 years, and two distinct types of change have been recognized: short-term fluctuations and long-term or secular variations.

Short-term periodic variations with durations of hours, days, or years can be explained by electric currents in the mantle and ionosphere; intensity variations are of the order of 10 to 100γ. Irregular, transient fluctuations which may exceed $10^3\gamma$, called magnetic storms, probably result from bombardment of the ionosphere by solar radiation because they usually occur at times of sunspot activity.

Secular changes include changes in declination, inclination, and intensity. Isoporic maps show the distribution of isopors or lines joining places with equal rates of changes of the magnetic elements during a given time interval. Secular variations in intensity are in the range of several to tens of gammas per year. There appear to be cyclic changes in the direction of the field with periods of several hundreds of years, which produce a range of field directions of 10 to 20° on either side of the average. The magnitude of the change varies from one magnetic observatory to another, which indicates that the cause of this secular variation is regional rather than planetary. In addition to these directional changes of the geomagnetic field, the patterns illustrated by isomagnetic maps drift westwards. This effect is produced because the nondipole field drifts westward over the regular dipole field at a rate of 0.18° per year. If this rate remains constant, the map of intensity anomalies shown in Figure 3-14b would travel right around the world in 2000 years. Another type of secular variation is shown by the decrease of the Earth's magnetic moment by 5% during the last 100 years. If this rate continues, then the Earth's magnetic field would vanish in about 2000 years. The evidence indicates, however, that the field is probably oscillating rather than reversing.

Paleomagnetism, the study of fossil or remanent magnetism in rocks, provides the means of extending observations farther back in time than the 400 years available in direct records from observatories. Paleomagnetic studies show that the Earth's magnetic field has vanished often in the geological time scale, on each occasion when the Earth's polarity underwent a complete reversal. Changes in the attitude of the geomagnetic axis, loosely termed polar wandering, have also been recognized from the paleomagnetic record. Polar wandering and the correlation of magnetic reversals with linear magnetic anomalies parallel to the oceanic rifts are largely responsible for the revival of the concept of continental drift. We shall return to these topics in Chapters 12 and 13.

The only plausible explanation for the origin of the Earth's magnetic field involves electric currents in the fluid core, which are probably induced and maintained by some magneto-hydrodynamic phenomenon. The speed of the secular variations supports this dynamo theory. It is hardly conceivable that such rapid changes could occur in the solid mantle, but the times appear reasonable for convective motions in a fluid core. The secular variations may be considered as resulting from large-scale eddies near the surface of the core, and the westward drift of the whole magnetic pattern could possibly be explained by variable coupling between the core and the mantle. The problem is extremely complex.

The Gravity Field

The computed variation of the acceleration due to gravity within the Earth was shown in Figure 3-11*b*, and in this section we will consider variations in gravity as measured at the surface. The major component of gravity is the attraction between the Earth and a body at the surface and this is dominated by the whole Earth. There are, however many effects modifying the measured values. If the Earth were a perfect sphere whose density varied only as a function of distance from its center, gravity at its surface would be uniform, as given by Equation 3-3. The force of gravity would be directed toward the center of the Earth and perpendicular to its surface, which is therefore an equipotential surface—one that is everywhere horizontal. The gravity potential at the surface would be $-GM/R$, where M and R are, respectively, the mass and radius of the Earth. For this hypothetical Earth the geoid would be the spherical equipotential surface.

Because of the Earth's daily rotation, however, there is a centrifugal force accelerating the matter of the Earth in opposition to gravity, and this force is greater nearer the equator. This accounts for the equatorial bulge and the flattening of the poles. The figure of the Earth is approximated by a spheroid or an ellipsoid, and the sea level value of gravity varies from 978.049 cm/sec^2 (gals) at the equator to 983.221 cm/sec^2 at the poles. The theoretical value of gravity as a function of latitude (θ) on the ellipsoid is given by an International Gravity Formula:

$$g = 978.049(1 + 5.2884 \times 10^{-3} \sin^2 \theta - 5.9 \times 10^{-6} \sin^2 2\theta) \text{ gal} \quad (3\text{-}7)$$

Satellite studies have provided more precise values than surface gravity surveys, and following is the revised geodetic standard:

$$g = (978.03090 + 5.18552 \sin^2\theta - 0.00570 \sin^2 2\theta) \text{ gal} \quad (3\text{-}8)$$

Variations or anomalies in gravity measurements are measured in milligals (1 mgal = 0.001 gal) or the gravity unit: 1 g.u. = 0.1 mgal.

The surface of the oceans is almost an equipotential surface, with minor departures arising from the attraction of the water toward the continents. The geoid is the equipotential surface which most nearly approximates mean sea level, and this is the reference level for all gravity measurements. Figure 3-15 shows that the geoid is slightly distorted compared to the equilibrium ellipsoid.

The value of gravity itself is not constant on the equipotential geoid, because it varies with latitude according to Equation 3-8. Data for the equilibrium figure of the Earth have been obtained from geodetic measurements, and location of the geoid by gravity measurements provides another method.

Gravity corrections and anomalies.

The results of gravity measurements vary not only with position on the geoid, but also because the surface of the Earth is not everywhere at the same altitude. Equation 3-3 gives the variation of gravity with distance from the center of the Earth and, using this equation, measured values of gravity can be corrected for the altitude effect by changing the result to what the value would have been if the measurement were taken at sea level, considered equivalent to the geoid. This free-air correction is about 0.3086 mgal/m, although it varies with latitude and altitude. This correction takes no account of the nature of the material between the geoid and the point of measurement. If the material were air, the free-air correction should correct the gravity measurement to the value appropriate for this point on the geoid. The difference between the theoretical value from Equation 3-7 and the corrected value is called the free-air anomaly.

A part of the free-air anomaly is caused by the attraction of land above sea level, or by the comparative lack of attraction of sea

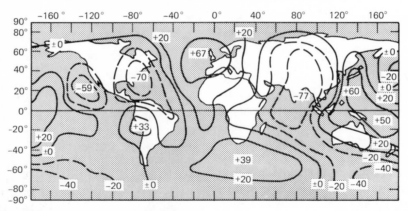

Figure 3-15. Satellite geoid map with contours in meters, representing the zonal and non-zonal gravity harmonics up to order 8. (From W. H. Guier and R. R. Newton, Jour. Geophys. Res., **70**, 4613, 1965, with permission; as presented by Schmucker, 1969.)

water in ocean basins. The Bouguer correction allows for the effect of the material between the elevation of the measuring station and the geoid, and additional topographic corrections can be made to allow for the effect of neighboring mountains and valleys. For rocks with average density 2.67 g/cm³, the Bouguer correction is 0.1119 mgal/m. The free-air and Bouguer corrections are always of opposite sign. The combined correction for an increase in elevation of 1 m is a reduction of the measured value by 0.1967 mgal.

The free-air anomaly is a measure of the effects of mass excesses and deficiencies within the Earth. Subtraction of the topographic correction and the Bouguer correction eliminates that part of the free-air anomaly due to elevation and topography. The remaining Bouguer anomaly, the difference between the corrected measurement of gravity and the value given by the International Gravity Formula (Equation 3-7), represents the effect of buried rocks of greater or less than average density, or of geological structures involving rocks of contrasting densities. Study of these anomalies has proved very useful in determining crustal structure, although no unique solution can be obtained from gravity data alone, because an infinite number of mass distributions at depth are capable of producing the same anomaly at the surface.

Consideration of Bouguer anomalies on a worldwide scale indicates that gravity is less than average in high areas and greater than average over oceanic areas, which indicates that rocks beneath high areas have relatively low density compared to rocks beneath oceanic areas. This relationship is explained by the law of isostasy, which can be stated thus: all large land masses on the Earth's surface tend to sink or rise so that, given time for adjustment to occur, their masses are hydrostatically supported from below, except where local stresses are acting to upset equilibrium. We can suppose that there is the same weight of material in each vertical column of rock above some level of compensation within the Earth. When the rock is lighter the surface is higher as for the continents, and when the rock is denser the surface is lower as in the ocean basins. Airy's theory of isostatic compensation is illustrated schematically in Figure 10-1, with the level of compensation situated at some depth beneath the deepest "crustal" column. Using either Airy's theory or Pratt's theory isostatic corrections for gravity measurements can be calculated. Subtraction of these from the Bouguer anomalies leaves a remainder which is called the isostatic anomaly; this is usually much smaller than the free-air and Bouguer anomalies. The assumption of hydrostatic equilibrium at some depth, of the order of 100 km, thus appears to be justified; but large isostatic anomalies do exist, indicating that portions of the Earth's crust are not in isostatic equilibrium.

The gravity anomalies discussed above are differences between corrected gravity measurements and the theoretical gravity calculated from the International Gravity Formula. This assumes that the geoid is a regular ellipsoid. In fact the geoid is slightly distorted. The amount of distortion is very small and it appears certain that the distortions are caused by density variations in the mantle rather than in the crust. Crustal effects have been compensated for by the free-air, Bouguer, and isostatic corrections. At present surface gravity measurements can tell us nothing about the deeper mantle.

Geoid heights from satellite data.
Since the first artificial satellite was launched in 1957 there have been many determinations of the Earth's geoid by satellite observations. A satellite samples all longitudes and a range of latitudes as the Earth rotates beneath it, and its path is affected by the irregularities in the gravity field produced by irregular mass distributions within the Earth. Satellite data provide no information about variations of gravity on the scale associated with physiographic features of the Earth's surface but they do detect large-scale anomalies in the Earth's gravity field.

The expression for the gravitational potential outside of the Earth is usually written in terms of spherical harmonics, and the equation includes coefficients describing the gravity field of the Earth. The satellite data reveal departures from the gravity field that would be associated with hydrostatic equilibrium within the Earth. It appears that the sources of some of the observed harmonics, and perhaps of most, must lie within the mantle.

The satellite data may be expressed interchangeably in terms of gravitational potential, surface gravity, or geoid shape. Figure 3-15

shows a convenient representation with the surface of the geoid contoured at intervals in meters above and below the standard ellipsoid, or theoretical geoid. Various authors have used different methods for determination of the geoid from satellite data; results have been obtained which differ by 40 m or more in some locations, but we can expect an accuracy to within 10 m in future work. Combinations of satellite and surface gravimetric data are now being used to obtain a more complete picture of the gravitational field of the Earth than has been obtained from satellite data alone.

Depressions of the geoid surface are regions of negative free-air gravity, and elevations of the geoid surface are regions of positive free-air gravity. These anomalies are probably produced by lateral density differences within the mantle; they bear no direct relation to the distribution of continents and oceans. One objective of the study of gravitational evidence is to set limits to the depth within the Earth that lateral density variations do occur. The density variations could result from lateral changes in composition or in temperature. Interpretations of the geoid distortions, and the probable density variations in the mantle, require comparison of the geoid with the results of other physical measurements of the Earth, such as heat flow, travel-time anomalies for seismic waves, and spherical harmonic analysis of major tectonic features.

Heat Flow

The present rate of heat loss from the Earth's surface is about 2.4×10^{20} cal/year. In terms of energy this outflow of heat is the most impressive of geophysical processes; the more spectacular energy loss involved in earthquakes and volcanic activity is orders of magnitude less.

Methods of measurement.
Despite its significance, satisfactory procedures for the measurement of heat flow have been developed only recently. In 1945 there were about

20 measurements, all on land. The design of apparatus for measuring heat flow through the ocean floor, together with the increased effort in marine geophysics, improved the situation greatly. By 1960 there were about 200 measurements, by the end of 1964 there were at least 1000 measurements published, and by July 1968 about 2600 results were available, with new values being reported at the rate of 500 per year. About 90% of all measurements were made at sea. Taking into consideration the areas of land and ocean (Table 2-1), this is equivalent to about three times more data per unit area of ocean than of land. Unfortunately the measurements on land are very unevenly distributed, and there are large gaps in continental and high-latitude regions.

The measurement of heat flow through the ocean floor, initiated by E. C. Bullard, is now a routine procedure, but the measurement of heat flow through land is still difficult. It is necessary to drill holes about 300 m deep, which is a very expensive business. Critical studies are required to find the optimum depths, hole size, horizontal spacing, instrumentation, and procedure for large-scale surveys of heat flow on continents. We can expect major efforts by groups such as the International Upper Mantle Committee to provide many more data in continental areas in the future, data which are critically needed to test geophysical hypotheses.

If the Earth's surface layer were an ideal uniform rock, the geothermal flux q would be provided by a measurement of the thermal gradient in the rock (two temperature measurements at distance z apart in a vertical direction), and the thermal conductivity K.

$$q = K \left(\frac{\partial T}{\partial z} \right) \qquad (3\text{-}9)$$

There are many factors that may affect underground temperatures: the circulation of underground water, including the migration of interstitial water in porous sediments on the ocean floor; vertical and lateral variations in thermal conductivity; the effects of topography, uplift and erosion, and recent glaciations; the effects of magmatic events; the effect of variable movements of water over the ocean floor; disturbances produced by drilling the hole on land or by penetration of the probe into ocean-floor sediments and by ventilation cooling in mines. Corrections may be applied for some of these effects.

On land the first measurements should be at a depth of at least 30 m in order to avoid the effect of annual temperature variations at the surface. The most difficult problem is determination of a good mean value for the thermal conductivity of the rocks within the vertical interval. The conductivity is low, and it may vary by tens of percent from one sample to another taken from an apparently uniform rock. Errors in individual heat flow measurements under favorable conditions at sea are usually no more than 10%, but they can reach 20%. The quality of heat-flow measurements on land varies from crude estimates to elaborate determinations. If the mean heat flow of a region is computed from more than 10 observations, it is believed that heat-flow variations among regions are significant if they exceed 0.2 μcal/cm²sec.

Review of data. The heat-flow data listed in Tables 3-2 and 3-3, and plotted in Figures 3-16, 3-17, 3-18, and 3-19 are derived largely from three reviews by W. H. K. Lee and S. Uyeda in 1965, R. W. Girdler in 1967, and R. P. Von Herzen and W. H. K. Lee in 1969. Table 3-2 gives the data for oceans and continents and shows that statistical analyses of 2584 selected heat-flow measurements in 1969 do not differ significantly from the analyses of fewer than half this number of measurements in 1965. Table 3-3 shows the values and statistics for major physiographic provinces.

Figure 3-16 compares the histograms of heat-flow measurements for continents and oceans. This demonstrates that the heat flows in these two environments are essentially the same (Table 3-2), with arithmetic means

TABLE 3-2 Selected Heat-Flow Values for Oceans and Continents: Individual Values and Grid Averages (9 × 10⁴ square nautical miles per grid element)[a]

Region	Lee and Uyeda, 1965			Von Herzen and Lee, 1969 (for July, 1968)		
	N	$\bar{q}$	S.D.	N	$\bar{q}$	S.D.
All continents	131	1.43	0.56	255	1.49	0.54
All oceans	913	1.60	1.18	2329	1.65	1.14
Atlantic	206	1.29	1.00	406	1.43	1.07
Indian	210	1.47	0.89	331	1.44	1.09
Pacific	497	1.79	1.31	1232	1.71	1.24
Arctic				29	1.23	0.33
Mediterranean seas	b			71	1.33	0.89
Marginal seas	b			260	2.13	0.63
				Girdler, 1967		
Africa	13	1.20	0.20	13	1.20	0.21
Asia[c]	37	1.49	0.58			
Japan				38	2.21	2.73
Australia	19	1.75	0.62	20	1.76	0.62
Europe	22	1.62	0.60	31	1.91	1.70
N. America	40	1.19	0.44	44	1.26	0.57
Grid Averages				Von Herzen and Lee, 1969		
All continents	51	1.41	0.52	79	1.45	0.47
All oceans	340	1.42	0.78	577	1.46	0.78
Atlantic	65[d]	1.21	0.64	127	1.32	0.56
Indian	94[d]	1.35	0.67	110	1.36	0.82
Pacific	181[d]	1.53	0.87	305	1.50	0.78

[a] N = number of observations. $\bar{q}$ = arithmetic mean of heat flow in $\mu cal/cm^2 sec$ (HFU). S.D. = standard deviation in HFU (heat-flow units).
[b] The few data available were included under oceans.
[c] Almost all data from Japan.
[d] Includes data from Mediterranean and marginal seas.

differing by less than 0.2 HFU (1 HFU = 1 heat flow unit = 1 $\mu cal/cm^2 sec$). Despite the absence of measurements from large areas, it appears that more than 99% of the Earth's surface can be considered normal, with heat flow varying around 1.5 HFU; the remainder represents thermal areas with heat flux many times greater. This is indicated by the positive skew distributions for the histograms in Figures 3-16a and c. The mode for each of these two distributions is 1.1 HFU. Mass transfer of material bringing energy from deep sources is required to explain the localized thermal areas. The arithmetic mean of the world heat-flow measurements is 1.58 HFU, with standard deviation 1.14, and a mode for the histogram of 1.1.

For comparing heat-flow over large regions it is useful to represent the measurements in such a way that the effect of localized high values is reduced. The logarithmic heat-flow values give distributions closer to normal frequency curves, as shown in Figures 3-16b and d. Girdler suggested that geometric

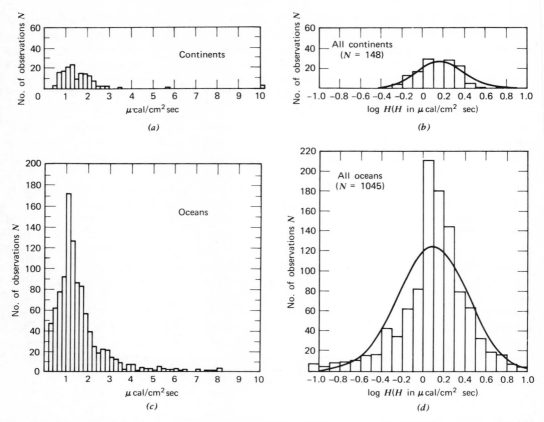

Figure 3-16. Histograms of heat flow observations for continents (*a*) and oceans (*c*) compared with histograms and normal frequency curves for the logarithms of heat flow values for all continental areas (*b*) and all oceanic areas (*d*) (after Girdler, 1967, with permission of Interscience).

means of heat-flow values should be considered as well as arithmetic means before making deductions concerning heat flow over large areas of the Earth's surface. The geometric means are smaller than the arithmetic means.

Lee and Uyeda introduced another method for reducing the bias of localized high heat-flow values, by working out the average heat flow for grid elements of equal area. Table 3-2 shows the statistics of averages for grid elements of 9×10^4 square nautical miles, or $5 \times 5°$ at the equator. The arithmetic mean and the standard deviation for the grid averages are smaller than for the original data, but the averages confirm the equality of heat flow through continent and ocean

floors. As noted in connection with Figure 3-12, this appears to require a different vertical distribution of radioactive materials within the mantle and crust beneath continents and oceans, and implies deep-seated differences in temperature between the two environments; unless convection in the mantle equalizes the temperature distributions.

Statistics of heat-flow values for the continents are listed in Table 3-2, but the small numbers and the irregular geographical distributions of the measurements permit no conclusions to be drawn. Similar statistics for the various oceans are also tabulated. The mean and scattering of the values increases from the Atlantic to the Indian to the Pacific, as shown by the histograms presented by Lee and Uyeda. These differences could be related to the different structures and histories of the oceans, but they could also be caused by nonuniform sampling of measurements.

Lee and Uyeda also presented histograms for heat-flow values associated with major geological features on land and beneath the ocean and found good correlation of heat flow with tectonic environments. The statistics are given in Table 3-3. Heat-flow values for shield areas are uniform and low, in accord with their long history of stability. The average heat flow in post-Precambrian orogenic and nonorogenic areas is essentially the same, but values in the orogenic areas are more scattered. If data from the Cenozoic volcanic areas were included with the orogenic areas, the average orogenic heat flow would be higher. Heat-flow values in the ocean basins are fairly uniform and low, with a mean of 1.28 HFU, whereas those in the ridges and other areas are higher and more scattered. The crestal zones of ridges are characterized by high heat flow and large variations, whereas relatively low values are observed over the flanks. This is shown in Figure 3-17a.

The high values for Japan represent the only extensive study among island arc areas, and the large standard deviation indicates the wide dispersion of values. The average heat flow in oceanic trenches is only 0.99 HFU, but the number of measurements is very small. The profiles of heat-flow values across island arc and ocean-trench systems are illustrated in Figures 3-17b and c. The lowest values are near the trench axis, and the high values are on the continental side of the arcs. These profiles correlate with the distribution of epicenters of earthquakes. The shallow focus earthquakes are concentrated in the belt of low heat flow, and the heat flow is high where the earthquakes have focal depths greater than about 150 km. This is coincident also with the oceanward limit of the active volcanoes.

Global analysis. A set of heat-flow data, in which each measurement point is expressed in terms of its latitude and longitude, can be fitted by a least squares method to a spherical harmonic expansion. This procedure provides an objective contour map of isoflux that best fits the observations. A global representation

TABLE 3-3 Statistics of Selected Heat-Flow Values for Major Physiographic Provinces.[a]
(Lee and Uyeda, 1965)

Province	N	$\bar{q}$	S.D.	S.E.	Modes
Ocean basins	273	1.28	0.53	0.03	1.1
Ocean ridges	338	1.82	1.56	0.09	1.1
Ocean trenches	21	0.99	0.61	0.13	1.1
Other seas	281	1.71	1.05	0.06	1.1
Precambrian shields	26	0.92	0.17	0.03	0.9
Phanerozoic nonorogenic	23	1.54	0.38	0.08	1.3
Phanerozoic orogenic areas[b]	68	1.48	0.56	0.07	1.1
Paleozoic orogenic areas	21	1.23	0.40	0.09	1.1
Mesozoic-Cenozoic orogenic	19	1.92	0.49	0.11	1.9, 2.1
Island arc areas	28	1.36	0.54	0.10	1.1
Cenozoic volcanic areas[c]	11	2.16	0.46	0.14	2.1

[a] N = number of observations. $\bar{q}$ = arithmetic mean of heat flow in μcal/cm²sec (HFU). S.D. = standard deviation in HFU (heat-flow units). S.E. = standard error in HFU.
[b] Excluding Cenozoic volcanic areas.
[c] Excluding geothermal areas.

of the heat-flow field up to third-order spherical harmonics is given in Figure 3-18. Contour lines over regions in which no data exist are dashed. The analysis averages out the small-scale variations in the heat-flow field, and for a third-order spherical harmonic analysis variations on a scale less than 2000 km do not appear. The contour map shows a general heat-flow pattern, with low values in the central Pacific and in the Atlantic and high values in the eastern Pacific and east Africa; the African high is uncertain because there are few measurements there. The heat-flow measurements are so un-

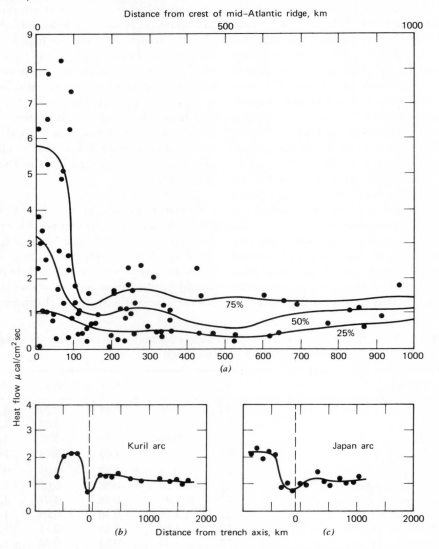

Figure 3-17. (*a*) Heat flow values versus distance from the crest of the mid-Atlantic Ridge; 75,- 50-, and 25-percentile lines are given. The 50-percentile line separates half the data points above and half the data points below it (after W. H. K. Lee and S. Uyeda, Geophysical Monograph **8**, 1965, with permission of Amer. Geophys. Union). (*b*) and (*c*) Profiles of heat flow values across island arcs, averaged in 100-km intervals (after Vacquier *et al.*, 1966; from Geophysical Monograph **12**, 349, 1968, with permission of Amer. Geophys. Union).

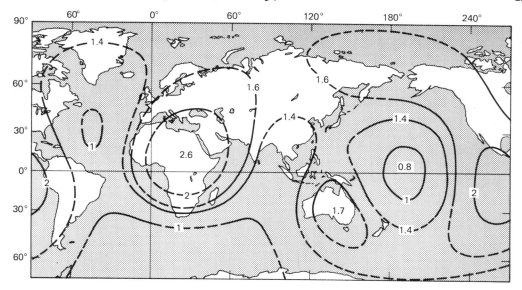

Figure 3-18. Orthogonal function representation (to third order spherical harmonics) of 987 heat flow values. Contour lines are in μ cal/cm² and are dashed over regions where no data exist (after W. H. K. Lee and S. Uyeda, Geophysical Monograph **8**, 1965, with permission of Amer. Geophys. Union; see Toksöz *et al.*, 1969, Figure 3*b*, for more recent map of heat flow variations).

evenly distributed across the Earth's surface that the use of spherical harmonic analysis could be misleading. There is a tendency for anomalies to be produced when there are no observations, as indicated by the large positive anomaly over Africa; much of this region is Precambrian shield, where low heat flow might be expected (Table 3-3).

The heat-flow map with all small-scale anomalies removed is related to properties within the mantle in a fashion similar to the map of the geoid in Figure 3-15. It has been proposed that a correlation exists between these two maps, with heat flow being low where gravity is high and vice versa. Girdler examined this proposition by assigning heat-flow measurements directly to regions of positive and negative gravity contoured on the geoid map (Figure 3-15). For each region he prepared a histogram of logarithmic heat-flow values to reduce the effects of local anomalies, and he compared the arithmetic and geometric means of each histogram with the world means. The overall results are plotted in Figure 3-19. The arithmetic mean heat flow for all four regions of negative

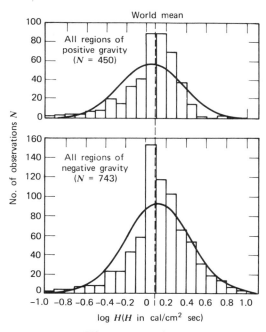

Figure 3-19. Histograms and normal frequency curves of the logarithmic values of heat flow for all regions of negative gravity and all regions of positive gravity for Guier and Newton's satellite geoid of figure 3-15 (after Girdler, 1967, with permission of Interscience).

geoid height is 1.67 HFU and that for all four regions of positive geoid height is 1.33 HFU. Girdler concluded that these results tend to support the hypothesis that undulations of the geoid are due to temperature differences in the mantle. The correlation is not well established, however, because of the uncertainty of our knowledge of both the heat-flow and gravity fields. Others have concluded that heat flow variations are not at all correlated with geopotential variations (e.g., Toksöz et al., 1969). We should also note that the gravity field represents the present mass distribution, whereas the surface heat-flow field may lag millions of years in indicating the temperature distribution at some depth in the Earth's interior because of slow thermal conduction. The thermal time scale for a layer 500 km thick is of the order of 10^9 years or longer. Convection in the mantle would reduce the time scale.

4. *Geological Processes and the Geological Time Scale*

INTRODUCTION

The aim of geology, and of the geophysical sciences in general, is to determine how and why the physical world became what it is today. We seek answers to questions such as these:

1. How did the continents form?
2. Why do they have their present distributions?
3. How is it that they persist?

In order to answer these questions we have to study the major features of the Earth's crust and to consider the geological processes involved in their formation and development. We have to examine carefully the rocks to see what clues they provide about their origins, and therefore we have to study the minerals composing the rocks. The study of minerals requires consideration of matter on an atomic scale. We discover that crustal processes have sources within the mantle, and that these processes can be related to mineralogical and phase changes occurring on an atomic scale and driven by thermal energy. Therefore, in order to comprehend the processes involved in large scale phenomena such as the formation of the great mountain ranges and the eruption of lavas from volcanoes and in floods, we have to study the materials of the Earth on all scales from the large dimensions considered in Chapter 2 down to atomic dimensions.

In Chapter 2 we examined the major features of the Earth as they exist at the present time. We know that despite their apparent permanence on the time scale of human life and civilizations, these are only transient features when considered in terms of the geological time scale. The mightiest mountains have been uplifted from beneath the oceans, and the low-lying continental shields were once traversed by rugged mountain ranges. Geology is a historical subject, and its study involves elucidation of the processes operating now, deduction of the processes operating during the past, determination of the effects of these processes operating over long periods of time, and their correlation with the stages of Earth history indicated in the stratigraphic record. This chapter is concerned with geological processes, and most of them are introduced and related to each other within the framework of the Rock Cycle, the Hydrologic Cycle, and the Tectonic Cycle. These cycles, combined within the dimensional framework of the geological time scale, provide an outline review of geology, and they introduce most of the subjects usually presented as separate courses in standard geological curricula. Subsequent chapters are more concerned with major physical units of the Earth or with major processes than they are concerned with conventional subjects.

GEOLOGICAL CYCLES

The major features of the Earth's surface are produced by the interactions of processes driven by two energy sources: the heat within the Earth is an internal source and the solar radiation is an external source. The internal source provides the energy for igneous, metamorphic, and tectonic activity, and the external source provides the energy dominating the circulation of the atmosphere and hydrosphere, the processes of weathering and erosion, and the formation of sedimentary rocks. These processes are connected through the Geological Cycle whose main features are summarized in Figure 4-1. Igneous, sedimentary, and metamorphic processes are reviewed in more detail in Figures 4-5, 4-6, and 4-7.

The Geological Cycle really consists of a number of interrelated cycles. (a) The Rock Cycle traces the relationships among igneous, sedimentary, and metamorphic rocks; but this relationship exists only because of the processes involved in (b) the Hydrologic Cycle which governs erosion and sedimentation, and those involved in (c) the Tectonic Cycle—specifically, subsidence and uplift. (d) The Geochemical Cycle is concerned with the migration of elements within the concentric shells of the Earth, from one shell to another and from one cycle to another.

Figure 4-1 shows that the Geological Cycle is not closed. New material is added by uprise of magma generated in the mantle at depths of 50 km or more. The additional material provides silicates for the Rock Cycle, as well as water and other volatile materials for the Hydrologic Cycle. There is evidence that the Earth's crust, hydrosphere, and atmosphere may have formed progressively, as a result of

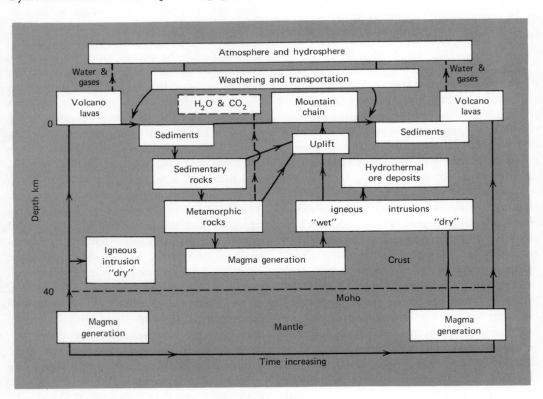

Figure 4-1. The geological cycle.

repeated magmatic activity, which leads to outward migration of the more fusible elements. The process is referred to as degassing or defluidization of the Earth.

We consider the Geological Cycle and the Rock Cycle to begin with the eruption of mantle-derived lava to form volcanoes, as shown in the left-hand side of Figure 4-1. Addition of water and other gases to the atmosphere and hydrosphere is indicated. Note also that some of the magma may be trapped in the crust to form a relatively dry igneous intrusion. Some of the water given off by volcanoes is undoubtedly juvenile, carried upward from the mantle, but at least a portion of it must be meteoritic, derived from water in the crust or from ocean water in the case of submarine eruptions and oceanic volcanoes.

As soon as a volcano emerges it is attacked by the atmosphere and hydrosphere. A series of physical and chemical processes disintegrates the exposed rocks, and the products of weathering are transported to lower altitudes where eventually they are deposited as sediments. Organisms and material of the biosphere may be involved in sediment formation in several different ways.

With continuation of the cycle sediments may become deeply buried and converted to sedimentary rocks by compaction, cementation, and recrystallization. If the Tectonic Cycle operates, the rocks are carried to deeper levels within the crust, where they are subjected to crushing, deformation, folding, recrystallization, and possibly reconstitution by metasomatism. Reactions occurring during progressive metamorphism cause evolution of carbon dioxide and water which migrate toward the surface. Regional metamorphism may culminate in partial fusion or anatexis of the rocks with the formation of migmatites and magmas. These magmas probably contain more dissolved water than mantle-derived magmas, and they are distinguished as "wet" intrusions in Figure 4-1. Crystallization of either "wet" magmas, or of relatively "dry" magmas from the mantle, leads eventually to the evolution of a dilute aqueous gas phase

which condenses at lower temperatures to a hydrothermal solution. The pneumatolytic gas phase and hydrothermal solution are products of an efficient geochemical process that concentrates certain elements. Precipitation of these elements in appropriate physicochemical environments may yield ore deposits.

The Tectonic Cycle continues with uplift and the formation of mountains composed of the folded and metamorphosed sediments and igneous intrusions. Figure 4-1 shows that the simple cycle of subsidence followed by uplift may be interrupted by periods of uplift and erosion at any stage in the cycle; these interruptions are represented in rock records by unconformities. The cycle is repeated with the deposition of sediments whenever significant uplift occurs, although the deposition of sediments is not always followed by deep burial and metamorphism. An Orogenic Cycle includes the whole sequence of cycles involved in the formation of a mountain chain (Chapter 9).

Normally the Hydrologic Cycle is considered with respect to the migration of meteoritic water through the atmosphere, the hydrosphere, and the uppermost part of the crust. A representation of the processes involved is shown in Figure 4-2. As shown in the figure, however, this cycle has direct connections with the deeper part of the Rock Cycle. Not only is juvenile water added from magmas of mantle origin but also meteoritic water is carried out of the main cycle for long periods of time, when sediments containing hydrous minerals and pore fluids are carried deep into the crust during a Tectonic Cycle. The water may return to the near-surface cycle via different routes. One route is simply by uplift and exposure through erosion of the overlying rocks. Progressive metamorphism of the buried sediment causes dehydration, and some of the released water migrates back to the surface. If anatexis occurs, the water dissolves in the magma and it may then be returned to the surface by volcanic activity or by upward migration of hydrothermal fluids emanating from the igneous intrusions.

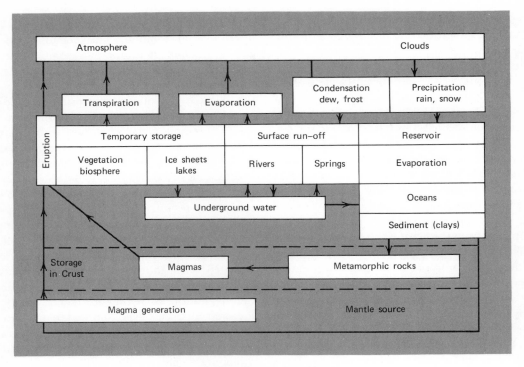

Figure 4-2. The hydrologic cycle.

GEOLOGICAL PROCESSES

Geological processes form rocks. Rocks are chemical systems adjusting to their physical environment in attempts to attain a state of thermodynamic equilibrium. As a basis for review of geological processes we should therefore examine the chemistry of the materials composing rock systems, the range of physical conditions to which they are subjected in the Geological Cycle, and the types of chemical reaction that result.

Rock Materials

Representatives of the main rock types involved at various stages of the Rock Cycle are listed in Table 4-1. The rocks are divided into four groups. The first constitutes the mantle material from which the original magma is derived, and the second includes the igneous rocks derived from the parental magma or magmas. Sediments constitute the third group,

and our main interest is in those occurring in orogenic environments, which subsequently become metamorphosed to form the fourth group. The compositions of metamorphic rocks cover the range of igneous and sedimentary rocks. Representative minerals composing these rocks are listed in the table.

For review of the types of chemical reaction experienced by rock systems we can group the minerals into three classes: (a) anhydrous silicate minerals stable only at high pressures (e.g. pyrope, jadeite); (b) anhydrous silicate minerals typical of crustal rocks (e.g. pyroxenes, feldspars, quartz); (c) hydrated and carbonated minerals (e.g. clay minerals, micas, hornblende, calcite).

Mineral Reactions

Magmas are generated in the mantle at depths to 100 km or more, corresponding to

TABLE 4-1   Selected Materials Involved in the Rock Cycle

	Rocks	Representative Minerals	
		Anhydrous	Hydrated and Carbonated
Upper mantle source	Peridotite Eclogite	Olivine Pyroxenes Garnet Spinel	Amphibole Phlogopite Clinohumite
Igneous rocks	Peridotite Gabbro Diorite Granite	Olivine Pyroxenes Feldspars Quartz	Amphibole Biotite Muscovite
Sedimentary rocks	Sandstone Arkose Greywacke Shale Limestones Evaporites	Quartz Feldspars Salts	Kaolin Clay minerals Chlorite Carbonates Salts
Metamorphic rocks	Schists Gneisses Marbles	Quartz Feldspars Garnet Pyroxenes Aluminosilicates Cordierite	Muscovite Chlorite Serpentine Biotite Epidote Staurolite Hornblende

pressures up to and greater than 30 kb and temperatures of the order of 1500°C. Conditions are less extreme in the Rock Cycle itself, which is confined to the range of crustal conditions with pressures up to 10 kb and temperatures rarely reaching 1000°C. There are three types of reactions illustrated in Figure 4-3 which occur within this range of pressures and temperatures: (a) polymorphic transitions and solid-solid reactions; (b) dissociation reactions: dehydration and decarbonation; (c) Melting reactions, either dry or in the presence of water under pressure.

Many simple mineral reactions are univariant, but in complex rock systems the reactions may be multivariant. These various types of reactions are represented on PT projections by curves or bands with different attitudes in terms of the P and T axes.

Figure 4-3a for the one-component system SiO_2 illustrates two kinds of polymorphic transition. The transition of quartz to coesite is pressure sensitive, being induced by increased pressure. In contrast the transition of α-quartz to β-quartz is temperature sensitive, the temperature of the transition being not greatly affected by change in pressure.

Pressure-sensitive reactions such as those illustrated in Figures 4-3b and 5-13 justify our division of anhydrous silicate minerals into groups (a) and (b) in the preceding sections. Plagioclase feldspar is a dominant constituent of crustal rocks, but Figures 4-3b and 5-11 show that it breaks down at high pressures as shown by univariant curves for the reactions:

$$\text{NaAlSi}_3\text{O}_8 \rightleftharpoons \text{NaAlSi}_2\text{O}_6 + \text{SiO}_2 \quad (4\text{-}1)$$
albite (Ab)   jadeite (Jd)   quartz (Qz)

$$3\text{CaAl}_2\text{Si}_2\text{O}_8 \rightleftharpoons \text{Ca}_3\text{Al}_2\text{Si}_3\text{O}_{12} + 2\text{Al}_2\text{SiO}_5$$
anorthite (An)   grossularite (Gr)   kyanite (Ky)

$$+ \text{SiO}_2 \quad (4\text{-}2)$$
quartz (Qz)

Figures 4-3*b* and 5-11 also show that the magnesian garnet, pyrope, is stable only at mantle pressures; the reaction plotted shows its formation from aluminous enstatite (En), sapphirine (Sa), and sillimanite (Si). In a complex rock system similar reactions occur through a pressure interval as shown in Figure 5-15, where the low pressure silicate assemblage of gabbro (augite + plagioclase) is converted to the high pressure assemblage of eclogite (omphacite + pyrope-rich garnet).

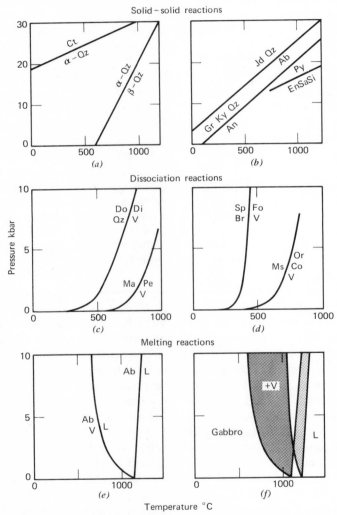

Figure 4-3. Types of reactions occurring in metamorphic and igneous processes. See text for details of reactions and abbreviations. Sources of data: (*a*) Coesite, Boettcher and Wyllie, 1968; β-quartz, Cohen and Klement, 1967. (*b*) Pyrope, Boyd and England, 1962; anorthite, Newton, 1966; jadeite, Boettcher and Wyllie, 1968. (*c*) Diopside, Turner, 1968, p. 144; periclase, Goldsmith and Heard, 1961. (*d*) Serpentine, Scarfe and Wyllie, 1967, and Johannes, 1968; muscovite, Evans, 1965, and Lambert *et al.*, 1969. (*e*) Boettcher and Wyllie, 1969, for data and review. (*f*) See Figure 6-12.

See also Figure 6-10 for transformations in peridotite.

The hydrated and carbonated minerals are abundant in sediments and many metamorphic rocks. During the progressive metamorphism of sedimentary rocks water and carbon dioxide are driven off by successive dehydration and decarbonation reactions as illustrated in Figures 4-3c and d. A dissociation reaction may involve the breakdown of a single mineral as illustrated by magnesite and muscovite:

$$\underset{\text{magnesite (Ma)}}{MgCO_3} \rightleftharpoons \underset{\text{periclase (Pe)}}{MgO} + \underset{\text{vapor (V)}}{CO_2} \quad (4\text{-}3)$$

$$\underset{\text{muscovite (Ms)}}{KAl_3Si_3O_{10}(OH)_2} \rightleftharpoons \underset{\text{orthoclase (Or)}}{KAlSi_3O_8}$$

$$+ \underset{\text{corundum (Co)}}{Al_2O_3} + \underset{\text{vapor (V)}}{H_2O} \quad (4\text{-}4)$$

In rock systems, however, it is more usual for reactions to involve at least one additional mineral among the reactants, as for the following reactions plotted in Figures 3-3c and d:

$$\underset{\text{dolomite (Do)}}{CaMg(CO_3)_2} + \underset{\text{quartz (Qz)}}{2SiO_2} \rightleftharpoons \underset{\text{diopside (Di)}}{CaMgSi_2O_6}$$

$$+ \underset{\text{vapor (V)}}{2CO_2} \quad (4\text{-}5)$$

$$\underset{\text{serpentine (Sp)}}{Mg_3Si_2O_5(OH)_4} + \underset{\text{brucite (Br)}}{Mg(OH)_2}$$

$$\rightleftharpoons \underset{\text{forsterite (Fo)}}{2Mg_2SiO_4} + \underset{\text{vapor (V)}}{3H_2O} \quad (4\text{-}6)$$

For similar reactions involving minerals that exhibit solid solution the univariant reactions could be replaced by reaction bands.

Other dissociation reactions may involve the evolution of both water and carbon dioxide as in the univariant reaction:

$$\underset{\text{tremolite (Tr)}}{Ca_2Mg_5Si_8O_{22}(OH)_2} + \underset{\text{calcite (Ca)}}{3CaCO_3} + \underset{\text{quartz (Qz)}}{2SiO_2}$$

$$\rightleftharpoons \underset{\text{diopside (Di)}}{5CaMgSi_2O_6} + \underset{\text{vapor (V)}}{3CO_2 + H_2O} \quad (4\text{-}7)$$

Such reactions may proceed through divariant intervals in rock systems, the precise nature of the reaction being controlled largely by the composition of the pore fluid in the rock and the ease with which the vapor can escape from the region of the reaction.

Figures 4-3c and d show that dissociation reactions tend to be temperature-sensitive, except at very low pressures, but at higher pressures corresponding to upper mantle conditions, hydrated minerals may break down by pressure sensitive reactions. This is illustrated in a general way for amphibole in Figures 6-11 and 6-12. At pressures above 15 kb the slope of the reaction (dP/dT) becomes negative.

Magma is formed by melting reactions, and the typical shape of a fusion curve for a silicate is shown in Figure 4-3e by:

$$\underset{\text{albite (Ab)}}{} \rightleftharpoons \underset{\text{liquid (L)}}{} \quad (4\text{-}8)$$

The corresponding melting interval for a mineral aggregate or rock is shown by the narrow shaded band in Figure 4-3f.

Solution of a small amount of water under pressure in the silicate liquid has a dramatic effect in lowering the melting temperatures of silicates, as shown in Figure 4-3e by the curve for:

$$\underset{\text{albite (Ab)}}{NaAlSi_3O_8} + \underset{\text{vapor (V)}}{H_2O} \rightleftharpoons \underset{\text{(L)}}{\text{liquid}} \quad (4\text{-}9)$$

The solubility of H_2O in the silicate liquid and the solubility of the solids in the gas phase increase with increasing pressure. The effect of water on a rock is to increase the melting interval as shown in Figure 4-3f for a gabbro, because the liquidus curve is depressed to a lesser extent than the solidus curve. In the presence of excess water the subsolidus gabbro is converted to amphibolite.

This pattern of melting reactions in the presence of excess water is modified when high pressure minerals are produced, as illustrated in Figures 6-12, 8-2 and 8-3. The slope of the solidus is reversed at pressures above about 15 kb, where the plagioclase of the gabbro breaks down to yield jadeite, grossularite, kyanite, and quartz (Figure 4-3b). The temperature of melting then increases with increasing pressure, and the solidus curve is then approximately parallel with the curve for the dry rock. The involvement of hydrous minerals in melting reactions is considered in Chapters 6 and 8.

The *PT* conditions for the occurrence of these various types of mineral reactions, involving materials of different composition, can be determined experimentally in the laboratory. The composition of rock material available for reaction varies, however, with depth and geological environment and the range of temperatures occurring within the Earth at any given depth (pressure) is also limited. For the experimentally determined mineral reactions to have relevance to geological processes, therefore, we must consider the appropriate combinations of rock composition, pressure, and temperature.

Figure 4-4 shows the pressure existing at each depth within the outer part of the Earth. The two dotted lines are geotherms showing estimates for the temperature distribution at depth in continental shield regions, and in oceanic regions (Figure 3-12*b*). The geotherm could rise to higher temperatures in active orogenic areas. The composition of the mantle

and crust will be considered in detail in Chapters 6 and 7, but for present purposes let us adopt a conventional model of a crust composed essentially of granitic rocks overlying gabbroic rocks, and a mantle composed of peridotite with layers and lenses of eclogite.

Figure 4-4 shows the range of *PT* conditions occupied by the facies of regional metamorphism. The facies are separated, for the most part, by dissociation reactions (Figures 4-3*c* and *d*), but the boundary between the eclogite facies and the other facies involves the formation of high-pressure phases (Figure 4-3*b*). See Figure 9-9 for details.

If water is available in the crust as is usually assumed, then magma generation can be expected in the crust if conditions correspond to the shaded bands shown in Figure 4-4 for granite-water and gabbro-water (Figure 4-3*f*). If water is available in the mantle, then traces of liquid may develop at temperatures down to the dashed line extending the solidus for the system gabbro-peridotite-water to pressures greater than 10 kb (see also Figures 4-3*f* and 6-13). Magma

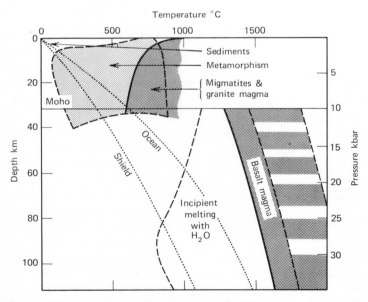

Figure 4-4. Depth-temperature ranges of major geological processes, compared with estimated temperatures beneath continental shields and oceans (see Figure 3-12). See Figure 9-9 for details of metamorphic facies.

generation in the mantle, however, is usually assumed to occur under essentially dry conditions, and the melting intervals for dry peridotite and eclogite are indicated by the alternating wide and narrow shaded bands.

Most reactions occurring in the formation of sedimentary rocks involve solution chemistry, and they occupy a very small range of pressures and temperatures near the origin of Figure 4-4. For discussion and illustration of sedimentary reactions, other diagrams, such as *Eh* versus *pH* diagrams, are more useful than the *PT* projections of Figures 4-3 and 4-4.

Figures 4-3 and 4-4 show the equilibrium positions of selected mineral reactions, as determined under static conditions in the laboratory. It does not necessarily follow that identical reactions will occur under identical pressure-temperature conditions in the Earth. Rock systems strive to attain a state of thermodynamic equilibrium but they do not always achieve this. Many rock materials include metastable ingredients, and these participate in metastable reactions. Many geological processes occur under irreversible conditions. Thus, although reaction curves determined experimentally in the laboratory provide a guide and possibly limits to the depths and temperatures of formation of rocks, they should be applied to rock systems only with due caution.

The occurrence of chemical or mineralogical reactions in rocks should not be considered separately from the many physical parameters that affect the rocks. In the formation of most sedimentary rocks, for example, the physical aspects of transportation and deposition of solid particles are at least as significant as the chemical reactions. In magmatic processes the temperature, pressure, and phase diagram suffice to define the conditions for the onset of partial fusion, but the extent of melting is controlled largely by the amount of heat available. This depends on the physical conditions in the crustal megasystem. Migration of magma from its place of origin introduces dynamic aspects which form no part of the static phase diagram. It is quite evident from their mineralogy and texture that many igneous rocks did not crystallize under equilibrium conditions.

The interrelationships of chemical reactions and physical controls are clearly displayed in metamorphic rocks. Metamorphic rocks can be formed at depth and subsequently uplifted to the surface with little or no mineralogical change; this fact proves that simply changing the pressure and temperature of a rock mass within the crust may not be sufficient to make a reaction proceed. Mineral reactions in metamorphic rocks are strongly influenced by the interstitial solutions filling pore spaces or absorbed as surface films on the mineral grains. The pore fluid also provides a medium for mass transport of material in response to temperature, pressure, or chemical potential gradients, and the metasomatism so produced is not represented on the equilibrium phase diagrams for closed systems. Equally important in the influence and control of chemical reactions in metamorphic rocks is the effect of stress. Folding, deformation, recrystallization, and mineral reactions in metamorphic rocks are all interdependent.

Igneous Processes

The location and source of igneous rocks within the mantle and crust, and with respect to the Rock Cycle, is illustrated in Figure 4-1. Figure 4-5 summarizes in schematic form the interrelationships among the major processes involved in the generation and emplacement of magmas. Some of these are discussed in more detail in Chapters 6 and 8.

Figure 4-5 shows the three types of rock material from which magma can be generated by partial fusion. These are (a) dry assemblages of anhydrous minerals, (b) dry assemblages of anhydrous and hydrous minerals, (c) assemblages of minerals with an aqueous pore fluid. The presence or absence of water, and its state of combination if present, are factors of fundamental importance for the

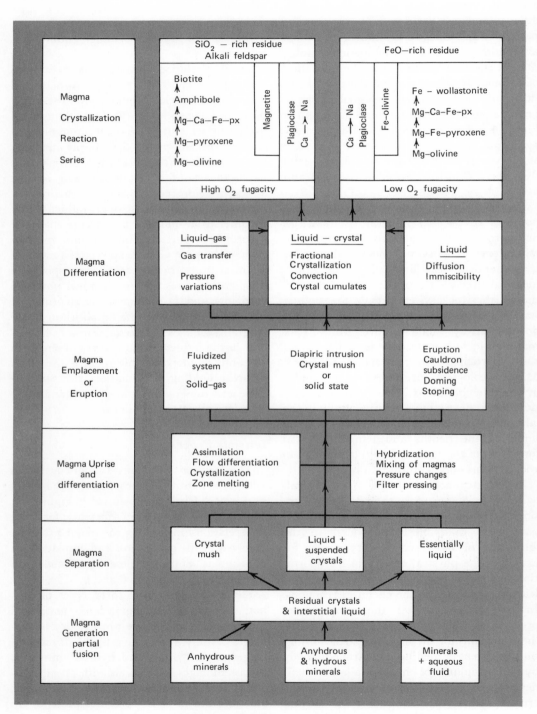

Figure 4-5. Schematic representation of igneous processes.

generation of magmas. The figure then outlines the major processes involved in (a) magma separation, (b) uprise of magma, (c) magma emplacement or eruption, (d) magma differentiation either during uprise or during solidification, and (e) the continuous and discontinuous reaction series which dominate differentiation in crystal-liquid systems.

Sedimentary Processes

The relationship of sedimentary rocks to others in the Rock Cycle was illustrated in Figure 4-1. The processes involved in their formation are summarized in schematic form in Figure 4-6. The main subdivision is into the processes of (a) weathering, which begins as soon as the rock is exposed to the hydrosphere, (b) transportation, (c) deposition, and (d) lithification. Extending below the part of the diagram showing transportation and deposition in the oceans is a path leading to a large box, which shows in more detail the various routes that materials derived by physical and chemical disintegration of exposed rocks can follow into the ocean. It also illustrates the processes leading to the formation of the various types of sediment occurring in different oceanic environments.

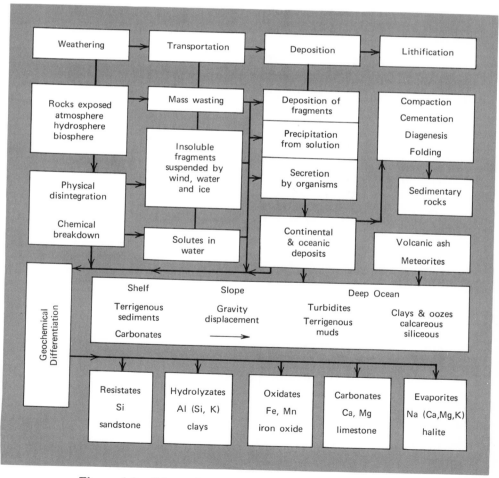

Figure 4-6. Schematic representation of sedimentary processes.

Extending from weathering boxes and transportation is a box labelled geochemical differentiation, which shows the chemical separation that occurs through sedimentary processes, and which classifies the sediments in these terms.

Metamorphic Processes

Figure 4-1 indicated that any rock could become metamorphosed if it were buried within the crust through subsidence. Subsidence is therefore labelled as process (1) in Figure 4-7, which summarizes in schematic form the major processes occurring during metamorphism. Deep burial need not be sufficient cause for regional metamorphism to occur, and box (2) indicates the possible effect of energy and material contributions from the mantle in active, orogenic belts. Up-

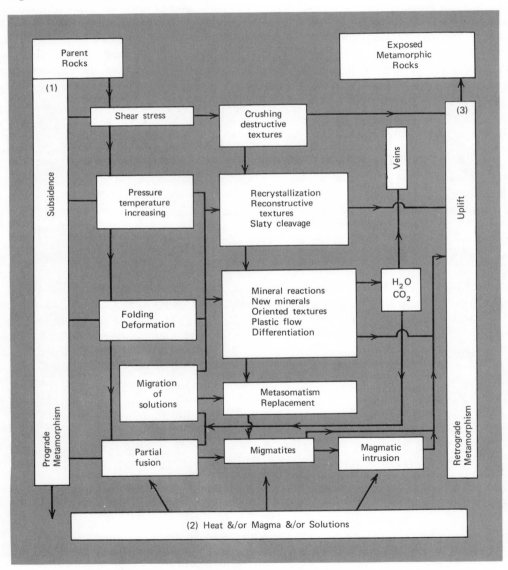

Figure 4-7. Schematic representation of metamorphic processes.

lift, which makes the examination of metamorphic rocks possible, is identified as major process (3). The main agents of metamorphism during subsidence are illustrated along with their effects upon the rocks, such as recrystallization, the formation of new minerals, and the development of textures peculiar to metamorphosed rocks.

GEOLOGICAL TIME SCALE

Geology is a historical subject, and its study can only be successful if sequences of geological events can be calibrated against a time scale. At any point on the surface of the Earth geological history is divided into (a) periods when rocks are forming either by sedimentation or by intrusion or extrusion of magma, and (b) periods of erosion when rocks are being destroyed and transported to other sites. The order and succession of rock units in time can be determined for a particular point or area, and this forms the stratigraphic column for that area. This column represents a finite time interval, including periods of rock formation, and unconformities representing periods of erosion. Lithological and biostratigraphical correlation of stratigraphic columns from different areas leads to the development of a generalized geological column representing the sequence of events in geological history without reference to any specific area. The subdivisions of this column, and the physical and biological events that they represent, provide an uncalibrated, relative time scale. Analysis of radioactive isotopes and their daughter elements permits calibration of the time scale.

Stratigraphic Classification

Two of the early attempts to subdivide the geological column into significant units are shown in Table 4-2. Charles Lyell's classification of 1833, incorporating the best of previous concepts, established the pattern to be followed for the next century. Many modified classifications were proposed. Continued study of the rocks in each group, their faunas, and the positions of major unconformities, led to the development of a catastrophic phi-losophy. Geological history was supposedly punctuated by repeated, synchronous changes in both physical and biological conditions. Diastrophism, represented by widespread unconformities, became the basis for correlation of stratigraphic columns and for the subdivision of the geological column. This was so in the text of T. C. Chamberlin and R. D. Salisbury in 1905, by which time classification had evolved into an essentially modern state (Table 4-2). The trend of the last fifty years has been toward a simplified classification, because detailed classifications developed to fit the stratigraphy of one area created additional problems in application to other areas. The idea of worldwide, synchronous orogeny and its relation to stratigraphy has undergone critical reexamination.

At the present time the science of stratigraphy is in a state of flux. The attention of stratigraphers in the petroleum industry has been directed by their search for oil and gas to tangible assemblages of strata that are potential structural traps, rather than to subdivisions representing portions of geological time. This led to the development of a dual classification, involving on the one hand mappable rock units, and on the other groupings of strata differentiated with respect to their position in time. This concept has now gained general acceptance, and Table 4-3 is the classification presented by W. C. Krumbein and L. L. Sloss in their text book of 1963.

Isotopic Age Determinations

All efforts to calibrate the geological time scale were failures until minerals and rocks were analyzed for radioactive isotopes and

TABLE 4-2 Stratigraphic Classification (modified from Krumbein and Sloss, 1963, Table 2-1)

Period	Lyell, 1833	Chamberlin and Salisbury, 1905 (Era)	Chamberlin and Salisbury, 1905	Present Usage (Era)	Present Usage	Age at Base of Period, M.Y.	Duration M.Y.
Recent / Tertiary Period	Newer Pliocene, Older Pliocene, Miocene, Eocene	Cenozoic	Present, Pleistocene	Cenozoic	Quaternary	1.5–2	1.5–2
			Pliocene, Miocene, Oligocene, Eocene		Tertiary — Neogene / Paleogene	65	63–63.5
Secondary Period	Cretaceous wealden	Mesozoic	Cretaceous Comanchean	Mesozoic	Cretaceous	136	71
	Oolite or Jura limestone group		Jurassic		Jurassic	190–195	54–59
	New red sandstone group		Triassic		Triassic	225	30–35
			Permian		Permian	280	55
	Carboniferous Group: Coal measures	Paleozoic	Coal measures or Pennsylvanian	Paleozoic	Pennsylvanian (U. Carboniferous)	345	65
	Mountain limestone		Subcarboniferous or Mississippian		Mississippian (L. Carboniferous)		
	Old red sandstone		Devonian		Devonian	395	50
	Grauwacke and Transition limestone		Silurian		Silurian	430–440	35–45
			Ordovician		Ordovician	~500	70
			Cambrian		Cambrian	570	70
Primary Period		Proterozoic	Keweenawan, Animikean, Huronian	Precambrian	Regionally defined systems	4600	85% of Earth's history
		Archeozoic	Archean complex				

TABLE 4-3   Classification of Time and Rock Units (from Krumbein and Sloss, 1963, Table 2-3)

A. Observable units
  1. *Rock-stratigraphic (lithostratigraphic) units*
     a. Formal rock units
        supergroup (rarely applied)
        group
        subgroup (rarely applied)
        formation
        member, tongue, lentil
        bed (rarely applied as formal unit)

     b. Informal rock units
        sequence
        bed (oil sands, quarry layers, key
          and marker beds)
        heavy mineral zone, insoluble
          residue zone
        electric-log zone, radioactivity
          zone, velocity zone
        "marker-defined" unit

  2. *Biostratigraphic units*
     a. Assemblage zone
        subzone
        zonule

     b. Range zone
        local range zone

     c. Concurrent-range zone

B. Inferential units
  1. *Time-stratigraphic (chronostratigraphic) units*
     _____
     _____
     system
     series
     stage

     Geologic time units
     eon
     era
     period
     epoch
     age

  2. *Ecostratigraphic units*
     a. Ecozone

     b. Geologic-climate unit
        glaciation
        stade
        interstade
        interglaciation

their daughter elements. Because radioactive isotopes decay at a constant rate, these analyses permit calculation of the time interval since the mineral or rock was formed, provided that the amount of daughter element was zero at the time of formation, and provided that elements were neither lost nor gained during this interval. If these conditions are satisfied the calculated time interval is the age of the mineral or rock. Alternately the calculated time may date an event such as metamorphism, during which the radioactive clock was reset by the migration of elements. If the assumptions and conditions are not satisfied the decay time calculated may have no geological significance. Since it is often difficult to decide whether or not the conditions were satisfied, it is usual to refer to the times determined as "apparent ages". Table 4-4 shows the isotopes most often used for age determinations of minerals and rocks, their half-lives, and the dating range for each.

A 1964 version of the ages of the boundaries between geological periods, based on radioactive isotopes, is given in Table 4-2. The age of the Earth is about 4.6×10^9 years. The Cambrian period began 570×10^6 years ago. Precambrian time thus spans more than 85% of the Earth's history. Figure 4-8 shows the durations of the geological periods plotted against a linear time scale. It also shows the extent of the time scale for calibrated magnetic reversals, compared with the extrapolated geomagnetic time scale based on magnetic anomalies in the oceanic regions (Chapters 12 and 13).

TABLE 4.4 Radioactive Decay Schemes and Age Determination Methods

Isotope		Half-life, years	Effective dating range
Parent	Daughter		
^{40}K	^{40}Ca	1.47×10^9	
	^{40}Ar	1.19×10^{10}	10^4 years to Earth formation
^{87}Rb	^{87}Sr	$4.7, 5.0 \times 10^{10}$	10^7 years to Earth formation
^{232}Th	^{208}Pb	1.39×10^{10}	10^7 years to Earth formation
^{235}U	^{207}Pb	7.13×10^8	10^7 years to Earth formation
^{238}U	^{206}Pb	4.51×10^9	10^7 years to Earth formation
^{14}C	^{14}N	5.57×10^3	0 to 5×10^4 years
Fission tracks from U decay			0 to 1×10^8 years

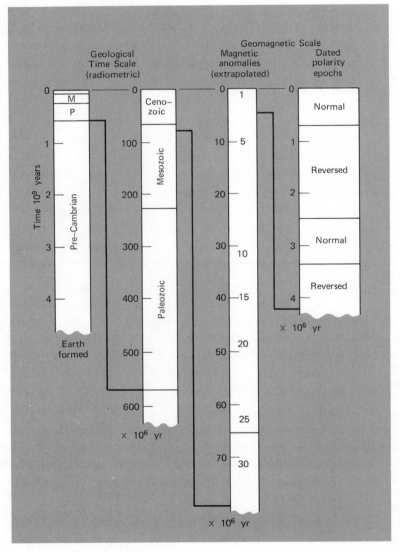

Figure 4-8. Radiometric geological time scale compared with geomagnetic time scale. The scale for each successive column is increased by a factor of 10. For the geomagnetic scales, see Figures 12-13, 12-21, and 13-15.

5. The Nature of the Crust-Mantle Boundary and the Gabbro-Eclogite Phase Transition

INTRODUCTION

The term "crust" has been used in various ways. It was for many years applied to the relatively strong, brittle, outermost shell of the Earth, about 100 km thick, which extended down to the asthenosphere or weaker shell of the interior. As more was learned about the nature of the outer layers of the Earth, the word crust became more widely used in a petrological sense, to refer to the rocks occurring at and near the surface. It was only in 1955 that the crust was limited specifically to the rock material occurring above the Mohorovicic discontinuity. The asthenosphere is now usually equated with the low-velocity zone, and the term lithosphere (formerly stereosphere) is used for the relatively rigid shell above the low-velocity zone, which includes the crust and part of the upper mantle. This can be treated as a single tectonic unit despite the abrupt change in seismic velocities occurring within it.

For fifty years since its discovery the Moho was identified as a rather sharp boundary at which the velocity of compressional seismic waves, P-waves, increased from the range 6.8-7.2 km/sec to the range 8.0-8.2 km/sec, characteristic of the upper mantle. It was considered to be a worldwide discontinuity as shown in Figure 5-1, a compilation by A. Holmes of data available to the end of 1962. Figure 5-1 shows the contrast between conti-

nents and oceans. The average depth to the Moho in stable continental regions is 35 km, whereas in the ocean basins the average depth is 11 km, with about 1 km of sediment above a crust 5.3 km thick and the rest ocean. The oceanic crust was thought to thicken beneath some, but not all, oceanic ridges and rises. Beneath the mid-Atlantic ridge, the Moho was reported at a depth of about 25 km, but the East Pacific Rise is underlain by crust of normal thickness.

Figures 5-1b and c show the Moho extending down to depths of more than 60 km beneath high mountain ranges forming mountain roots. In general higher continental elevations are accompanied by thicker crusts in agreement with Airy's principle of isostasy, but Holmes mentioned exceptions like the Colorado Plateau and some parts of the Andes. The cross sections through island arcs and ocean trenches in Figures 5-1b and c show that the Moho is lowered to depths of about 30 km, producing considerable crustal thickening in a narrow belt.

This simple picture has been revised since 1960, largely through the results of explosion seismology, because it is now known that the upper mantle is heterogeneous and P_n is variable immediately beneath the crust (Figure 3-7a). There are also lateral variations in P for rocks in the lower crust (Figure

3-7*b*). Figure 3-7*a* shows that in the eastern part of United States P_n exceeds 8.3 km/sec, but beneath the mountainous part of western United States P_n falls as low as 7.7 km/sec. Low values of P_n have also been measured beneath other tectonically active regions, including island arcs and midoceanic ridges. The picture is further complicated in some tectonically unstable environments by the occurrence of layers of rock with P-wave velocities between 7.2 and 7.7 km/sec, which are intermediate between normal values for lower crust and upper mantle. In these environments the Moho ceases to give clear refractions and the terms Moho and P_n are not unambiguously defined. This led to some discussion in the early 1960's as to whether

the material with intermediate P-wave velocities should be called anomalous lower crust, anomalous upper mantle, or mantle-crust mix.

At the present time the crust is usually defined as the outer shell of the Earth above the level where the P-wave velocity increases to more than 7.7 km/sec, and the Moho is then defined as the layer within which the P-wave velocities increase rapidly or discontinuously from crustal values to values above 7.7 km/sec. The sharpness of this layer is poorly defined, with estimates ranging from 0.1 km below parts of the Pacific Ocean through about 0.5 km in some stable continental regions, to several kilometers in other regions.

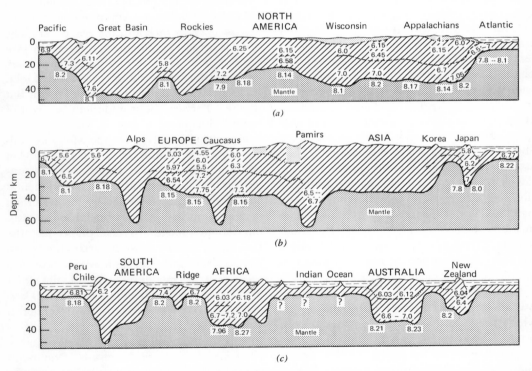

Figure 5-1. Composite seismic sections through characteristic parts of the continental and oceanic crusts. Figures in italics are the velocities of P-waves in km/sec. The levels of discontinuities are indicated where they have been recorded. Certain thick layers of sediments are shown by fine dots. Compiled from the data available to the end of 1962. (A. Holmes, *Principles of Physical Geology*, 2nd edition, The Ronald Press Company, New York, Thomas Nelson, London. Copyright 1965.)

CRUST-MANTLE BOUNDARY IN TECTONICALLY ACTIVE REGIONS

The layers of rock with *P*-wave velocities between 7.2 and 7.7 km/sec were earlier called anomalous, but it has now been established that they are a constant feature of tectonically active regions. For this reason, and because both crust and mantle are finely stratified, Kosminskaya and Zverev (1968) drew attention to the need for determining not only the depth to the Moho, but also the deep structure in the transition from the crust to the mantle and the layers within the mantle itself. A summary and an averaged representation of many Soviet results obtained by deep seismic soundings (DSS) are given in Figure 5-2. Crustal and upper mantle structures are subdivided into four types: continental and oceanic, with the continental margins including subcontinental and suboceanic types. The seismic velocity sections illustrated represent average characteristics for the four types. The Moho is readily identified in all sections except the subcontinental, which includes a great thickness of material with seismic velocity between 7.6 and 7.8 km/sec. Material with velocities in this range does not occur in the other three sections.

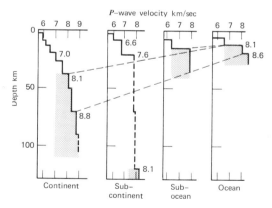

Figure 5-2. Velocity sections of the crust and upper mantle for different types of crust, averaged from data collected from many locations (after I. P. Kosminskaya and S. M. Zverev, Geophysical Monograph **12**, 122, 1968, with permission of Amer. Geophys. Union).

Figures 5-3 and 5-4 show similar data plotted in different ways by Drake and Nafe (1968) who have examined the variation of *P*-wave velocities with depth in different geological environments throughout the world. Figure 5-3*a* shows all available *P*-wave velocities from the Atlantic Ocean basins. The velocities are plotted against the depth of their measurement, depth below the solid surface being recorded rather than depth below sea level in order to reduce the scatter. The data points for mantle and crustal velocities are rather narrowly confined to the boxes in Figure 5-3*a*, although the data for the oceanic basement, layer 2, are more scattered compared to the box which encloses the most typical values. The fourth box shows the range of velocities recorded for the ocean basin sediments.

Similar diagrams were prepared for velocity data available in the various geological environments shown in Figure 5-4, and results are compared in Figure 5-3*b* for the stable ocean basins, the stable Precambrian shields with elevations of less than 500 m, and continental margins. The data for the Pacific and Indian Ocean basins fit into the boxes of Figure 5-3*a* for the Atlantic Ocean basin, so the same boxes represent all ocean basins in Figure 5-3*b*; there are distinct velocity gaps between the four boxes representing mantle, crust, layer 2, and oceanic sediments. Data for the low-lying continental shields occupy the three boxes shown; there is a distinct velocity gap between the boxes representing mantle and crust, but there is an overlap in velocities for the boxes representing lower and upper crust; velocities in the lower crust occupy the same range as those in the oceanic crust. Velocities measured beneath continental margins occupy the whole range from sediments to mantle, with a general trend of increasing velocity with increasing depth; there is no gap between the mantle velocities and the velocities characteristic of the crust beneath the stable continental

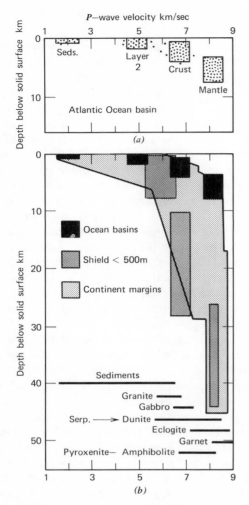

Figure 5-3. Seismic *P*-wave velocities located with respect to their depth beneath the solid surface. (*a*) Data points for the Atlantic ocean basin are concentrated into four distinct ranges, for mantle, crust, layer 2, and sediments. (*b*) Results based on world wide data are shown for three types of crust: data points for all ocean basins occupy the same ranges as in *a*; data points for the shields occupy three distinct ranges in terms of velocities and depths; data points for continental margins overlap with the specific ranges for other environments and occupy the values between as well. Velocities for rocks measured in the laboratory are plotted (after C. L. Drake and J. E. Nafe, Geophysical Monograph **12**, 174, 1968, with permission of Amer. Geophys. Union).

regions and ocean basins. Figures 5-3 and 5-2 both show that material with *P*-wave velocities between 7.2 and 7.7 km/sec is present beneath continental margins and absent beneath stable continents and ocean basins.

Figure 5-4 summarizes the data of Figure 5-3 in a different way and includes results from additional geological environments. For each environment there are four columns. Each column represents a certain range of *P*-wave velocities; the ranges 4.5–6.5 km/sec and 6.5–7.2 km/sec correspond to the crustal boxes in Figure 5-3. The range 7.2–7.7 km/sec corresponds to the velocity gap between crust and mantle boxes in Figure 5-3, and the range 7.8–8.5 km/sec corresponds to the mantle boxes in Figure 5-3. The vertical extent of the columns in Figure 5-4 shows the depth interval below the solid surface within which these velocities are encountered. This figure shows ranges of velocity in various environments, and it must not be confused with specific velocity profiles like those given in Figure 3-5*b*.

The central part of Figure 5-4 shows a progression from stable ocean basins, across continental margins and orogenic belts, to the low-lying, stable continental shields. The minimum crustal thickness, indicated by the top of the mantle column, indicates a depth to the Moho increasing from ocean basins to older orogenic belts and decreasing for the shields. The maximum crustal thickness occurs beneath the younger orogenic belts. Layers with the anomalous seismic velocities, represented by the black column, are restricted to continental margins, orogenic belts, and to the environments on the left and right of Figure 5-4, which correspond to anomalous oceanic and continental regions respectively.

Drake and Nafe discussed the possibility that the rock material with anomalous velocity is of a transient nature, representing additions to the crust from the mantle during orogenic processes. Production or disappearance of this material in response to changing conditions at depth could be responsible for up-

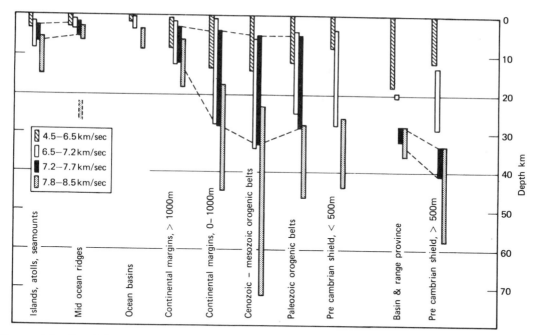

Figure 5-4. Depth intervals for *P*-wave velocities in four ranges, as they are encountered in various structural provinces; this is a summary of data gathered on a world-wide basis. The four velocity ranges correspond to (1) the upper crust, (2) the stable lower crust, (3) anomalous crust or mantle in tectonically active regions, and (4) upper mantle. See Figure 5-3 (after C. L. Drake and J. E. Nafe, Geophysical Monograph **12**, 174, 1968, with permission of Amer. Geophys. Union).

lift or subsidence and for changes in crustal thickness. Figure 5-3 includes the known seismic velocity ranges of a number of rock types and of particular interest are those for gabbro, eclogite, amphibolite, and dunite (or peridotite). These are the rock types with velocities appropriate for those measured in the lower crust and upper mantle. The most likely interpretation of the transient material discussed by Drake and Nafe is that this is related in some way to the gabbro-eclogite phase transition.

Conventionally the Moho has been re-garded as a level marking a chemical discontinuity between two different types of rock. In more recent years the hypothesis that the Moho is a phase boundary between gabbro of the crust and chemically equivalent eclogite of the upper mantle has received serious consideration, and the consequences of the hypothesis have been examined in detail because of their tectonic implications. Changes of pressure or temperature at depth in the region of such a phase transition would result in uplift or subsidence at the surface of the Earth.

CHEMICAL DISCONTINUITY OR PHASE TRANSITION?

The rock materials comprising the upper mantle include peridotite, eclogite, and dunite as described in Chapters 3 and 6. Models for the lower continental crust reviewed in Chapter 7 involve gabbro, amphibolite, or rocks of intermediate composition in the high pressure granulite facies. The upper continental crust corresponds in composition to a granodiorite. In oceanic regions the crust is conventionally interpreted as basalt, but serpentine is an alternative. Arrangements of these materials providing a chemical discontinuity with physical properties appropriate for the Moho are summarized in Figure 5-5, and arrangements providing a phase transition appropriate for the Moho are summarized in Figure 5-6.

The dashed lines in Figures 5-5 and 5-6 represent chemical discontinuities, and the heavy solid lines represent the Moho. The chemical discontinuities indicated in these figures are schematic; although a boundary between upper and lower crust is detectable in many localities by geophysical methods, it is not universally present and the generalized model of a two-layer crust is passing from favor. There is also considerable uncertainty about, and possibly variation in, the thickness of the Moho. The Moho line in the figures is 1 km thick.

The Moho as a Chemical Discontinuity

Figure 5-5 shows four models for the Moho as a chemical discontinuity based on current information about the upper mantle and crust. These are:

1. Peridotite-gabbro (or basalt).

2. Peridotite-amphibolite.

3. Peridotite-granulite of intermediate composition.

4. Eclogite-granulite of intermediate composition.

In model (4) the upper mantle of gabbroic composition is composed of eclogite, because the overlying crustal material is in the eclogite facies (although of different composition). Note that the eclogite of the upper mantle does not extend indefinitely downwards, but it is terminated eventually by a chemical discontinuity against peridotite.

The conventional view that the Moho is a chemical discontinuity between basic crustal material and peridotite of the upper mantle provides reasonable explanations for petrological processes. Reasons for concluding that the upper mantle is of ultrabasic composition are discussed in Chapters 3 and 6. Petrogenesis demands that mantle material be capable of supplying the basaltic magma which has been erupted so frequently at the Earth's surface throughout the span of geological time. Furthermore it is likely that this magma is produced by partial melting of crystalline material rather than by complete melting, which is what would be required if the upper mantle were composed mainly of basaltic material in the form of eclogite.

There are difficulties facing this model, and these were discussed in 1959 by G. C. Kennedy with reference to major geological problems including the differences between continents and ocean basins, the permanence of continents, the elevation of large plateaus, the formation of geosynclines, and the origin of mountain belts. The present rate of heat loss from the Earth requires that radioactive material be concentrated in the upper mantle and crust. Because of the high concentration of heat-producing radioactive elements in rocks of the continental crust, and their low concentration in known ultrabasic rocks, we might expect higher values for heat flow from the continents than from the thin crust of the oceanic regions. The average heat flow from the continents, however, appears to be approximately the same as the average heat flow from the ocean basins (Chapter 3). This is more readily explained if the upper mantle beneath the oceans is of basaltic composition than if it is a peridotite. Finally there is the question of whether or not a chemical discontinuity between basalt and peridotite could persist for periods of the order of 10^9 years without being smoothed out by geological processes.

The Moho as a Phase Transition

Figure 5-6 illustrates two models for interpretation of the Moho as a phase transition

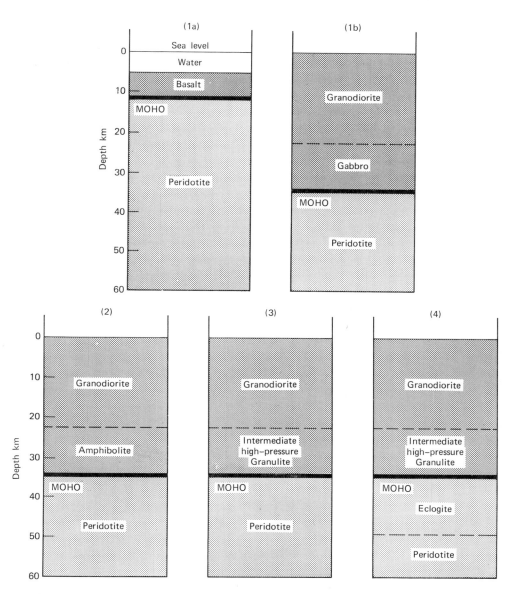

Figure 5-5. The Mohorovicic discontinuity as a chemical boundary; four models.

These are (5) peridotite-serpentinite in oceanic regions, and (6) eclogite-gabbro (basalt). Note that the eclogitic upper mantle is replaced at deeper levels by a peridotite. The Moho in model (5) is represented by a reaction involving the hydration of peridotite. Migration of water is therefore involved if it changes its position and the reaction is not an isochemical phase transition. Model (6) does involve an isochemical phase transition; the term eclogite has been used for rocks in the eclogite facies with a fairly wide range of compositions, but in this context eclogite is considered to be the chemical equivalent of the overlying basalt. Although the phase transitions are indicated in Figure 5-6 as sharp boundaries, following most models in the literature, everyone has recognized the fact that the transitions in multicomponent rocks would occur through a finite thickness, which could extend through several kilometers of rock.

Before we can decide whether or not a phase transition occurs at the depths corresponding to the Moho, in different tectonic environments, we must know the following:

1. The composition of the rock in the region of the Moho. Figure 5-5 indicates that there are different interpretations available.

2. The distribution of temperature with depth in any locality under consideration. The uncertainties here have been discussed in Chapter 3, and Figure 5-10 will illustrate the kind of problem encountered because of these uncertainties. In the following figures we will use the geothermal curves for different tectonic environments which were plotted in Figure 3-12b.

3. The nature of the phase transitions occurring in the rocks of (1) and their positions in terms of pressure (depth) and temperature. These have to be determined experimentally and critical evaluation of the phase transition hypothesis depends upon review of the experimental data.

If it appears that a phase transition layer does correspond to the Moho, then the kinetic response of the transition to changes in pressure and temperature or the availability of water (for serpentinite) must be determined for evaluation of the tectonic influence of the phase transition layer. We might expect that phase transitions near the base of the continental crust would migrate readily in response to temperature and pressure perturbations but, even with geological time available, similar migrations near the base of the oceanic crust might be far more sluggish. Whatever the rate of reaction peridotite can not be serpentinized if there is no water available.

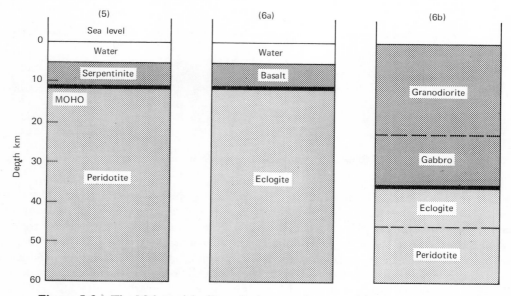

Figure 5-6. The Mohorovicic discontinuity as a phase transition zone; two models.

THE PERIDOTITE-SERPENTINITE MODEL OF H. H. HESS

Hess proposed in 1959 that layer 3 of the oceanic crust (Figure 7-8) might be serpentinite rather than the generally accepted basalt, and he has developed the idea in several papers (Chapter 12). Model (5) of Figure 5-6 is illustrated in more detail in Figure 5-7. Hess had earlier discussed the tectonic implications of serpentinization of peridotite in the upper mantle beneath the continents and Figure 5-7 provides the background for this process. Detailed examination of the process, however, will be reserved for our discussion of Figure 10-1.

Figure 5-7 is made up of three diagrams. The P-T diagram shows the experimentally determined univariant reactions involving the formation of serpentine from olivine:

$$Mg_3Si_2O_5(OH)_4 \quad + \quad Mg(OH)_2$$
serpentinite (Sp)   brucite (Br)

$$\rightleftharpoons 2Mg_2SiO_4 \quad + \quad 3H_2O \quad (5\text{-}1)$$
forsterite (Fo)   vapor (V)

$$5Mg_3Si_2O_5(OH)_4 \rightleftharpoons 6Mg_2SiO_4$$
serpentinite (Sp)   forsterite (Fo)

$$+ Mg_3Si_4O_{10}(OH)_2 + 9H_2O \quad (5\text{-}2)$$
talc (Ta)   vapor (V)

We will let reaction (5-2) represent the transition of serpentinite to peridotite by dehydration. The other two diagrams correspond to crustal models (5) in Figure 5-6 for the oceanic environment and (1b) in Figure 5-5 for an average continental crust.

The geotherm for continental regions on the P-T diagram intersects reaction (5-2) at point A, and that for oceanic regions intersects it at point B. The intersection A corresponds to a depth of about 55 km in the continental model, in the peridotite of the upper mantle. Given equilibrium conditions in the system peridotite-H_2O, and sufficient water, the upper mantle between the Moho and a depth of 55 km should be composed of serpentinite, with peridotite occurring only below 55 km. If this situation existed the abrupt change in physical properties corresponding to the Moho would occur at the depth of 55 km instead of the measured depth of 35 km. This indicates that a

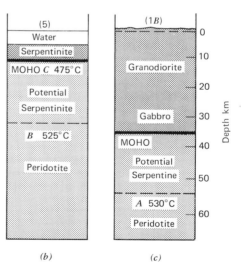

Figure 5-7. Mantle peridotite and the stability of serpentinite. (5-1) $Sp + Br = Fo + V$; (5-2) $Sp = Fo + Ta + V$ (Kitahara *et al.*, 1966; Scarfe and Wyllie, 1967). Hess' proposal that the oceanic crust is composed of serpentinite, or hydrated mantle, requires that at one time the geotherm was in position D. Note the extent of mantle peridotite that is in the stability field of serpentinite, should water become available. See Figure 10-2.

significant layer of the upper mantle in stable continental areas (20 km in model (1*B*)) is potential serpentinite, should water become available to hydrate the mantle peridotite.

The intersection *B* in Figure 5-7 corresponds to a depth of about 32 km, 20 km below the Moho in the oceanic model (5). Therefore the Moho is definitely not an equilibrium reaction boundary between serpentinite and peridotite. An equilibrium reaction between peridotite and serpentinite would correspond to conditions given by point *C* on reaction curve (5-2) in Figure 5-7, and this requires that the geotherm passes through *C* following a curve such as *CD*. Hess proposed that generation of new ocean basin floor at the midocean ridges was accompanied by a high geothermal gradient such as *CD*. This produced an isothermal surface for 475°C at a depth of about 12 km. Water emanating from the Earth, or percolating downward from the oceans, hydrated that portion of the

mantle above the isotherm, converting it to serpentinite. After the emanation of water from the mantle ceased the geotherm migrated toward the present oceanic geotherm in Figure 5-7 and the 475°C isotherm moved to deeper levels in the Earth. The boundary between serpentinite and peridotite remained to mark the position of a fossil 475°C isotherm.

This hypothesis modifies the conventional model of the oceanic basaltic crust extending beneath the continents to form the lower continental crust (models (1*a*) and (1*b*) in Figure 5-5). The 5 km-thick layer 3 of the oceanic crust is here regarded simply as hydrated mantle, with only a very thin veneer of basalt and deep-sea sediments representing crust in the normal sense, that is material above the mantle. This is an integral part of Hess' concept of sea-floor spreading, and the proposal is reviewed in some detail in Chapter 12.

THE GABBRO-ECLOGITE TRANSITION AS THE MOHO

The suggestion that the upper mantle is composed of eclogite appears to have originated with L. L. Fermor in 1913. The idea was explored in the 1920's by V. M. Goldschmidt and A. Holmes, but the concept of a peridotite mantle overlain by a basaltic layer was preferred by most petrologists. Following theoretical and experimental studies in the 1950's, this conventional view was challenged by proposals that the Moho represent a phase transition from gabbro to eclogite, its dense chemical equivalent. The first detailed discussions of the geological and geophysical implications of this hypothesis appear to be those of J. F. Lovering in 1958 and G. C. Kennedy in 1959.

Lovering compared the Earth with achondrite meteorites concluding that a considerable portion of the Earth's upper mantle is of basaltic composition and that the Moho is therefore a phase change from basalt to eclogite. Kennedy cited several major geological observations which have been inadequately explained by the traditional view that the Moho is a chemical discontinuity. He then proceeded to develop plausible explanations for these observations by assuming that the Moho is represented by a phase transition from gabbro to eclogite. These two papers have been superseded by more recent work, but their historical significance and the lessons to be learned from them justify a rather detailed examination and explanation.

The Meteorite Analogy of J. F. Lovering

In 1958 Lovering outlined evidence supporting his model of a differentiated parent meteorite body, with 95% by volume of its silicate mantle being composed of achondrite material. He compared this model with a layered Earth, and concluded by analogy that at least the outer 100 km of the Earth is of basaltic composition. Extraterrestrial evi-

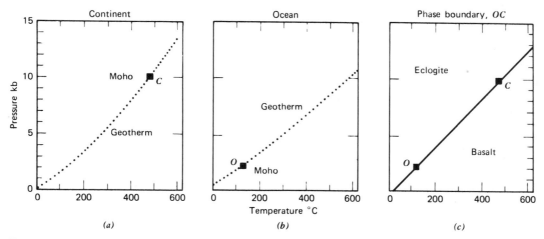

Figure 5-8. J. F. Lovering's hypothesis that the Moho is a phase transition boundary. (*a*) Given the calculated geotherm and the known depth of the Moho beneath continents, point *C* lies on the basalt-eclogite transition boundary. (*b*) Given the calculated geotherm and the known depth of the Moho beneath oceans, point *O* lies on the basalt-eclogite transition boundary. (*c*) According to the assumptions made, the basalt-eclogite transition boundary is the line through *OC*. (After J. F. Lovering, Trans. Amer. Geophys. Union, **39**, 947, 1958, with permission.)

dence for the composition and mineralogy of the Earth's mantle is reviewed in Chapter 6, and the analogy with chondrites is generally preferred over the analogy with achondrites. Lovering (1962) himself has since modified his assumptions about the relationships among chondrites and achondrites, but there does remain significant geochemical evidence that the Earth's chemistry may be closer to that of basaltic achondrites than to chondrites.

If the upper mantle is of basaltic composition then the Moho can be explained only as a phase change from basalt to eclogite. Therefore, Lovering reasoned, if the pressure and temperature conditions were known for the base of the crust beneath the oceans and the continents, this would provide two points on the basalt-eclogite phase transition curve. Temperatures in these two environments were estimated from surface heat flux and an assumed typical distribution of radioactivity and resulting geotherms are plotted in Figure 5-8. They are very similar to the curves that we have adopted in this text. A depth of 35 km to the Moho in continental regions corre-

sponds to a pressure of about 10 kb. This provides the point *C* in Figure 5-8*a* as a point on the basalt-eclogite phase transition curve at 480°C. Similarly the point *O* represents a depth of 11 km below sea level in oceanic regions, where the temperature is 123°C according to the calculated geotherm and this is a second point on the phase transition curve. In Figure 5-8*c* the phase transition curve has been drawn through these two points. Lovering noted that the position of this estimated curve was in reasonable accord with the available theoretical and experimental data, and he pointed out that the use of a univariant line for the complex rock reaction is a simplification.

Given the estimated phase boundary of Figure 5-8, and assuming that the Moho is everywhere a transition from basalt to eclogite, a knowledge of the depth of the Moho in any locality would be sufficient to provide the temperature at the Moho. Similarly for a given temperature distribution with depth, the position of basalt and eclogite layers within the Earth and the boundaries between them can be read directly from diagrams like those in Figure 5-8. When this is done for the conditions used in Figure 5-8 to derive the phase transition, the system works well for the continental region but it provides incorrect results for the oceanic region as shown in Figure 5-9.

Figure 5-9a reproduces from Figure 5-8 the continental geotherm and the phase transition curve. Figure 5-9b illustrates the crustal section used by Lovering for the temperature distribution calculation, with about 10 km of granodiorite overlying basalt, the Moho, marking the transformation of basalt to eclogite, occurs at the depth provided by the intersection point *C*, and the resulting section corresponds to that illustrated in (6b) of Figure 5-6. The oceanic geotherm and the

phase transition curve from Figure 5-8 are reproduced together in Figure 5-9c, with the intersection point *O* occurring at a pressure corresponding to the depth of the Moho, beneath 5 km of ocean and 6 km of crust. The relative positions of the geotherm and the transition boundary in Figure 5-9c, however, indicate that the material above the Moho is eclogite and that below the Moho is basalt, and this, of course, is the wrong way round (see section (6a) in Figure 5-6). This anomaly in Lovering's derivation was pointed out two years later by Harris and Rowell (1960), who noted that only if the continental thermal gradient is steeper than the oceanic one can the Moho coincide with the eclogite transformation in both environments. This is the model used by Kennedy as illustrated in Figure 5-10(a).

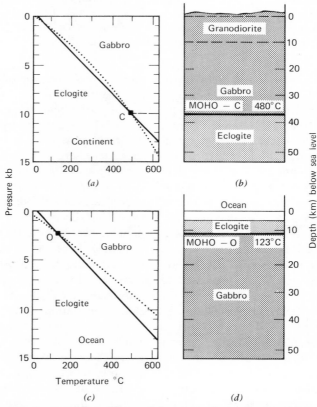

Figure 5-9. J. F. Lovering's model from Figure 5-8 expressed as crustal sections. (*a*) Combination of Figure 5-8a and 5-8c for the continents. (*b*) Crustal section for continents derived from *a*; 10 kms granodiorite assumed by Lovering in heat flow calculations. (*c*) Combination of Figure 5-8b and 5-8c for the oceans. (*d*) Crustal section for oceans derived from *c*.

G. C. Kennedy's Model

Lovering adopted a model for the mantle requiring the existence of a phase transition, and deduced where the phase transition curve would be on the basis of the known Moho depths and calculated temperature distributions. Kennedy (1959) adopted the phase transition hypothesis and selected a phase transition curve and geotherms which provided intersections at pressures corresponding to the Moho depths in different environments. His scheme is presented in Figure 5-10a, and what the intersections mean in terms of crustal sections are illustrated in Figures 5-10b, c, and d. The oceanic geotherm intersects the phase transition curve at point A, providing the crustal section shown in Figure 5-10b, and similarly the intersections B and C in Figure 5-10a provide the crustal sections in

Figure 5-10c and 5-10d respectively. These correspond to sections (6a) and (6b) in Figure 5-6.

The main difference between the models of Lovering and Kennedy is in the positions of the geotherms for oceanic and continental regions. Kennedy's model certainly explains the variation of depth to the Moho in different tectonic environments, but unfortunately it does not fit with current opinions for temperature distributions at depth. In Figure 5-10e the same phase transition curve is compared with the two geotherms adopted in Figure 3-12b. Intersection of the oceanic geotherm at the point E in Figure 5-10e provides the crustal section shown in Figure 5-10f, with a depth to the phase transition (the Moho) of about 25 km, which is too deep to satisfy geophysical measurements (Figure 5-5). Similarly the intersection of the

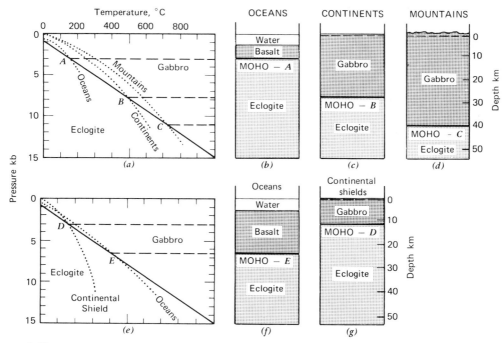

Figure 5-10. G. C. Kennedy's hypothesis that the Moho is a phase transition boundary. (a) Assumed gabbro-eclogite transition boundary, and geotherms for different environments. (After Kennedy, 1959, with permission, *American Scientist*, Journal of the Society of the Sigma Xi). (b), (c), and (d). Crustal sections derived from a. (e) Assumed gabbro-eclogite transition boundary from a with geotherms from Figure 3-12. (f) and (g) Crustal sections derived from e.

continental shield geotherm at the point *D* provides the crustal section in Figure 5-10*g* with the phase transition occurring at a depth of only 12 km, which is too shallow for continental regions. It is also too shallow for the occurrence of a basaltic layer according to most crustal models (Figures 5-5 and 5-6). Thus the arrangements of geotherms following the current best estimates do not support the occurrence of a basalt-eclogite phase transition beneath both continents and oceans, using Kennedy's estimate for the phase transition boundary as a basis for analysis.

Figure 5-10 shows that quite small changes in the positions of either the phase transition boundary or of the geotherms would cause significant variation in the depths to the phase transition in crustal sections. In order to evaluate the hypothesis, therefore, it is most important to know precisely the temperature distributions and position of the phase transition. The problems in connection with the geotherms have already been discussed so let us now examine the experimental evidence for the phase transition. More is available than when Lovering and Kennedy presented their papers.

EXPERIMENTAL STUDIES ON THE GABBRO-ECLOGITE PHASE TRANSITION

In this context eclogite is considered to be the high pressure, dense chemical equivalent of a gabbro. The olivine, pyroxene, and plagioclase of the gabbro are replaced by omphacite and pyrope-rich garnet. The transformation of gabbro to eclogite involves a series of complex reactions which can be oversimplified as follows:

$$3Mg_2SiO_4 \quad + \quad 3CaMgSi_2O_6$$
olivine (Ol)   pyroxene (Px)

$$+ \ [3CaAl_2Si_2O_8 \quad + \quad 2NaAlSi_3O_8]$$
plagioclase (Pl)

$$\rightleftharpoons 3CaMg_2Al_2Si_3O_{12}$$
garnet (Ga)

$$+ \ [2NaAlSi_2O_6 + 3CaMgSi_2O_6]$$
omphacite (Om)

$$+ \ 2SiO_2 \quad (5\text{-}3)$$
quartz (Qz)

In addition to the formation of pyrope and jadeite there are reactions involving increased solubility in pyroxenes of Al_2O_3 from the anorthite component of the plagioclase and the formation of almandine and grossular components in garnet at lower pressures than the pyrope-rich garnet which characterizes eclogite.

There are three experimental approaches providing information about the location of the gabbro-eclogite phase transition in terms of pressure and temperature:

1. Locate experimentally the positions of the reaction curves for the conditions of breakdown or formation of individual minerals, and of more complex reactions involving the stability of solid solutions.

2. Locate the positions of univariant reactions involving mineral assemblages with bulk compositions approaching that of the rock.

3. Study the transition interval in natural or synthetic rock samples.

We will examine these experiments in some detail to provide examples of the difficulties involved in obtaining conclusive results in laboratory phase studies of complex systems.

Instability of Crustal Silicates at High Pressures

The breakdown of plagioclase to yield jadeite is essential for the formation of omphacite, and the two appropriate univariant reactions are plotted in Figure 5-11:

$$NaAlSi_3O_8 \quad + \quad NaAlSiO_4$$
albite (Ab)   nepheline (Ne)

$$\rightleftharpoons 2NaAlSi_2O_6 \quad (5\text{-}4)$$
jadeite (Jd)

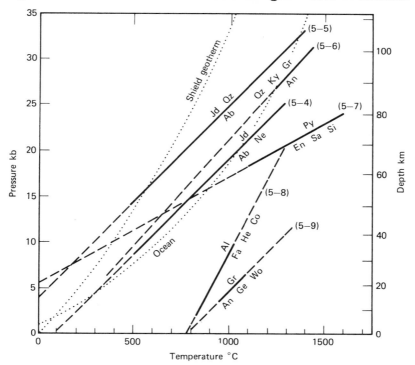

Figure 5-11. Mineral reactions related to the gabbro-eclogite transition. Solid lines show the limits of experimental measurement; dashed lines are extrapolations. (5-4) $Ab + Ne \rightleftharpoons Jd$; (5-5) $Ab \rightleftharpoons Jd + Qz$ (Boettcher and Wyllie, 1968); (5-6) $An = Gr + Ky + Qz$ (Newton, 1966, Boettcher, 1970); (5-7) $En + Sa + Si \rightleftharpoons Py$ (Boyd and England, 1962); (5-8) $Al \rightleftharpoons Co + He + Fa$ (Yoder, 1955, Hsu, 1968); (5-9) $Gr \rightleftharpoons An + Ge + Wo$ (Newton, 1966, Boettcher, 1970).

$$NaAlSi_3O_8 \rightleftharpoons NaAlSi_2O_6 + SiO_2 \quad (5\text{-}5)$$
$$\text{albite (Ab)} \qquad \text{jadeite (Jd)} \qquad \text{quartz (Qz)}$$

In their reports of the first experimental studies of portions of these reaction curves in 1957 and 1960 Robertson, Birch, MacDonald, and LeComte noted with due caution that, although the results give an indication of the position and slope of the basalt-eclogite transition, they are insufficient to determine whether or not the Moho was a phase change.

The anorthite component of plagioclase also breaks down under similar pressure conditions, yielding grossular garnet and other minerals as shown in Figure 5-11:

$$3CaAl_2Si_2O_8 \rightleftharpoons Ca_3Al_2Si_3O_{12}$$
$$\text{anorthite (An)} \qquad \text{grossular (Gr)}$$

$$2Al_2SiO_5 + SiO_2 \quad (5\text{-}6)$$
$$\text{kyanite (Ky)} \qquad \text{Quartz (Qz)}$$

The stability field of pyrope, bounded on its low pressure side by dense silicate minerals, was located by Boyd and England as shown in Figure 5-11:

Aluminous enstatite (En) + sapphirine (Sa)

+ sillimanite (Si) $\rightleftharpoons$ pyrope (Py)   (5-7)

They suggested that the pyrope and jadeite stability curves could together be taken as an approximation to the basalt-eclogite transition. Yoder's preliminary curve for the stability of almandine in Figure 5-11:

$$5Fe_3Al_2Si_3O_{12} \rightleftharpoons 2Fe_2Al_4Si_5O_{18}$$
$$\text{almandine (Al)} \qquad \text{Fe-cordierite (Co)}$$

$$+ FeAl_2O_4 + 5Fe_2SiO_4 \quad (5\text{-}8)$$
$$\text{hercynite (He)} \quad \text{fayalite (Fa)}$$

shows that the Fe-bearing garnet in eclogite

would be stable at somewhat lower pressures than pure pyrope, and this is confirmed by the schematic curves in Figure 5-12 showing the pressure conditions required for the formation of garnets with given pyrope contents. The stability conditions of garnet are further clouded by the effect of CaO. Figure 5-11 shows that at high pressures grossular is produced along with other phases by the breakdown of anorthite, but that grossular also has a stability field similar to that of almandine; it decomposes according to the reaction:

$$2Ca_3Al_2Si_3O_{12} \rightleftharpoons CaAl_2Si_2O_8$$
grossular (Gr)   anorthite (An)

$$+ Ca_2Al_2SiO_7 + 3CaSiO_3 \quad (5\text{-}9)$$
gehlenite (Ge)   wollastonite (Wo)

The effect of solid solution on the stability of jadeite is illustrated in Figure 5-12, where a set of curves shows the pressure required to produce an omphacite with a specified percentage of jadeite in solid solution according to the reaction:

$$\underset{\text{diopside (Di)}}{CaMgSi_2O_6} + \underset{\text{albite (Ab)}}{NaAlSi_3O_8}$$

$$\rightleftharpoons \underset{\text{omphacite (Om)}}{[CaMgSi_2O_6 + NaAlSi_2O_6]} + \underset{\text{quartz (Qz)}}{SiO_2}$$

$$(5\text{-}10)$$

These results indicate that considerable solid solution of diopside in jadeite is required in order to reduce the pressure of formation by more than a kilobar or two and the same is true for the solution of acmite in jadeite.

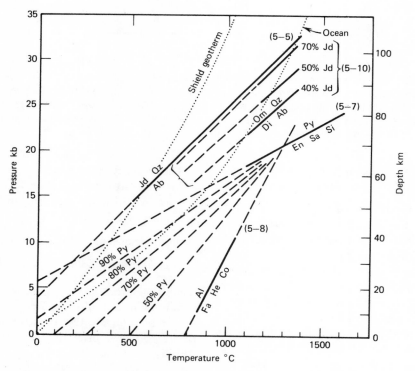

Figure 5-12.   Mineral reactions related to the stability of omphacite and pyrope solid solutions. Solid lines show the limits of experimental measurement; dashed lines are extrapolations or estimates. See Figure 5-11 for reactions (5-5), (5-7), and (5-8). (5-10) Di+Ab ⇌ Om+Qz (Kushiro, 1969); the lines show pressure required for formation of omphacite with specific jadeite content. The dashed lines between reactions (5-5) and (5-8) show the pressure required to form a garnet with specific pyrope content (based on *PX* and *TX* diagrams of Yoder and Chinner, 1960).

The intersections of the two geotherms with the reaction curves in Figures 5-11 and 5-12 indicate that crystalline gabbro would be converted to eclogite at no great depth within the Earth, but there are large margins of uncertainty regarding the depth at which the change occurs. The depth would obviously be strongly dependent upon the composition of the gabbro and in particular upon the degree of silica saturation as indicated by the two curves for jadeite in Figure 5-11. Figure 5-12 demonstrates the depth dependence imposed by other compositional variables.

Reactions Among Mineral Assemblages

Experimental results for the stability of mineral assemblages are plotted in Figure 5-13. The reaction:

$$4MgSiO_3 \quad + \quad MgAl_2O_4$$
$$\text{enstatite (En)} \qquad \text{spinel (Sp)}$$
$$\rightleftharpoons Mg_3Al_2Si_3O_{12} + \quad Mg_2SiO_4 \quad (5\text{-}11)$$
$$\text{pyrope (Py)} \qquad \text{forsterite (Fo)}$$

indicates that in a silica-undersaturated environment, a higher pressure is required to produce pyrope along with forsterite than to produce pyrope alone (compare Figure 5-12). The other reactions plotted in Figure 5-13, most of which involve the formation of a garnet, indicate the complexities involved in a four-component system, CaO-MgO-Al$_2$O$_3$-SiO$_2$, which is simple compared with a whole rock composition.

Kushiro and Yoder (1966) studied reactions occurring between anorthite and forsterite and between anorthite and enstatite. The mixtures used approached gabbro in composition. Univariant reactions encountered

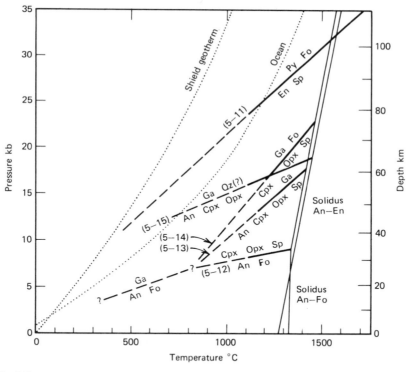

Figure 5-13. Mineral reactions in the system CaO–MgO–Al$_2$O$_3$–SiO$_2$ after Kushiro and Yoder (1966): (5-12) An + Fo $\rightleftharpoons$ Cpx + Opx + Sp; (5-13) An + Cpx + Opx + Sp $\rightleftharpoons$ Ga; (5-14) Cpx + Opx + Sp $\rightleftharpoons$ Ga + Fo; (5-15) An + Cpx + Opx $\rightleftharpoons$ Ga + Qz (?). Reaction (5-11) is from MacGregor (1964): En + Sp $\rightleftharpoons$ Py + Fo. Solid lines show the limits of experimental measurements; dashed lines are extrapolations. (With permission from Journal of Petrology, Clarendon Press, Oxford.)

involve the formation of aluminous pyroxenes and of garnets with compositions represented by the formulas $CaMgSi_2O_6.xCaAl_2SiO_6$, $MgSiO_3.xMgAl_2SiO_6$, and $CaMg_2Al_2Si_3O_{12}$:

anorthite (An) + forsterite (Fo)
$$\rightleftharpoons \text{aluminous diopside (Cpx)}$$
$$+ \text{aluminous enstatite (Opx)} + \text{spinel (Sp)}$$

$$(5\text{-}12)$$

and at a higher pressure any remaining anorthite is used up by the reaction:

anorthite (An) + aluminous diopside (Cpx)
$$+ \text{aluminous enstatite (Opx)} + \text{spinel (Sp)}$$
$$\rightleftharpoons \text{garnet (Ga)} \quad (5\text{-}13)$$

the garnet having a composition near 75% pyrope, 25% grossular (Py_3Gr_1). Additional garnet is produced at a higher pressure by reaction among the pyroxenes and spinel:

aluminous diopside + aluminous enstatite
$$+ \text{spinel} \rightleftharpoons \text{garnet} + \text{forsterite} \quad (5\text{-}14)$$

Reaction between anorthite and enstatite to produce aluminous pyroxenes is rather similar to reaction (5-12), except that quartz is produced instead of spinel. At a higher pressure anorthite reacts with the aluminous pyroxenes to yield garnet:

anorthite + aluminous diopside
$$+ \text{aluminous enstatite} = \text{garnet}$$
$$+ \text{quartz(?)} \quad (5\text{-}15)$$

This reaction occurs at a higher pressure than the similar reaction (5-13), indicating that the formation of garnet with composition close to Py_2Gr_1 takes place at higher pressures in silica-saturated environments than in silica-undersaturated environments. A similar statement is true for the formation of jadeite as shown by reactions (5-4) and (5-5) in Figure 5-11. Note also that the slope of the garnet-producing reactions (5-13) and (5-15) changes markedly with silica-saturation.

It appears from these results that mineral assemblages of anorthite with enstatite and forsterite are transformed through a transition interval characterized by aluminous pyroxenes, with garnet becoming stable on the high pressure side. Extrapolation of reactions (5-12), (5-13), and (5-14) to lower temperatures would probably produce an invariant point pinching out the aluminous pyroxene field, with a reaction involving the direct formation of a garnet from anorthite and forsterite at temperatures below about 750°C. None of the reactions in Figure 5-13 were determined at temperatures below 1100°C and it is a long extrapolation from the determined curves to the geotherms. The dependence of the positions and slopes of the garnet-producing reactions upon the chemical composition of the reacting system confirms the necessity for examining the gabbro-eclogite phase transition in systems approaching more closely the compositions of the rocks themselves. The fact that one can write a balanced chemical equation for a group of minerals is no guarantee that the reaction will occur, and even less of a guarantee that the reaction will occur in a complex rock system containing additional components.

Stability Limits of Gabbro and Eclogite

Figure 5-14a shows the preliminary diagram published by Yoder and Tilley in 1962 together with the definitive runs on samples of basalt and eclogite. Their results just below the solidus at 10 and 20 kb provide a reversed bracket for the gabbro-eclogite transition at high temperatures but, as the plotted runs show, the transition is not closely defined. At 20 kb and 1200°C, an olivine tholeiite was converted completely to an eclogitic assemblage (clinopyroxene + garnet) and at 10 kb, plagioclase feldspar was produced from partial reaction of natural eclogite in runs at 900°C and above.

The slope of the interval is based on the results of Kennedy (1959) at 500°C. Using a basaltic glass (fused olivine tholeiite) as starting material, he reported that plagioclase grew at 10 kb, decreased in amount with increasing pressure, and in later

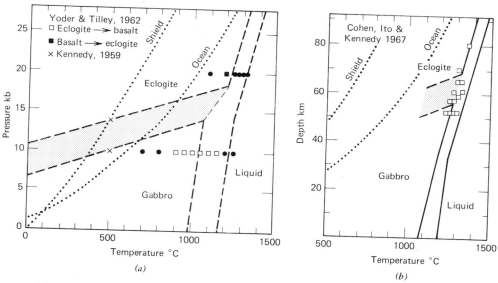

Figure 5-14. The gabbro-eclogite phase transition. Solid lines show the limit of experimental determination; dashed lines are preliminary estimates. Compare Figure 5-15. (*a*) After Yoder and Tilley (1962). Runs at 10 and 20 kbar were made with natural rock samples. Kennedy's preliminary runs were made with a basaltic glass. (With permission from Journal of Petrology, Clarendon Press, Oxford.) (*b*) After Cohen *et al.* (1967). Runs were made with a basaltic glass. (With permission from American Journal of Science, Yale University.)

experiments reported to Yoder and Tilley (pp. 472 and 504) was replaced by jadeitic pyroxene by 14 kb. Yoder and Tilley retained reservations about these results because garnet, essential for eclogite, was not produced. In a more recent paper Kennedy emphasized the problem of nucleation of garnet (Cohen et al., 1967, p. 514). In the absence of other information, however, Figure 5-14a was presented as the best estimate available for the gabbro-eclogite transition. This preliminary phase diagram has been widely reproduced in the literature, and applied to problems related to the crust and upper mantle. It is testimony to the difficulty of obtaining satisfactory results in subsolidus mineral reactions that, despite the great significance of this transition, it was not until five years later that any more results were published in detail. A preliminary report by Ringwood and Green (1964) on the recrystallization of quartz tholeiite glasses was

followed in 1966 by a comprehensive discussion of the geophysical applications of their results, and the detailed experimental data were published in 1967 (Green and Ringwood). In the same year Cohen, Ito, and Kennedy (1967) described results obtained with the olivine tholeiite glass used by Kennedy in his earlier experiments at 500°C.

Cohen, Ito, and Kennedy completed determination of the melting relationships of their basalt up to 40 kb pressure, with the results shown in Figure 5-14b; the only runs plotted are those related to the gabbro-eclogite phase transition. Their results show that the phase transition at the basalt solidus, in the temperature interval 1250 to 1300°C, lies between 16 and 21 kb with the minimum possible width being between 17.2 and 19.9 kb. Ten exploratory subsolidus runs at temperatures between 1000 and 1150°C were made in an attempt to locate the slope of the transition interval. The samples were usually inhomogeneous and results were inconsistent. The authors concluded that development of other techniques would be necessary for location of the gabbro-eclogite phase transition at lower temperatures. The difficulty of garnet nucleation, noted by all previous investigators of reactions involving pyrope, becomes an increasingly serious problem with decreasing temperature and pressure. Details of Kennedy's earlier experimental data at 500°C do not appear to have been published and his results were not used by Cohen, Ito, and Kennedy to estimate the position of the transition zone at lower temperatures. They did, however, discuss the gradient required for extrapolation of their results in such a way that the transition would occur in the region of the continental Moho, and they noted that relevant reactions such as the formation of pyrope, reaction (5-7) in Figure 5-11, and the reaction between anorthite and forsterite, reaction (5-12) in Figure 5-13, do have appropriate gradients. Such an extrapolation would be consistent with Kennedy's 500°C results and would thus be

rather similar to the zone indicated by Yoder and Tilley.

Green and Ringwood (1967) used a glass prepared from a natural quartz tholeiite as their main starting material, and by adding oxides and minerals in required proportions they obtained seven other glasses representing a range of basaltic types. Their results for quartz tholeiite B are shown in Figure 5-15a. Only the definitive runs are plotted, those which locate the solid lines bounding the transition interval between gabbro and eclogite. The lower line represents the appearance of garnet, and the upper line represents the disappearance of plagioclase; the transition interval was located between 1000°C and the solidus at about 1250°C. The transitional rocks were called garnet granulites in contrast to the gabbros or pyroxene granulites. The corresponding diagram of Green and Ringwood (Figure 7) includes additional runs on other quartz tholeiite compositions, and it shows the extent of the shifts produced in the boundary positions by minor changes in rock composition.

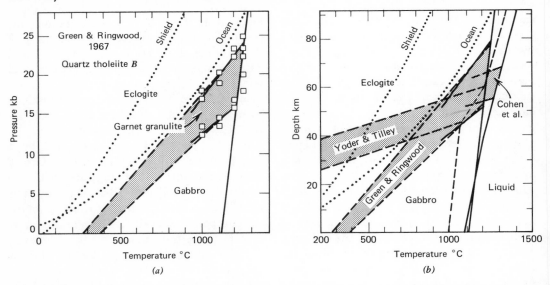

Figure 5-15. The gabbro-eclogite phase transition. Solid lines show the limit of experimental determination; dashed lines are extrapolations or estimates. Compare Figure 5-14. (*a*) After Green and Ringwood (1967). Runs were made with basaltic glass. (From Geochimica Cosmochimica Acta, **31**, 1967, with permission. Copyright Pergamon Press.) (*b*). Comparison of transition intervals from *a* and Figures 5-14*a* and *b*.

Most runs using the other basaltic compositions were limited to 1100°C. Each basalt glass crystallized in the same three principal mineral stability fields and at 1100°C the transition interval between the low pressure and high pressure assemblages varied in width from 3.5 to 12 kb. The positions of the transition intervals for two of the basalts are shown in Figure 5-16.

As Ringwood and Green pointed out (1966, p. 399) extrapolation of the boundaries of the garnet granulite field to lower temperatures has an appreciable uncertainty. The average gradient of the two experimental boundaries in Figure 5-15a is 21 bar/°C, which is similar to the slope of the jadeite reactions and several others plotted in Figures 5-11, 5-12, and 5-13. This slope was used for the extrapolation shown by the dashed lines in Figure 5-15a, with the width of the transitional zone made proportional to the absolute temperature.

The three interpretations for the position of the basalt-eclogite transition zone, based on experimental studies with natural rock compositions, are compared in Figure 5-15b. The subsolidus runs of Yoder and Tilley, the reversals with crystalline rock samples plotted in Figure 5-14a, are consistent with the results obtained by Cohen et al. with olivine tholeiite glass (Figure 5-14b). Also the subsolidus results of Green and Ringwood for the beginning of the transition zone marked by the appearance of garnet, using quartz tholeiite glass (Figure 5-15a), are consistent with the other two sets of results. There is a significant difference in the pressure required for the elimination of plagioclase, the curve of Green and Ringwood indicating higher pressures at temperatures near the solidus. With this exception, which could be the effect of compositional differences among the starting materials, the results are in reasonable agreement at temperatures near the solidus. The reaction curves in Figure 5-13 for the appearance of garnet and the disappearance of anorthite in the system CaO-MgO-Al_2O_3-SiO_2 are also consistent with the rock results at near-solidus temperatures.

The gradient for the transition interval presented tentatively by Yoder and Tilley depended upon Kennedy's observation that basaltic glass yielded plagioclase feldspar at 10 kb and 500°C, and this result conflicts with the extrapolation of Green and Ringwood's results also obtained with glass starting material. The pressure difference of more than 5 kb at 500°C is too large to be explained by uncertainties in pressure measurement in different apparatus and the discrepancy probably relates to metastable crystallization.

Many experimenters have found that subsolidus runs of short duration using glass as a starting material have yielded metastable products (Fyfe, 1960) and it is well known that metastable plagioclase is produced readily from glass, whereas magnesian garnet is extremely reluctant to nucleate from glass. Another complication is the tendency of metastable pyroxene to form first in basaltic glasses. Cohen, Ito, and Kennedy did not report the results of their subsolidus experiments at temperatures below 1200°C because they obtained inconsistent results. Green and Ringwood described experiments at 1100°C using the quartz tholeiite composition which were designed to investigate the attainment of equilibrium and reversibility of the plagioclase boundary, and they concluded that "there is no evidence that the use of glassy starting material causes growth of long-persisting metastable phases or nonnucleation of other phases at least for the compositions and experimental procedures used in this study".

The work of Green and Ringwood (Figure 5-15a) on subsolidus relations in basaltic compositions is the most detailed study available, but I would be surprised if the last word has been written on the position of the gabbro-eclogite phase transition at low temperatures. The difficulty of locating subsolidus reactions is indicated by the history of successive reports for the experimental determination of transitions among the polymorphs of Al_2SiO_5, most reports differing from the others and many with confident claims that reversibility was achieved. Location of the curve bounding the transition interval on its high pressure side (Figure 5-15a) is based on the limit of x-ray detection of plagioclase. In these multiphase assemblages such a procedure is not likely to provide a reliable slope for long extrapolation to low temperatures. Furthermore the effect of bulk composition on the slope of the transition zone remains to be determined. Figure 5-11 illustrates the effect of silica saturation on the pressure of formation of one eclogitic component, jadeite. The positions of reactions (5-13) and (5-15) in Figure 5-13 demonstrate that the degree of silica saturation may have a marked effect on the other eclogitic component, pyrope-rich garnet, not only with respect to the pressure of formation but also with respect to the slope of the garnet-producing reaction.

The Transition of Gabbro to Eclogite

Green and Ringwood studied the phase relationships for a number of basaltic compositions at 1100°C, and illustrated in diagrammatic form the nature of the mineralogical changes occurring through the transition zone for quartz tholeiite *B* and for alkali olivine basalt. The proportions of minerals present were estimated by comparison of powder patterns and diffractometer records of runs with specially prepared standards. Their figure incorporated the results of runs and interpretations of inferred reactions occurring among minerals; a somewhat simplified version is given in Figure 5-16. Note that the transition zone for the alkali olivine basalt is wider than that for the quartz tholeiite and it occurs at a lower pressure. With increasing pressure the anor-

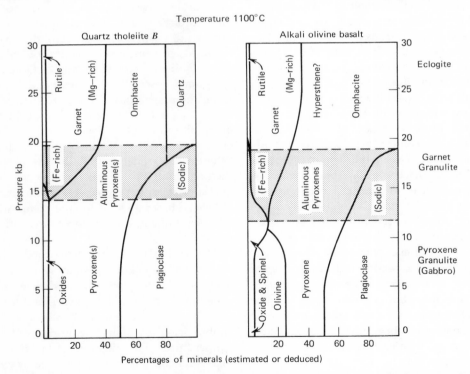

Figure 5-16. The mineralogical changes occurring within the basalt-eclogite transition for two different rocks at 1100°C, after Green and Ringwood (1967) and Ringwood and Green (1966). This is a simplified version. Note that the pressure of the phase transition interval varies with rock composition. (From Geochimica Cosmochimica Acta, **31**, 1967, with permission. Copyright Pergamon Press.)

thite component of the plagioclase reacts with the pyroxenes and olivine to yield aluminous pyroxene—see reaction (5-12)—with garnet being formed at a slightly higher pressure as in reactions (5-13) and (5-15). The pyroxenes probably contain Al_2O_3 in the form of Tschermak's molecule with the jadeite content remaining low until the upper part of the transition zone. Within the transition interval the mineralogical changes are gradual with steady decrease in plagioclase and increase in garnet. The garnet becomes richer in pyrope through the interval, and the plagioclase becomes richer in albite until this breaks down to yield the jadeite component for omphacite.

Other inferred mineralogical changes were discussed in detail by Green and Ringwood.

Ringwood and Green discussed the transition through garnet granulite in some detail in their 1966 paper with specific reference to Figure 5-16. They concluded that to a first approximation, but beyond reasonable doubt, both density and seismic velocity change regularly throughout the garnet granulite interval, from gabbro with density of 3.0 g/cm^3 to eclogite with average density 3.5 g/cm^3. Their measured results for the alkali olivine basalt at 1100°C are given in Figure 5-17. Note the regular density varia-

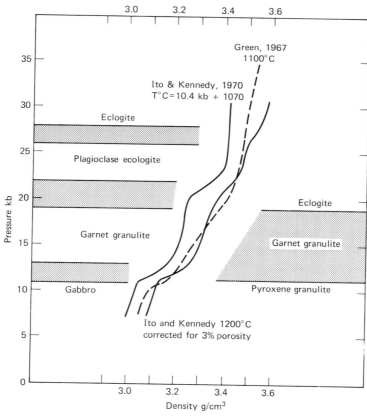

Figure 5-17. Experimentally measured variation in density as gabbro is transformed to eclogite (see transition interval in Figure 5-16). According to Green (1967) the density changes continuously through the transformation interval, but according to Ito and Kennedy (1970 and personal communication) there are two distinct density changes at the beginning and near the end of the transformation interval. (Based on preprint figure kindly supplied by K. Ito and G. C. Kennedy.)

tion through the garnet granulite interval after the first change.

In 1970 Ito and Kennedy reported experiments showing sharp density changes within the transition interval:

1. From gabbro with density near 3.0 g/cm^3 to garnet granulite with density near 3.2 g/cm^3.

2. From garnet granulite to eclogite with density 3.4–3.5 g/cm^3.

They used the same olivine tholeiite glass previously studied with Cohen (Figure 5-14*b*). Runs were made between 6 and 35 kb at points on a line believed to be 20–50°C below the solidus. The density variation curve through their data points is reproduced in Figure 5-17. The change for the incoming of garnet occurs between 12 and 14 kb, or 5 kb lower than indicated in Figure 5-14*b*. The earlier results were too high because garnet nucleates with difficulty; the 1970 results were obtained by seeding the glass starting material with powdered garnet granulite of the same bulk composition.

Ito has since informed me that the runs for the 1970 results were actually just above the solidus, so they repeated the work at 1200°C with no glass present. Figure 5-17 compares their revised density curve, corrected for 3% porosity in the run charges. Two sharp density changes near 12 and 21 kb are still present, and there is also a less marked density change near 27 kb where plagioclase disappears. According to these results it appears that the gabbro-eclogite transition could be considered as composed of three transitions; two of these are associated with density changes sufficient to cause seismic discontinuities.

Plagioclase feldspar persists in plagioclase-eclogite for a pressure interval above the plagioclase-out boundary determined by Ringwood and Green for the alkali olivine basalt. Ito and Kennedy determined the boundary of garnet appearance as P = 0.014T − 5.4, and that of plagioclase disappearance as P = 0.020T + 4.0, with reactions being reversed at 800°C and higher. These boundaries should be added to Figure 5-15*b*: an exercise for students.

DEVELOPMENT OF THE PHASE TRANSITION HYPOTHESIS

Within the space of a few years the status of the phase transition hypothesis as an explanation for the Moho has changed several times. It is clear from the preceding review that neither the structure of the crust-mantle boundary, nor the fine structure of the gabbro-eclogite phase transition is as simple as it appeared to be in the late 1950's when the hypothesis was developed by Lovering and Kennedy.

We can review the changing status of the phase transition hypothesis and the development of ideas about the nature of the Moho, against a series of historical markers.

1. The revival of the hypothesis in the late 1950's by Lovering and Kennedy.

2. The preliminary experimental work on the transition by Yoder and Tilley in 1962.

3. More detailed experimental studies on gabbroic rocks published between 1966 and 1968 by Ringwood and Green; and by Cohen, Ito, and Kennedy.

4. Experimental determination in 1970 of the fine structure of the gabbro-eclogite transition by Ito and Kennedy, which evoked a rejoinder from Green and Ringwood in 1971.

Revival by J. F. Lovering and G. C. Kennedy

The models presented by Lovering and Kennedy have been described and reviewed in Figures 5-8, 5-9, and 5-10. Figure 5-9 shows that Lovering's model is internally inconsistent, and Figure 5-10 shows that Kennedy's model is inconsistent if currently accepted geotherms are employed. It works only if temperatures beneath the oceans are lower than those beneath the continents at moderate depths. Several papers published during the following years pointed out that the Moho beneath both continents and oceans could not readily be explained as the result of the same phase transition.

One of the difficulties facing the phase transition hypothesis relates to surface heat flow. If the Moho is a phase transition variations in surface heat flow, which reflect variations in the positions of geotherms at depth, should be accompanied by changes in the thickness of the crust. Wide variations in heat flow measurements from the Pacific Ocean are not accompanied by changes in crustal thickness. Observations of this kind were cited as evidence against the phase transition hypothesis in oceanic regions. Hess's proposal that the Moho beneath the oceanic crust represents a fossil isotherm for the reaction between serpentinite and peridotite, according to model (5) in Figure 5-6, is independent of present values of heat flow.

Modifications Based on Results of H. S. Yoder and C. E. Tilley

The phase diagram published by Yoder and Tilley in 1962, and illustrated in Figure 5-14a, supported many proposals that the Moho beneath the continents, at least in some environments, could be caused by the gabbro-eclogite phase transition, whereas the Moho beneath the oceans could be caused either by a chemical change from basalt to peridotite or by some other phase transition, such as serpentinite-peridotite (models 5 and 6b in Figure 5-6).

Observations on surface heat flow and crustal thickness do not conflict with the phase transition hypothesis if the model presented by Yoder and Tilley (Figure 5-14a) is correct, because of the shallow gradient of the phase transition, about 7 bar/°C. Cohen, Ito, and Kennedy also

seem prepared to accept a shallow gradient of about 10 bar/°C. Green and Ringwood, on the other hand, presented experimental evidence that the slope of the phase transition zone was much steeper, 21 bar/°C (Figure 5-15a), which cannot be reconciled with the phase transition hypothesis because it would produce large variations in crustal thickness with surface heat flow.

Another difficulty introduced by the work of Yoder and Tilley is the indication that the gabbro-eclogite phase transition is spread through a width of several kilobars (Figure 5-14a), which represents even more kilometers within the Earth. We know that the Moho discontinuity itself may be of appreciable width, but it is not as wide as the interval in the phase diagram. Yoder and Tilley suggested that the part of the reaction producing garnet might make the largest contribution to the seismic velocity change, so that seismic techniques would sense an effective change through a much smaller depth interval. This was denied by Ringwood and Green who concluded from their experiments (Figure 5-16) that garnet increases gradually through the transition interval. They concluded further that if the Moho were a gabbro-eclogite phase transition, then the seismic velocity distribution in the transformation zone would be smeared out to such an extent that there would be no discontinuity. The more recent data of Ito and Kennedy, however, suggest that the formation of garnet does make a large contribution to the density (Figure 5-17) and therefore to the seismic velocity.

The preliminary phase diagram of Yoder and Tilley (Figure 5-14a) was widely used and applied to many petrological and geophysical problems; for example Pakiser (1965) presented a model for the crustal structure of the United States based on geothermal gradients, observed seismic discontinuities, and the basalt-eclogite transformation similar to that given in Figure 5-14a. He made excellent use of all available data, and concluded that a wide transformation zone

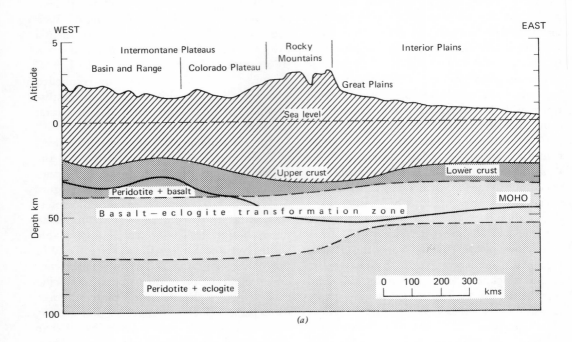

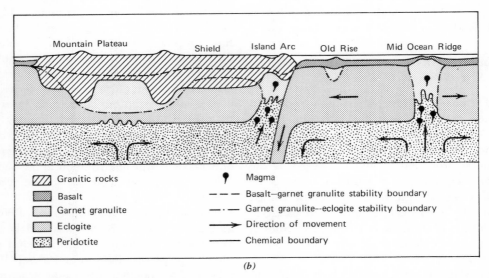

Figure 5-18. (*a*) Section through the crust from central Nevada to southeastern Nebraska showing the crustal and upper mantle structure from the Basin and Range province to the Great Plains. The inferred basalt-eclogite transformation is based on the data available in 1965 Figure 5-14*a* (after Pakiser, 1965 survey. Reprinted with permission from U.S. Geological Professional Paper 525–B). (*b*) Hypothetical crust-mantle structure based on a gabbro-eclogite phase transformation with two density contrasts bordering the garnet granulite as in Figure 5-17 (after Ito and Kennedy, 1970, with permission of Amer. Mineral. Society).

occurred mainly above the Moho in much of the eastern United States and entirely beneath the Moho in much of the western United States, as shown in Figure 5-18a. This model requires that the Moho is a chemical discontinuity separating basic crustal rocks from ultrabasic mantle rocks. It also requires considerable modification if the more recent results of Green and Ringwood are used (Figure 5-15a).

Rejection by A. E. Ringwood and D. H. Green

The hypothesis that the continental Moho is caused by an isochemical transformation from gabbro to eclogite was rejected by Ringwood and Green in 1966 on the basis of their experimental results illustrated in Figures 5-15a, 5-16, and 5-17. According to these results, and their extrapolation to lower temperatures, eclogite is thermodynamically stable through most of the normal continental crust (Figure 5-15a). This conclusion invalidates the gabbro-eclogite phase transition as an explanation of the Moho in normal continental regions. It also requires that the base of the continental crust is not of basaltic composition, but presumably of an intermediate composition occurring in the eclogite mineral facies. This produces a crustal model of type (3) or (4) in Figure 5-5, model (3) being favored.

Extrapolation of the experimental results to low temperatures is subject to uncertainties, and the position and slope of the phase boundaries may vary considerably with composition. Alternatives to Figure 5-15a therefore should not be dismissed until more data are available. The phase transition could be located at depths appropriate for the Moho in stable continental regions if extrapolations of high temperature results to low temperatures adopted a more gentle gradient for the transition as indicated by Yoder and Tilley's zone in Figure 5-15b and as discussed by Cohen, Ito, and Kennedy. This, however, seems unlikely.

Ringwood and Green reviewed the relationship between temperature at the Moho beneath the continents and the thickness of the crust, and they noted that although the surface heat flux varies by a factor of two or more in stable continental regions there is no apparent correlation between heat flow and crustal thickness.

They also argued that an eclogite layer (density 3.5 g/cm^3) immediately below the continental crust passing downwards into peridotite (density 3.3 g/cm^3) would be gravitationally unstable; furthermore the density of the upper mantle was generally believed at that time to lie between 3.3 and 3.4 g/cm^3. The gravitational instability of eclogite overlying peridotite had been mentioned in 1960 by Harris and Rowell, who stated that the phase transition hypothesis might therefore endear itself to those seeking a new cause for mantle convection. By 1966 many geophysicists were actively seeking mechanisms for mantle convection, and Ringwood and Green did develop this theme of gravitational instability in a tectonic scheme for the orogenic cycle.

Despite the arguments listed by Ringwood and Green against the phase transition hypothesis for the Moho beneath normal continental areas, they allowed the possibility that conditions for the phase transition might occur *within the crust* in tectonically active areas characterized by high heat flow. Such areas include midoceanic ridges, island arcs, continental rifts, and continental areas that have undergone recent mountain building. These are the areas where the *M* discontinuity beneath the crust is poorly defined, and where there appears to be a layer with seismic velocities intermediate between those regarded as characteristic of the crust and those characteristic of the upper mantle (Figure 5-4). Figure 5-15a shows that a geotherm tracing higher temperatures than that beneath the oceanic regions would intersect the phase transition zone; and that a phase transition possibly from gabbro to eclogite, and certainly from garnet granulite to eclogite, could occur within the crust. The similar gradients of the geotherms and the phase transition zone would cause the transition phase to be spread out through several kilometers, which is an argument against the occurrence of the phase transition in continental regions where the Moho is sharply defined, but which is quite consistent with the occurrence of the phase transition in regions characterized by intermediate crust-mantle seismic velocities or by abnormally low upper mantle velocities.

Revitalization by F. Press, K. Ito, and G. C. Kennedy

One of the major points put forward by Ringwood and Green against the existence of an eclogite layer in the upper mantle is its high density. Despite the apparent gravitational instability of eclogite overlying peridotite, however, this is the structure of the suboceanic mantle deduced by Press in 1969 from independent geophysical evidence. Press introduced the concept of gabbro crystallizing from magma beneath the midoceanic ridges and becoming converted to eclogite as a result of cooling upon lateral migration away from the ridge (Figure 14-11).

Ringwood and Green also concluded that the seismic velocity distribution within the gabbro-eclogite transition interval would be smeared out to such an extent that occurrence of the transition within the Earth could not produce a seismic discontinuity. The results of Ito and Kennedy (Figure 5-17) introduce two abrupt changes in density associated with the transition, each of which would be capable of producing a relatively sharp discontinuity in seismic velocity. They related the garnet granulite interval to the layers with P-wave velocities in the range 7.4–7.8 km/sec which occur beneath some tectonically active areas (Figure 5-4). In some of these areas there are double discontinuities associated with the crust-mantle boundary, which could correspond to (a) the phase boundaries above and below the garnet granulite zone, or to (b) a combination of a chemical change from crust of acid or intermediate composition to underlying garnet granulite, and a phase change where the garnet granulite is replaced at a deeper level by eclogite.

Ito and Kennedy suggested that eclogite could be stable at the Moho beneath shield areas where the temperature is about 400°C at 35 km depth, but that in active regions with higher heat flow garnet granulite would be stable beneath the crust. There is no evidence for a garnet granulite layer where the temperature is low as at the suboceanic Moho and they discussed various possibilities to account for this. Great flexibility is introduced by the combination of two discontinuities associated with the gabbro-eclogite phase transition and the various chemical discontinuities possible among granitic rocks, peridotite, and rocks of gabbroic composition. Their sketch of hypothetical crust-mantle structures demonstrating some of these combinations is shown in Figure 5-18*b*.

Rejoinder by D. H. Green and A. E. Ringwood

This story of the development of a hypothesis is not closed. In the 1971 preprint from which Figure 5-17 was taken, Ito and Kennedy concluded that their results on the stability field of eclogite measured at temperatures down to 800°C contrast with those published by Green and Ringwood (Figure 5-15*a*). Green and Ringwood respond that this statement is not true. Green kindly gave me a 1971 preprint in which he and Ringwood concluded that the new work by Ito and Kennedy is in excellent agreement with their own earlier work and conclusions. They described the contribution as an improved determination of the gabbro-eclogite transition for a specific rock composition which adds little information not already demonstrated in their earlier work (Figures 5-15*a* and 5-16).

Green and Ringwood reviewed the experimental data and the range of uncertainty in the determination of densities for the Ito and Kennedy results in Figure 5-17, and they concluded that there is no experimental justification for the stepped density versus pressure curve. In their view, as stated in 1966, the gabbro-eclogite phase transition may certainly be involved in active regions with high heat flow, but it is not the cause of seismic discontinuities. Teachers will continue to find excellent material for seminars in future publications related to this controversy.

6. *The Composition and Mineralogy of the Earth's Mantle*

INTRODUCTION

The core and mantle together essentially determine the bulk composition of the Earth. The mantle constitutes 67.2% of the total mass of the Earth and 90% of its volume. The crust, hydrosphere, and atmosphere, which are so significant in geological and geochemical theories, together amount to less than 0.5% by weight of the whole Earth. Geophysical measurements have provided reasonable physical models for the Earth (Chapter 3), but the development of chemical models which correlate with the variation of physical properties within the Earth is hindered by lack of direct information about the chemistry of the Earth's interior.

We have four main approaches for estimating the composition and mineralogy of material within the earth:

1. The use of data from extraterrestrial sources and the formulation of physical and geochemical models for the origin of the solar system and the Earth. Sources include the stellar and solar abundances of the elements and the compositions of meteorites. The lunar rocks have been strongly fractionated, so they provide little direct information about early planetary processes and compositions.

2. The study of ultramafic and basaltic igneous rocks derived from the mantle. The rocks selected as representatives of the upper mantle must have concentrations of K, Th, and U adequate to account for surface heat flow.

3. Comparison of the variations of density, elastic properties, and other properties in the mantle measured by geophysical methods, with experimentally determined values for appropriate materials at high pressures and temperatures. This involves the results of static compression and shock wave experiments.

4. Given chemical compositions deduced from the above approaches the mineral phase assemblages and phase transitions at various depths are estimated by phase equilibrium studies at high pressures and temperatures using whole rocks, silicate systems, or germanate model systems.

We cannot infer the bulk composition of the Earth from terrestrial rocks, because the rocks exposed in the crust have been too strongly fractionated by various geological processes, but examination of igneous rocks derived from the mantle does place limits on the composition of the uppermost mantle. For the Earth as a whole we have to rely on interpretation of the chemistry of extraterrestrial bodies.

EXTRATERRESTRIAL EVIDENCE FOR THE COMPOSITION OF THE EARTH

In this section we consider only enough of the voluminous literature on the chemistry of extraterrestrial materials to permit comprehension of the basis for estimates of core and mantle composition. We start with the cosmic abundances of the elements which are generally assumed to be approximated by the solar abundances and stellar abundances, neither of which have been adequately determined. Meteorites, the oldest rocks known to us, provide samples of extraterrestrial material that can be analyzed in exhaustive detail. Much effort has been directed to recognition among the meteorites of the most primitive, least differentiated material. Comparison of the chemistry, mineralogy, and petrology of the meteorites with terrestrial igneous rocks should eventually yield an acceptable model for Earth formation and differentiation, although at present the application of meteorite analogies to quantitative calculations of Earth compositions must be regarded as conjectural.

Cosmic and Solar Abundances of the Elements

The term "cosmic" abundances refers to elements in the solar system and the stars of our galaxy. The composition of the sun is central to all discussions relevant to the bulk composition of the Earth, but even during the past decade there have been large changes in the estimates of the solar abundances of some elements. Stellar spectroscopic abundance measurements extend the range of elements measured compared to those measured in the sun, without improving the precision of abundances. Table 6-1 compares estimates of cosmic abundances of the elements with the solar abundances and the abundances of the elements in selected meteorites. The values are listed in cosmic abundance units (cau), which give the number of atoms present per 10^6 Si atoms.

Suess and Urey refined the subject of cos-mic abundances in the 1950's and their interpretive approach used data on meteoritic and terrestrial abundances as well as solar and stellar abundances. Table 6-1 lists their estimates published in 1956, and compares them with the solar abundances adopted by Goles after a careful review of available data in 1969 and with abundance data for two types of carbonaceous chondrites obtained for the most part after 1956. The fifth column gives a 1967 estimate of cosmic abundances obtained by normalizing the recent solar and meteoritic abundances and by averaging the abundance ratios of 10 nonvolatile elements. The sixth column gives Goles' adopted abundances for the solar system and this bears comparison with the solar abundances. Comparison of the meteorite data with the solar and cosmic abundances indicates that there are significant differences among them. Let us therefore examine meteorites in somewhat more detail.

Classification and Chemical Composition of Meteorites

Meteorites differ from each other in terms of chemistry, mineralogy, structure, and color, and all of these properties have been used as criteria for classification. Figure 6-1 is a classification scheme illustrating the main chemical and mineralogical features. The four major groups in the left portion of the diagram are the irons, the stony-irons, and the stony meteorites, the stones comprising the chondrites and the achondrites. These are distinguished from each other on the basis of their proportions of metal (iron-nickel alloy) to silicate, which is shown by the horizontal axis. Troilite, FeS, occurs in all except the achondrites. The right-hand portion of Figure 6-1 shows the subclassification of the major classes.

Of all meteorites known, 61% are stones, 35% are irons, and only 4% are stony-irons; but of all the meteorites that were actually

TABLE 6-1   Comparisons of Cosmic, Solar, and Meteoritic Abundances of the Elements (from Goles, 1969) in Cosmic Abundance Units (cau, atoms per 10^6 Si atoms; to two significant figures)

Element	Cosmic (Suess and Urey, 1956)	Solar (Goles, 1969)	Carbonaceous chondrites		Cosmic (Cameron, 1968)	Solar System (Goles, 1969)
			Type I	Type II		
H	4.0×10^{10}	4.8×10^{10}	5.5×10^6	3.0×10^6	2.6×10^{10}	4.8×10^{10}
Li	100	1.7	50	16	45	16
Be	20	11	—	0.81	0.69	0.81
C	3.5×10^6	1.7×10^7	8.2×10^5	4.5×10^5	1.4×10^7	1.7×10^7
N	6.6×10^6	4.6×10^6	4.9×10^4	2.6×10^4	2.4×10^6	4.6×10^6
O	2.2×10^7	4.4×10^7	7.7×10^6	5.5×10^6	2.4×10^7	4.4×10^7
Na	4.4×10^4	9.1×10^4	6.0×10^4	3.5×10^4	6.3×10^4	3.5×10^4
Mg	9.1×10^5	7.4×10^5	1.1×10^6	1.0×10^6	1.1×10^6	1.0×10^6
Al	9.5×10^4	6.9×10^4	8.5×10^4	8.4×10^4	8.5×10^4	8.4×10^4
Si	$\equiv 1.0 \times 10^6$	$\equiv 1.0 \times 10^6$	$\equiv 1.0 \times 10^6$	$\equiv 1.0 \times 10^6$	$\equiv 1.0 \times 10^6$	$\equiv 1.0 \times 10^6$
P	1.0×10^4	1.9×10^4	1.3×10^4	8100	1.3×10^4	8100
S	3.8×10^5	8.0×10^5	5.1×10^5	2.3×10^5	5.1×10^5	8.0×10^5
K	3100	2200	3200	2100	3200	2100
Ca	4.90×10^4	6.0×10^4	7.2×10^4	7.2×10^4	7.4×10^4	7.2×10^4
Sc	28	30	31	35	33	35
Ti	2200	1800	2300	2400	2300	2400
V	220	630	300	590	900	590
Cr	7800	3800	1.3×10^4	1.2×10^4	1.2×10^4	1.2×10^4
Mn	6900	3000	9300	6200	8800	6200
Fe	6.0×10^5	2.5×10^5	9.0×10^5	8.3×10^5	8.9×10^5	2.5×10^5
Co	1800	2400	2200	1900	2300	1900
Ni	2.7×10^4	2.3×10^4	4.9×10^4	4.5×10^4	4.6×10^4	4.5×10^4
Cu	210	160	590	420	920	420
Zn	490	250	1500	630	1500	630
Ga	11	20	46	28	46	28
Ge	50	16	130	76	130	76
Rb	6.5	10	6.0	4.1	6.0	4.1
Sr	19	25	24	25	58	25
Y	8.9	80(?)	4.6	4.7	4.6	4.7
Zr	55	20	32	23	30	23
Nb	1.0	10	—	—	1.2	0.90
Mo	2.4	10	—	—	2.5	2.5
Ru	1.5	3	1.9	1.8	1.6	1.8
Rh	0.21	1	—	—	0.33	0.33
Pd	0.68	1	1.3	1.3	1.5	1.3
Ag	0.26	0.4	0.95	0.33	0.5	0.33
Cd	0.89	3	2.1	1.2	2.1	1.2
In	0.11	1	0.22	0.10	0.22	0.10
Sn	1.3	6(?)	4.2	1.7	4.2	1.7
Sb	0.25	0.1(?)	0.40	0.20	0.38	0.20
Ba	3.7	16	4.7	5.0	4.7	5.0
Yb	0.22	8(?)	0.21	0.22	0.21	0.22
Pb	0.47	4	2.9	1.3	2.9	1.3

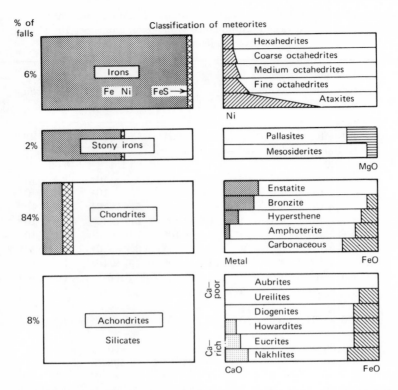

Figure 6-1. Meteorites are divided into four broad classes (left) on the basis of their ratio of metal (shaded) to silicate (white). Further subclassification (right) is based on various compositional or structural criteria (after E. Anders. Reprinted from Accounts of Chemical Research, October 1968, p. 289. Copyright American Chemical Society, reprinted by permission).

observed to fall, over 92% are stones and of these 84% are chrondrites. This has led many investigators to conclude that outer space is populated preponderately by chondritic meteorites. A disproportionate number of iron meteorites has been collected, which certainly results from their unusual and distinctive appearance. Stony meteorites are easily overlooked among terrestrial materials unless they are observed to fall.

Iron meteorites. The iron meteorites consist essentially of iron-nickel alloy distributed between two phases, kamacite with composition $Fe_{93}Ni_7$ and taenite with composition $Fe_{65}Ni_{35}$, together with troilite. The absolute value for the nickel content of all iron meteorites is about 11% by weight; no

irons contain less than 4% Ni, and a very few contain more than 20% Ni. There are small included crystals containing P, S, Cr, and C and, rarely, magnesian olivine and enstatite. The average composition of the metal is listed in Table 6-2.

Figure 6-1 shows the three main types of iron meteorites, the hexahedrites, the octahedrites, and the ataxites which are classified on the basis of their nickel content as shown and on their structure. Octahedrites are the commonest class and only one ataxite fall has been observed.

Iron meteorites are extremely fractionated compared with other meteorites and the sun (Tables 6-1 and 6-2). Such an efficient fractionation could have been achieved by a process of melting and separation of the

TABLE 6-2   Compositions of Meteorites and Meteoritic Material (weight percent)

	Iron meteorites		Average Bronzite Chondrite				Average Carbonaceous Chondrite	
	(1) Average Metal	(2) Metal	(3) Silicate	(4) Chondrite	Oxide	(5) Chondrite	(6) Type I	(7) C-, S- and H_2O-free
O			43.7	33.24	SiO_2	36.57	22.63	33.32
Fe	90.78	90.72	9.88	27.79	MgO	23.69	15.96	23.50
Si			22.5	17.10	FeO	9.67	10.42	35.47
Mg			18.8	14.29	Al_2O_3	2.30	1.64	2.41
S			—	1.93	CaO	1.77	1.56	2.30
Ni	8.59	8.80	—	1.64	Na_2O	0.86	0.74	1.10
Ca			1.67	1.27	Cr_2O_3	0.42	0.34	—
Al			1.60	1.22	MnO	0.33	0.22	—
Na			0.84	0.64	P_2O_5	0.26	0.35	—
Cr			0.38	0.29	K_2O	0.10	0.07	—
Mn			0.33	0.25	TiO_2	0.10	0.07	—
P			0.14	0.11	CoO	—	0.06	—
Co			—	0.09	NiO	—	1.29	1.90
K			0.11	0.08	H_2O	—	19.29	—
Ti			0.08	0.06	C	—	3.97	—
FeS			(5.3)	—	Organ.	—	5.53	—
	100.00	100.00	100.03	100.00	Fe	16.90	—	100.00
					Ni	1.64	—	
					Co	0.09	—	
					FeS	5.30	16.73	
						100.00	100.87	

(1) Average composition of metal from iron meteorites. Brown and Paterson, 1947.
(2)–(5) Average bronzite chondrite. B. Mason, 1965.
(2) Average metal from bronzite chondrite.
(3) Average silicate from bronzite chondrite.
(4) Average bronzite chondrite in terms of elements.
(5) Average bronzite chondrite in terms of oxides.
(6) Average carbonaceous chondrite Type I. Wilk, 1956.
(7) Analysis (6) recalculated on C-, S-, and H_2O-free basis. A. E. Ringwood, 1966.

molten alloy, presumably by gravity settling, from a complementary silicate rock or immiscible silicate liquid. Study of the textures and phase relationships of the kamacite and taenite indicate that the cooling rate between 600 and 400°C was very slow, in the range 1 to 10°C per million years, which is consistent with the rates expected if the material were situated near the center of an asteroidal body with radius 70 to 200 km.

Achondrites. The achondrites are stony meteorites without chondrules amounting to only 8% of observed falls. They are very similar to terrestrial igneous rocks in texture and mineralogy and they exhibit a wide range of chemical composition and mineralogy. Nickel-iron alloy is rarely present. Figure 6-1 shows their subclassification into calcium-rich and calcium-poor types, and the positive correlation between calcium content and the ratio of FeO to MgO. These are usually regarded as differentiated materials which have crystallized from a silicate melt in a meteorite parent body. Measured ratios of K/Rb, Rb/Sr, $^{87}Sr/^{86}Sr$, and K/U in meteorites and terrestrial rocks have been used to support the thesis that the composition of the mantle is closer to that of Ca-rich achondrites than to that of chondrites. Implications for interpretation of the Moho, given an Earth model with outer layers composed of achondritic material, were reviewed in Chapter 5.

Chondrites. Chondrites are characterized by the presence of chondrules, small rounded bodies consisting largely of olivine or pyroxene or both, with an average diameter of about 1 mm. The textures of chondrules indicate an igneous origin: they crystallized from drops of silicate liquid. Because it is generally agreed that of all meteoritic material the chondrites represent most closely the composition of the primordial dust of the solar nebula, they have been subjected to intensive chemical and mineralogical study. This has established the existence of significant variations among them.

In terms of their elemental compositions chondrites are very similar to one another, but Figure 6-1 shows a subdivision into five classes on the basis of degree of oxidation of their iron. The more oxidized chondrites contain less metal and FeS, and more FeO in the silicates. There are also differences in the absolute abundances of iron. There are distinct chemical differences between each group of chondrites, as shown in Figure 6-2, which plots both the oxidation state of the iron and the abundance of iron. The dashed lines with 45° slope represent constant total iron contents, and the areas covering analyses for the different groups lie on at least two lines.

The carbonaceous chondrites are the most primitive types among the chondrites. Their mineralogy is not well known because of their fine grain size and the presence of opaque

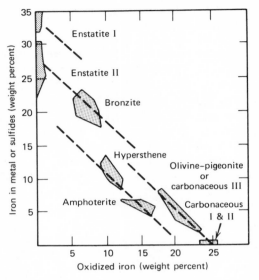

Figure 6-2. Relationship between oxidized iron and iron as metal or sulfide in chondritic meteorites. Dashed lines are lines of constant total iron content. Based on data from B. Mason and H. Craig (after Kaula, 1968, with permission of John Wiley, New York).

carbonaceous material which hinders microscopic examination. They are characterized by the presence of hydrated silicate material, including minerals such as serpentine and organic compounds in amounts up to 10%. There is little or no nickel-iron alloy because the iron is usually fully oxidized. Types I, II, and III are recognized on the basis of volatile element content: Type I does not contain chondrules. These volatile-rich chondrites obviously formed at low temperatures and they cannot have been subjected to temperatures any higher than a few hundred degrees. It has been suggested that they may be more abundant in space than the ordinary chondrites (bronzite and hypersthene chondrites), despite the fact that the latter constitute more than 90% of chondrite falls. The carbonaceous chondrites are so friable that most of them may break up when they enter the atmosphere. Analysis of the first sample of lunar soil returned from the Apollo 11 voyage in 1969 supports this suggestion; the chemistry of the soil indicates that it contains an admixture of about 2 wt% of carbonaceous chondrites (Keays, *et al.*, 1970).

The other classes of chondrites have mineralogy consistent with the attainment of high temperatures during at least one stage of their history. The mineralogy of the various classes varies around an average of 40% olivine, 30% pyroxene, 10% oligoclase, 10–20% nickel iron alloy, and 5–15% troilite. Note the sodic character of the plagioclase; in terrestrial igneous rocks containing 70% of magnesian olivine and pyroxene, plagioclase would be much richer in the anorthite component. Table 6-2 lists the average compositions of the metal and silicate fractions in chondrites. The metal fraction is very similar to the average composition of the metal in iron meteorites. The textures of many chondrites indicate that they have been heated and metamorphosed. Those with sharply defined chondrules have not been metamorphosed and the mineral compositions indicate lack of equilibrium. In others recrystallization has caused blurring of the chondrule boundaries, and the minerals approach equilibrium compositions appropriate for the temperature experienced. The degree of metamorphism thus provides another criterion for the classification of chondrites.

The chemical compositions of carbonaceous chondrites are compared with the cosmic and solar element abundances in Table 6-1 and with the bronzite chondrites in Table 6-2. The abundances of most of the nonvolatile elements in the chondrites agree with those of the sun, which lends support to the development of a chondritic Earth model. It has also been shown that the rate of heat production in an Earth of chondritic composition is consistent with the known heat flow through the surface of the Earth, provided that the radioactive elements are strongly concentrated towards the surface. This is not a unique argument for a chondritic model, however, because the same heat flow can be produced using other models.

Abundance patterns in chondrites. The detailed chemical study of meteorites has revealed the pattern of element abundances depicted in Figures 6-3 and 6-4. The abundances of 48 elements, mostly in the transition groups of the periodic table, remain similar to solar and cosmic abundances in all classes of meteorites, but the shaded elements in Figure 6-3 are depleted in many meteorites.

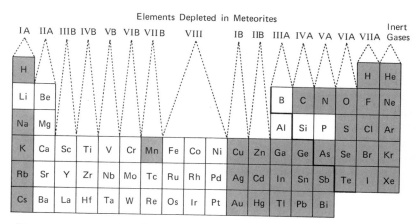

Figure 6-3. Elements in meteorites (after Anders, 1968). With the exception of Mn, all depleted elements (shaded) are situated in the main groups of the periodic table. Their only common property is volatility. (Reprinted, by permission, from Accounts of Chemical Research, October 1968, p. 289. Copyright American Chemical Society.)

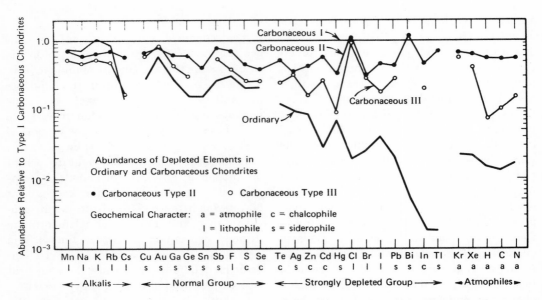

Figure 6-4. Elements in meteorites (after Larimer and Anders, 1967). Regardless of geochemical character all depleted elements are underabundant in Type II and III carbonaceous chondrites by nearly constant factors: 0.6 and 0.3 respectively. Apparent exceptions occur at Cl and Bi, but the parallelism of the curves suggests that the fault lies with the Type I data used for normalization. In ordinary chondites elements of the "normal" group are likewise depleted by a constant factor of ~0.25, while elements of the "strongly depleted" and "atmophile" groups are depleted by progressively smaller factors. Alkalis other than Cs are not depleted. (From Geochimica Cosmochimica Acta, **31**, 1967, with permission. Copyright Pergamon Press.)

Type I carbonaceous chondrites approximate most nearly the solar abundances (Table 6-1), and elements which are strongly depleted in ordinary chondrites may be only slightly depleted in other carbonaceous chondrites and enstatite chondrites. Figure 6-4 compares the depletion in the various chondrites. The more volatile, depleted elements decrease in abundance consistently from carbonaceous chondrites Types I, II and III in ratios of 1.0 to 0.6 to 0.3, and enstatite chondrites of Type I fit into the same pattern.

It has been shown by Larimer and Anders (1967) and Anders (1968) that the abundance patterns of trace elements in chondrites can only be explained if the meteorites are a mixture of two types of materials:

1. Material that lost most of its volatile components during a high-temperature event.

2. A low-temperature material that retained its volatile elements.

The observed trace element depletion factors can be explained if we postulate that carbonaceous chondrites of Type I contain 100% of the low-temperature material, and Type III contains only 30% of the low-temperature material. There is a second fractionation pattern evident in the ordinary chondrites and Type II enstatite chondrites in which some elements show even further depletion.

The trace element fractionation patterns shown in Figure 6-4 strongly support a two-component origin for the chondrites. Loss of the depleted elements from the high-temperature fraction could possibly have occurred (a) during accretion of the meteorite parent bodies from the solar nebula, or (b) during a later reheating of these bodies. Larimer and

Anders concluded that the fractionations must have occurred in the solar nebula as it cooled from high temperatures, and that they could not have occurred in meteorite parent bodies. Some other properties of meteorites probably were established in parent bodies. The evidence suggests that most meteorites accreted at a temperature of 520–680°K and carbonaceous chondrites at ≤ 400°K.

Genetic Relationships Among Meteorites

One aim of meteorite studies is to determine the petrogenetic relationships among them and to apply these relationships to the history of planetary origin and evolution. The meteorites were classified in Figure 6-1 and their chemistry has been illustrated in Figures 6-2 and 6-4. Carbonaceous chondrites of Type I are of special interest because their element abundances agree so well with the solar abundances. Ringwood (1966) pointed out that, with very few exceptions, the compositions of all chondrites could be derived solely by the removal of appropriate

amounts of trace and minor elements from the Type I carbonaceous chondrites. Figure 6-5 is Ringwood's interpretation of the genetic relationships among the principal groups of meteorites, showing some of the major chemical variations and the processes required for their derivation from the primitive carbonaceous chondrite.

There are three main theories for the origin of meteorites, each with many variants. They begin with the condensation from the primeval cloud, the solar nebula, of material similar to chondrites. Planetary models assume that all meteorites are fragments from a single disrupted planet. Serious objections to this theory are overcome in the second theory by the assumption that meteorites are derived from two successive generations of bodies; the primary objects of lunar dimensions which were subsequently fragmented by mutual collisions and the secondary objects of asteroidal size which accumulated from the debris about 4.3×10^9 years ago and which were in turn broken into fragments to produce the meteorites. This is a complex theory and other investigators prefer

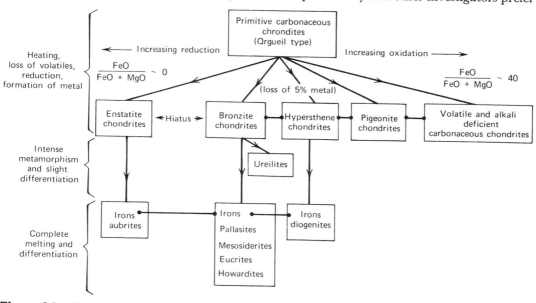

Figure 6-5. Genetic relationships among principal groups of meteorites (after Ringwood, 1961, see also Ringwood 1966. From Geochimica Cosmochimica Acta, **24**, 1961, with permission. Copyright Pergamon Press).

a third theory proposing that meteorites were formed in asteroidal bodies or planetismals which are presumably easier to break up than a single body of planetary size.

Whatever the precise sequence of events Figure 6-5 indicates that all of the chondrites could be derived from carbonaceous chondrite, Type I, by selective loss of volatile elements (Figure 6-4) and by varying degrees of reduction (Figure 6-2). Figure 6-2 also shows that hypersthene chondrites must also be depleted in about 5% of metallic iron, if derived from carbonaceous chondrites. The iron meteorites and the achondrites can then be derived from specific chondrites by melting and differentiation as depicted in Figure 6-5.

Larimer and Anders (1967) emphasized that the observed uniform depletion factors (Figure 6-4) cannot be attained in a simple heating process, and they concluded that the only feasible alternative was a two-component model with chondrites being mixtures of an undepleted fraction (A) and a depleted fraction (B). They identified fraction A as ^{18}O-rich matrix, and fraction B as ^{18}O-poor high-temperature material comprising the chondrules-plus-metal.

Apparent ages reported for meteorites lie consistently within the range 4.4–4.7×10^9 years, and these date the event resulting in the differentiation of meteoritic material into irons and achondrites. Gas retention ages give a wider spread, including an apparent age at 600×10^6 years, which could possibly represent a catastrophic collision in space between parent bodies. Studies of excess ^{129}Xe concentrations in meteorites, derived from the extinct radionuclide ^{129}I, indicate that the history of the solar system extended no further back than 200×10^6 years beyond the time of cooling the meteorite parent bodies about 4.5×10^9 years ago.

Origin and Differentiation of the Earth

Theories for the origin of the Earth begin with nucleosynthesis in the Galaxy, about 10×10^9 years ago, when various nuclear processes formed more complex elements from the primeval hydrogen. Matter initially dispersed through the rotating galaxy gathered together into more dense clouds, and the increased gravitational attraction caused these to collapse inward to form protostars. Gravitational potential energy was converted to heat and the gas pressure increased. The solar nebula produced in this way consisted of a large protosun, at high temperature, and a thin disc-shaped nebula of gas and dust particles, with temperature decreasing away from the protosun. Several lines of evidence suggest that the solar system was formed about 4.6×10^9 years ago, either by the action of some external force which produced the planets catastrophically from the sun, or more likely by the local condensation of gas and aggregation of particles within the rotating nebula to form small solid bodies which accreted to form protoplanets and planets. It is usually assumed that meteoritic material was formed during this stage of accretion of the tenuous solar nebula into solid bodies about 4.55×10^9 years ago.

The sun contains over 99.6% of the mass of the solar system and its composition is thus effectively that of the whole system, both now and probably when the solar nebula was forming. An estimate of its composition is compared in Table 6-1 with the compositions of carbonaceous chondrites, and it is clear that these meteorites have been depleted in volatile elements compared to the sun; H, C, N, O, and the noble gases are depleted by factors of 10 to 100. The achondrites and other chondrites are even more deficient in volatile elements as shown by Figure 6-4. Some of the more volatile elements that were lost during accretion of the parent meteorite bodies were presumably lost from the Earth as well. On the other hand the relative abundances of many elements within the sun differ only by small factors from those in the chondrites. The abundances of the non-volatile lithophile elements Mg, Al, Ca, Sc, Ti, Sr, and Ba are about the same in the sun and carbonaceous chondrites as are those of Co and Ni. It is therefore assumed that the abundances of the nonvolatile elements in the sun and carbonaceous chondrites provide a good estimate of element abundances in the Earth. There is one possible exception: Table 6-1 shows a considerable excess of Fe in the meteorites compared to the sun. Ringwood maintains, however, that when realistic estimates of uncertainties are made, it is not justifiable to conclude that solar Fe abundance is significantly less than that in meteorites.

According to most theories the Earth was formed by the cold accretion of particles of metal, troilite, and silicates with bulk composition approximated by chondrites (Table 6-2). Mason, Anders, and especially Urey have developed models involving several stages. Fractionation of iron with respect to silicate may have occurred in the solar nebula. At least one period of high temperature is required during the preterrestrial stage, and this may have developed in the dust cloud itself, or within earlier generations of bodies of asteroidal or lunar size that were subsequently fragmented. A rather complex series of events is required, causing the loss of volatile elements from the early accreted material believed to be similar to carbonaceous chondrites (Figure 6-4), the reduction of iron oxides to yield metal (Figure 6-2), and the formation of chondrules. Once the Earth had formed, slowly enough that conversion of gravitational potential energy to heat did not cause melting, then heating by short-lived radioactive elements would cause melting of the dispersed iron within about 600×10^6 years. The heavy molten iron, incorporating nickel and troilite (Table 6-2), would become concentrated towards the center of the Earth, while the lighter silicate fraction would be displaced outward. According to some models the silicate fraction also was partially fused, and in other models the fractionation was achieved by convection without melting of the silicate phases. The mantle and core material remain in chemical equilibrium with each other during the differentiation.

Ringwood concluded that these multistage theories lacked plausibility mainly because of their complexity, and he proposed a single-stage theory involving the processes shown in Figure 6-5. The Earth formed directly by accretion from the primitive oxidized dust in the solar system, which he considered chemically equivalent to carbonaceous chondrites Type I with a slightly higher H/C ratio (Tables 6-1 and 6-2). As a direct result of the accretion process, and simultaneously with it, there occurred (a) the loss of volatile elements (Figure 6-4), (b) reduction to metal or iron oxides (Figure 6-2), nickel oxide, and some silica, (c) melting of both metal and silicate phases, and (d) a major differentiation of the Earth into metallic core and silicate mantle, followed by (e) further differentiation within the silicate mantle. Ringwood argued against an earlier silicate-iron fractionation within the nebula, claiming in effect that the solar Fe/Si ratio is in error (Table 6-1). He maintained also that although the mantle and core are separately in equilibrium, they are not in chemical equilibrium with each other as required in the multistage theories.

Compositions of the Core and Mantle

The masses of the core and mantle are known from geophysical measurements, and the bulk composition of the Earth is estimated by assigning compositions to the core and mantle and combining these in the appropriate proportions. The results obtained thus depend upon the model adopted for the origin and differentiation of the Earth. An early estimate by Washington is compared in Table 6-3 with recent estimates derived from the multistage and single-stage theories.

Washington's estimate is based on a model with a central core composed of material equivalent to iron meteorites, subsequent layers composed of stony-iron and chondritic material, and a dominant peridotite shell

TABLE 6-3   Estimated Compositions of the Whole Earth, the Core, and Mantle Using Extraterrestrial Information (weight percent)

	Earth			Core		Mantle			
	Washington 1925	Mason 1966	Ringwood 1966	Mason 1966	Ringwood 1966	Mason 1966	Oxide	Mason 1966	Ringwood 1966
O	28	30	30			44	SiO_2	48	43
Fe	40	35	31	86	84	9.9	MgO	31	38
Si	15	15	18	—	11	23	FeO	13	9.3
Mg	8.7	13	16	6.0		19	Fe_2O_3	—	—
S	0.64	1.9	—	7.4			Al_2O_3	3.0	3.9
Ni	3.2	2.4	1.7		5.3		CaO	2.3	3.7
Ca	2.5	1.1	1.8			1.7	Na_2O	1.1	1.8
Al	1.8	1.1	1.4			1.6	Cr_2O_3	0.55	—
Na	0.39	0.57	0.9			0.84	MnO	0.43	
Cr	0.20	0.26	—			0.38	P_2O_5	0.34	
Mn	0.07	0.22	—			0.33	K_2O	0.13	
P	0.11	0.10	—			0.14	TiO_2	0.13	
Co	0.23	0.13	—	0.40	—				
K	0.14	0.07	—			0.11			
Ti	0.02	0.05	—			0.08			

with the composition of achondrites; the outer shells of basalt and granite amount to a small fraction.

The second column of Table 6-3 gives Mason's estimate of the Earth's composition based on the average composition of the bronzite chondrites as listed in Table 6-2. He used Birch's model of a mantle plus crust making up 67.6% and a core 32.4% the mass of the Earth. The calculation assumes that the mantle and crust together have the same composition as the silicate portion of the bronzite chondrites, and that the 5.3% of troilite in the bronzite chondrites is incorporated in the core, the remaining 27.1% of which is nickel-iron of the same composition as in bronzite chondrites. The calculated compositions of the core and of the mantle plus crust are also listed in the table, the mantle plus crust composition being given in terms of elements and also of oxides. Mason emphasized that the assumptions that he made are oversimplifications, but he con-

sidered the results to be encouraging when compared with other estimates (Tables 6-3 and 6-6).

Ringwood adopted a single-stage model for the origin of the Earth mainly because of the complexity of the multistage models. But in order to estimate the composition of the Earth and its major parts from meteoritic compositions he had to make a more complex series of assumptions than did Mason. Ringwood took as his starting material the carbonaceous chondrite, Type I, which is listed in Table 6-2 and the last column shows the principal components recalculated on a C-, S-, and H_2O-free basis. He assumed that this composition approximates closely to the primitive, accreting, oxidized Earth in terms of nonvolatile elements. He concluded that geochemical and geophysical evidence indicates that the FeO/(FeO + MgO) molecular ratio of the mantle is between 0.1 and 0.2, and he adopted a value of 0.12. The chemical analysis in Table 6-2 was rearranged to give this ratio by reducing all of the nickel and an appropriate amount of iron, and by transferring the reduced nickel and iron to the core. This provided a core amounting to only 26.48% of the mass of the Earth, and in order to increase the mass of the core to its correct value of 31% more material must be transferred from the silicate portion of the devolatilized and partly reduced carbonaceous chondrite. Ringwood concluded that SiO_2 is the common oxide that would be most readily reduced to metal, and in order to obtain an Earth model from the primitive composition, he therefore transferred 11 wt/% of silicon (20 atomic %) from the mantle to the core; this provided the required mantle/core mass ratio. The resulting estimate of the Earth's composition is shown in the third column of Table 6-3.

The core composition estimates by Mason and Ringwood are compared in Table 6-3. The density of the outer core shown in Figure 2-5 is at least 10% lower than the probable density of iron at these conditions, which indicates that the core contains a low density element in addition to iron. The only likely elements appear to be sulfur or silicon. Mason prefers an Earth model with core and mantle in equilibrium, which denies the existence of free silicon in the core if the mantle contains appreciable amounts of FeO. He concluded that the addition of sulfur to the core satisfied both geochemical and geophysical requirements. Ringwood, on the other hand, concluded that most of the primordial sulfur was lost from the Earth by

volatilization either before or during accretion, and that the separation of the mantle from the core occurred in such a manner that the compositions of the core and mantle are not in chemical equilibrium. Silicon is then considered to be the most likely extra component of the Earth's core.

Table 6-3 also compares the estimated compositions of mantle plus crust according to Mason and Ringwood. These analyses are reproduced in Table 6-6 along with estimates of mantle composition derived from terrestrial evidence, and these will be discussed together in a later section.

There is geochemical evidence that the abundances of many nonvolatile trace elements, including U, Th, Ba, Sr, Ta, Zr, Hf, and the rare earths, are strongly concentrated in the upper mantle and crust. Ringwood concluded that their upward concentration occurred at a very early stage in the Earth's history, about 4.55×10^9 years ago, when all or most of the mantle was subjected to partial or complete melting. This coincides with the single decisive high-temperature event which caused loss of volatile elements, reduction of oxides, and formation of the core. These trace elements are incompatible with the major mineral phases of the mantle because of their sizes and ionic charges, and therefore they would have migrated upward with the silicate liquid, forming a protocontinental layer. The continents, hydrosphere, and atmosphere were then derived later by fractional melting of the protocontinent and subsequent magmatic processes.

ULTRAMAFIC ROCKS AND THE UPPER MANTLE

The estimates of mantle composition in Table 6-3 are equivalent to ultramafic rocks, and we turn therefore to examination of ultramafic rocks exposed at the Earth's surface. Ultramafic rocks could possibly have been derived directly from the mantle or indirectly from other mantle-derived material such as basaltic magma. The petrological problem is to determine which rocks, if any, represent mantle

material and to deduce the processes which led to their emplacement into the crust.

Peridotite Mineralogy and Field Associations

There are many different rock types in the ultramafic clan and these occur in a variety of field and petrographic associations. The most abundant minerals in rocks with compositions similar to those in Table 6-3 are olivine and pyroxenes, and combinations of these minerals in different proportions form the various peridotites. The alumina is distributed among the pyroxenes and accessory minerals such as plagioclase, spinel, and garnet. Small amounts of water are likely to be accommodated in amphibole, phlogopite, or serpentine, and given sufficient water at crustal temperatures, peridotite may be converted to serpentinite.

For petrogenetic discussions relevant to the composition of the upper mantle it is convenient to consider four associations of ultramafic rocks which are summarized in Table 6-4.

1. The layered, stratiform, and other intrusions involving gabbro or diabase together with accumulations or concentrations of mafic minerals. These occur in varied tectonic environments, but they are rarely affected by contemporaneous orogeny.

2. The alkalic rocks of stable continental regions including kimberlites, mica peridotites, members of ring complexes, and ultrabasic lava flows. These usually occur in stable or fractured continental regions, and their distribution in belts appears to be controlled by deep-seated tectonics with linear trends.

3. The several serpentinite-peridotite associations of the orogenic belts that have been classified together as alpine-type intrusions. These include large and small bodies distributed along deformed mountain chains and island arcs, usually along with gabbro or basic volcanic rocks. Relationships among rocks in this association are often complicated by metamorphism and metasomatism. The orogenic peridotites are subdivided into the high-level ophiolites, and the "root-zone" peridotites which occur in crystalline rocks.

4. Serpentinites and peridotites of the oceanic regions. These have been identified on midoceanic ridges, within deep ocean trenches, and where small scarps expose material below the cover of basalts forming the ocean floor.

5. In the present context we must also consider peridotite and eclogite nodules occurring in alkali olivine-basalts and kimberlites.

Petrogenesis of Ultramafic Rocks

Important factors to be considered in petrogenesis include the source of the material, its variation in physical state and temperature between its source and its posi-

TABLE 6-4   Classification of Main Ultramafic Rock Associations

Ultramafic Associations	Tectonic Environments	Source of Rock
1. Layered intrusions	Nonorogenic Varied	Basalt magma from mantle
2. Alkalic complexes, Kimberlites	Cratonic	Magmas from mantle
3. Alpine-type intrusions (a) ophiolites (b) root-zone	Orogenic Island arcs	Magmas and Solids from mantle
4. Oceanic rocks	Ocean trenches Ocean floors Oceanic ridges	Mantle or oceanic crust
5. Nodules (a) alkali olivine-basalts (b) kimberlites	Varied	Mantle, crust, or magma concentrates

tion of intrusion, and its postintrusion history. The more complex the postintrusion history of an ultramafic rock the more difficult is its petrogenetic interpretation. The following discussion is a brief outline of a very complicated subject which I have reviewed elsewhere in some detail (Wyllie, 1967, 1969, 1970). Figures 6-6 and 6-7 provide a schematic representation of the processes involved in the petrogenesis of ultramafic rocks.

Layered intrusions. There is general agreement that the parent for association (1) in Table 6-4 is basaltic magma derived by partial fusion of the upper mantle, and that gravity settling of early-formed crystals is the dominant process. Figure 6-6 depicts magma generation beginning with diapiric uprise of mantle peridotite to a level where partial melting produces interstitial basaltic

liquid. The crystal mush continues to rise, and in the left-hand diagram basaltic magma separates from the mush within the mantle and rises into a large reservoir or magma chamber within the crust. Crystal settling from this basic magma produces a series of ultramafic cumulates. The magma flows over the cumulates and upwards out of the chamber, possibly producing dykes, sills, volcanic necks, volcanoes, and lava flows. Crystal settling, flow differentiation, and other processes of crystal concentration may yield ultramafic rocks in the general locations shown in the figure.

Alkalic complexes, and kimberlites. The parent magmas proposed for the ultrabasic rocks of group (2) in Table 6-4 include undersaturated alkalic basalts and alkalic ultrabasic liquid magmas, evidence for the

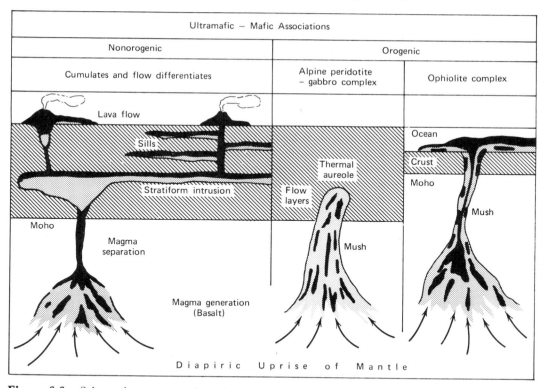

Figure 6-6. Schematic representation of the processes involved in the origin and emplacement of associations of ultramafic and mafic rocks. See text for explanation. Stipple—crystalline ultramafic material; black associated with stipple—interstitial basic liquid or crystalline basalt or gabbro (after Wyllie, 1970, with permission, Amer. Mineral. Society).

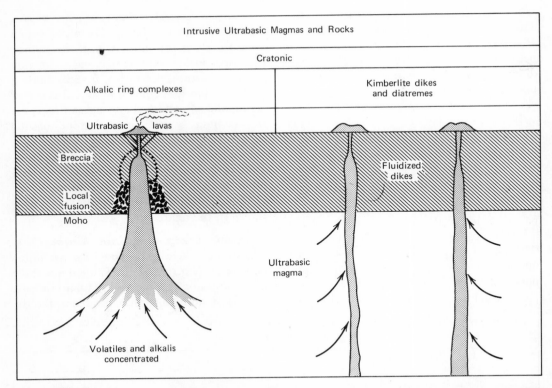

Figure 6-7. Schematic representation of the processes involved in the origin and emplacement of intrusive ultrabasic magmas and rocks. See text for explanation. Stipple—ultrabasic material (after Wyllie, 1970, with permission, Amer. Mineral. Society).

latter including the existence of alkalic ultrabasic lavas. The upper mantle is usually considered to be the source of these magmas, as illustrated schematically in Figure 6-7, but contamination with crustal material is often invoked to explain their unusual chemistry. Alkalis and volatiles are concentrated with the generation of ultrabasic magma which rises in dyke or pipe-like form into the crust and, eventually, ultrabasic lavas are erupted at the surface. The alkalis and volatiles become concentrated in residual liquids in the intrusions, facilitating differentiation. Exsolution of the volatiles causes brecciation, and explosive eruption of fragmental volcanic rocks. This effect is well illustrated by kimberlites: the deep-seated massive rock crystallizes from an ultrabasic magma, and higher level rocks are emplaced in diatremes as fluidized solid-gas systems, xenocrysts, and xenoliths of wall rocks being mixed with fragments of altered kimberlite. The explosive effects and the metasomatic effects of alkalis and volatiles in the residual magmas and exsolved gases and solutions, tend to obscure the early stages of petrogenesis.

Ultramafic rocks of the orogenic belts (alpine-type). There is no general agreement about the petrogenesis of these rocks, and there is little doubt that the association includes rocks with a variety of origins and histories. If we are to extrapolate down to the upper mantle from the study of rocks now exposed at the surface, it is necessary to decipher these histories. Figure 6-6 illustrates schematically some of the processes involved in the emplacement of these rocks.

The middle diagram represents the root-zone peridotites, group (3b) in Table 6.4. The ultrabasic crystal mush rises into the crust under conditions permitting little physical separation of basaltic magma. Some gravity layering may develop within the mush within the upper mantle, and this is further complicated by the development of flow layering during emplacement, with characteristics different from that of ultramafic cumulates in stratiform intrusions. A similar diagram would represent another process often invoked, the diapiric intrusion of solid mantle material. A major question is the extent to which ultramafic rocks of cumulate origin, depicted in the left-hand diagram of Figure 6-6, have become involved in the orogenic process and thus converted into rocks which merit the appellation "alpine-type". The origin of ultramafic rocks enclosed in crystalline rocks of the orogenic belts has been explained by five types of interpretation: (a) intrusion of ultrabasic magma, (b) intrusion of crystal mush of partially melted mantle, (c) diapiric or tectonic intrusion of solid mantle peridotite or of peridotite plus gabbro, (d) reintrusion of peridotite or serpentinite from any source, and (e) metasomatism and metamorphism of basic lavas and dolomites.

The right-hand diagram of Figure 6-6 represents the processes involved in the petrogenesis of the ophiolite suite, with the crystal mush breaking through the oceanic crust. There are four main types of interpretation for the origin of the associated ultramafic and mafic rocks. The first three involve the extrusion of very large volumes of material in submarine eruptions, forming a thick peridotite-gabbro-basalt complex, with later gravity slides occurring on regional scales to produce chaotic structures.

1. Extrusion of essentially liquid basaltic magma which undergoes differentiation on the sea floor,

2. Extrusion of a partially molten diapir of rising mantle with partial separation of basic magma during and after uprise and continued differentiation of the massive pile on the sea floor (Figure 6-6),

3. A two-stage process with solid intrusion of peridotite and gabbro from the mantle during a pregeosynclinal period being followed by intrusion and extrusion of basaltic magma at intervals through the whole geosynclinal cycle,

4. Basaltic lavas and sediments are metasomatically transformed into peridotites and serpentinites.

Ultramafic rocks of the oceanic regions. During the last few years serpentinites and peridotites have been dredged from many sites in the Atlantic and Indian Oceans especially near fracture zones although very few have been found in the Pacific Ocean. Several recent interpretations of ultramafic rocks of the ophiolite suite involve the supposition that these represent ocean floor or suboceanic mantle in eugeosynclines, which has been incorporated into the orogenic belt by thrusting or gravity sliding. The structure of midoceanic ridges appears to be similar to that of some chromitite-bearing alpine peridotite-gabbro complexes of continental environments, which implies that gabbro interlayered with peridotite is an essential constituent of the upper mantle. According to Hess's sea-floor spreading hypothesis (Chapter 12) the oceanic crust is composed of serpentinized mantle peridotite, migrating laterally away from midoceanic ridges. The global scheme of plate tectonics (Chapter 14) introduces the prospect that a midoceanic ridge may be overridden by a continent. These ideas imply that the ultramafic rocks of the orogenic belts, island arcs and ocean trenches, the midoceanic ridges, and the ocean floor between all have common petrogenetic links with the upper mantle.

Ultramafic nodules. The processes involved in the formation of the nodules of group (5) in Table 6-4 can be related to the left-hand diagram of Figure 6-6 and to Figure 6-7. Nodules in basalts may represent fragments of the mantle or fragments of crystal cumulates or flow concentrates from a magma chamber or the walls of a conduit or volcanic neck (Figure 6-6). The nodules in kimberlites may represent fragments carried upward from various levels in the mantle or from metamorphic rocks of the deep crust (Figure 6-7). There is good evidence that in a single locality, for either basalt or kimberlite host, the nodules present may be of several types with distinct origins. Pyroxene peridotite nodules of uniform composition and mineralogy occur in both basalts and kimberlites, and garnet peridotites with a wide variation in mineralogy occur only in kimberlites. Eclogitic nodules, ranging in composition from basalt to picrite basalt, may accompany peridotite nodules in kimberlites and rarely in basalts. For interpretation of nodule suites it is necessary to determine both the depth of origin and process of formation: processes may be elucidated by recognition of cumulus, tectonite, or metasomatic textures in the rocks.

Extrapolation to the Upper Mantle

Review of the petrogenesis of the ultramafic rocks in groups (3), (4), and (5) of Table 6-3 indicates that some of these rocks represent mantle material, but it also posts clear warnings that extrapolation to the upper mantle should be made with caution. Rocks of group (1), in stratiform or layered intrusions or extrusions, are produced by the concentration of mafic minerals crystallized from basaltic magma. They can provide only indirect evidence of mantle chemistry. The alkalic rocks of the cratonic belts in group (2) do not provide direct representatives of the mantle because they have been strongly fractionated. The high concentrations of volatile components and alkalis obscure many details of their petrogenesis, and their chemical relationships to the upper mantle.

The ultramafic rocks of the orogenic belts probably include representative mantle material as depicted in Figure 6-6, and these

rocks have often been cited as guides to mantle composition. Unfortunately the effects of metamorphism and reintrusion tend to blur the petrogenesis of these associations, and peridotites of any origin that have become involved in the orogenic process may be difficult to distinguish from those of alpine-type that are contemporaneous with the orogenic cycle. Certainly petrologists must exercise careful judgement before equating any specific orogenic peridotite with the upper mantle.

The ultramafic rocks of the oceanic regions, especially those near the center of the mid-oceanic ridges, may provide samples of the upper mantle, but more data are needed about the rock association and the geological structures before extrapolation can be made with confidence.

The ultramafic nodules may provide the most direct evidence of mantle chemistry and mineralogy, assuming that distinction can be made among those that are crystal cumulates or concentrates. The problems of interpretation were summed up by Jackson in 1969:

"Trying to reconstruct the depth of origin and process of formation of xenoliths in any one individual tuff or flow is akin to trying to map the geology of an inaccessible highland area by looking at the boulders in the bed of an emergent stream. Multiple origins and differences in distances of transport of fragments, even those deposited side by side, might as well be assumed in both cases, and generalizations ought to be made with extreme care."

Once the petrogenesis of an ultramafic rock (or of an ultramafic-mafic rock association) has been unravelled and an ultramafic rock has been traced back to a mantle source, its composition and mineralogy have still to be related to the history of that portion of the mantle from which it was derived. Petrologic considerations support the geophysical evidence that the upper mantle is heterogeneous (Chapter 3). The upper mantle will include material ranging in composition from original, primitive, undifferentiated peridotite to residual peridotite or dunite from which a basaltic or picritic liquid fraction has been partially or completely removed. It will include peridotite precipitated from picritic or basaltic magma as it fractionated en route to the surface (Chapter 8). It will include layers and lenses of eclogite crystallized from a basaltic or picritic liquid that failed to reach the crust. It may also include portions of peridotite somewhat enriched in eclogitic components by percolation of basaltic liquid and its local concentration in dispersed form in favorable sites.

Ultramafic rocks, properly selected, should thus provide information about the composition and mineralogy of the upper mantle, but the petrogenetic processes involved between mantle source and crustal exposure are varied and complex. For satisfactory estimates of mantle composition based on terrestrial ultramafic rocks the field association and the petrogenesis of the rocks should be thoroughly investigated before extrapolation to the mantle.

COMPOSITION OF THE MANTLE

The general idea of a peridotitic mantle as a source for basaltic magmas by partial melting dates at least as far back as Bowen's treatise on the igneous rocks in 1928, but it was Ringwood who formalized the concept in 1962. Ringwood is largely responsible for directing the attention of petrologists to model compositions for the upper mantle. He postulated the existence of a primitive mantle material which was "defined by the property that on fractional melting it would yield a typical basaltic magma and leave behind a residual refractory dunite-peridotite of alpine type." This he called "pyrolite" (pyroxene-olivine-rock). He referred to the valid objections to inventing new names for hypothetical materials and defended this one mainly on the grounds that the trace element chemistry of "pyrolite" does not match that of natural peridotites. Rocks are named on the basis of their mineralogy, however, not on their trace element content; and I see no need to add yet another to the names that already exist. Most petrologists concerned with upper mantle constitution refer to known varieties of peridotites although geophysicists tend to use "pyrolite".

Among the first attempts to correlate the composition of the upper mantle with specific mantle-derived ultramafic rocks are those of Hess and Green both in 1964. Hess reviewed the chemistry of oceanic serpentinites, and Green proposed that the average composition of the orogenic peridotite at Lizard, England represented undepleted upper mantle material. Since 1967 many accounts of the chemistry of ultramafic rocks in a variety of environments have included discussions of mantle chemistry. Petrologists have been concerned with identification of ultramafic rocks representing (1) undepleted mantle, and (2) residual mantle after removal of a basic liquid fraction; they have also attempted to identify (3) the rock produced by crystallization of the liquid fraction (Chapter 8). In this section we examine first (Table 6-5) the compositions of selected ultramafic rocks which have been cited as representatives of the upper mantle and then the calculated compositions of hypothetical mantle peridotites. Estimated mantle compositions based on extraterrestrial evidence are then compared with estimates based on terrestrial rocks and the hypothetical rocks (Table 6-6). Some aspects of the chemistry are compared in chemical variation diagrams, which include not only the rock compositions cited in this chapter but many others.

Estimates Based on Terrestrial Rocks

The Lizard peridotite of Cornwall, England, is one of a distinctive group of orogenic ultramafic intrusions, which Green interprets as a portion of the upper mantle mobilized by partial melting (middle diagram of Figure 6-6). The average composition of this intrusion, listed in column (1) of Table 6-5, was presented by Green (1964) as equivalent to undepleted mantle peridotite.

Garnet peridotites occur in the root-zones of orogenic belts and as nodules in kimberlites. Carswell (1968a) studied the garnet peridotites enclosed in gneiss at Ugelvik, Norway, and concluded that they represent primary mantle material; their average composition is listed

in column (2) of Table 6-5. He also reviewed other occurrences of garnet peridotite in basement rocks of Czechoslovakia and Switzerland and in kimberlites. All of these rocks prove to be similar mineralogically, occupying restricted chemical ranges, and he therefore concluded that they too probably represent primary upper mantle material. Points for these rocks are plotted in Figure 6-8.

Hess (1964) reviewed the mineralogy and chemistry of the serpentinite cores from the drill hole near Mayaguez, Puerto Rico where

TABLE 6-5   Compositions of Mantle-Derived Ultramafic Rocks

Weight percent	Continents		Oceans		Nodules		Hypothetical	
	(1)	(2)	(3)	(4)	(5)	(6)	(7)	(8)
SiO$_2$	44.77	44.65	39.82	43.56	44.18	41.10	42.71	40.3
MgO	39.22	41.66	48.60	41.53	40.95	46.33	41.41	32.7
FeO	8.21	6.81	7.86[a]	7.77[a]	7.34	9.31	6.51	7.1
Fe$_2$O$_3$	—	—	1.00[a]	1.00[a]	1.16	1.24	1.57	1.8
Al$_2$O$_3$	4.16	3.50	0.87	2.36	2.81	0.56	3.30	3.7
CaO	2.42	2.02	0.37	2.51	2.49	0.17	2.11	2.1
Na$_2$O	0.22	0.23	0.37	0.32	0.22	0.23	0.49	0.5
K$_2$O	0.05	0.04	[b]	[b]	0.04	0.03	0.18	0.0(2)
Cr$_2$O$_3$	0.40	0.59	0.46	0.40	0.3	0.35	0.45	0.3
NiO	0.24	0.29	0.46	0.34	0.27	0.44	0.42	0.2
CoO	—	—	—	—	—	—	0.02	—
MnO	0.11	0.14	0.10	0.10	0.14	0.15	0.13	0.1
P$_2$O$_5$	—	—	0.08	0.07	—	—	0.06	0.1
TiO$_2$	0.19	0.08	0.01	0.04	0.09	0.08	0.47	0.4
H$_2$O +	—	—	—	—	—	—	0.17	9.7
CO$_2$	—	—	—	—	—	—	—	0.8
Cl	—	—	—	—	—	—	—	0.2
Total	99.99	100.01	100.00	100.00	99.99	99.99	100.00	100.00

[a]Ferrous ferric ratio adjusted so that Fe$_2$O$_3$ is 1%.
[b]means less than 0.005.
(1) Green (1964). Average composition for the Lizard peridotite.
(2) Carswell (1968a). Mean of three garnet peridotites from Ugelvik, Norway.
(3) Hess and Otalora (1964). Average (D + E)-type serpentinite, recalculated water free, residual type.
(4) Hess and Otalora (1964). Average C-type serpentinite, recalculated water-free.
(5) Harris et al. (1967). Mean of five high calcium, high aluminum olivine nodules.
(6) Harris et al. (1967). Average of three olivine nodules with CaO and Al$_2$O$_3$ contents less than 1%, residual nodules.
(7) Green and Ringwood (1963). Pyrolite with 4:1 of, respectively, average anhydrous dunite and the mean of Nockold's (1954) average normal tholeiite and normal alkali basalt.
(8) Nicholls (1967). Composition of volatile-rich parts of the upper mantle, such as may occur beneath the midoceanic ridges (Nicholls, analysis 3, table 9).

the exposed serpentinite appears to be similar to rocks dredged from the ocean bottom near the island. The serpentinite forms the basement beneath Puerto Rico, and is considered by Hess to represent oceanic crust or hydrated mantle. Several distinct types of serpentinite are represented, and columns (3) and (4) in Table 6-5 give average compositions involving three types recalculated on an anhydrous basis. Hess proposed that the average in column (3) might be the residue after extraction of basalt from original mantle material with average composition (4). These analyses are similar to those of serpentinized peridotites from midoceanic ridges as shown by the recent review of 23 specimens by Miyashiro, Shido, and Ewing (1969). The peridotite mylonites from St. Peter and St. Paul Rocks in the mid-Atlantic Ocean include several varieties, and they appear to represent rocks which crystallized at different levels within the mantle. They therefore have bearing on the heterogeneity of the upper mantle according to Melson, Jarosewich, Bowen, and Thompson (1967). Individually the rocks analyzed are not considered equivalent to the mantle, but an estimated average composition of St. Paul's Rocks which is plotted in Figure 6-8 is in accord with suggested mantle compositions.

Figure 6-8 includes seven compositions specified as upper mantle material on the basis of study of peridotite nodules in kimberlites and basalts. Two of these are listed in Table 6-5. Harris, Reay, and White (1967) derived columns (5) and (6) from the chemistry of 27 peridotite and three garnet-peridotite nodules. The average of five selected analyses of peridotite nodules with CaO and Al_2O_2 contents between 2 and 3% is given in column (5), and this they considered to represent mantle material that has undergone only limited or no depletion in fusible components. Column (6) is the average of three peridotite nodules with CaO and Al_2O_3 each less than 1%. This they considered to represent samples of residual mantle left after extraction of basaltic magma

or samples of cumulate olivine precipitated from magma rising through the mantle.

Hypothetical Peridotites

The definition of Ringwood's "pyrolite" was given above. Ringwood and Green have emphasized flexibility in their model, and between 1963 and 1967 they have presented four pyrolite analyses. All four are plotted in Figure 6-8 and two of them are listed in column (7) of Table 6-5 and column (5) of Table 6-6. The ratio of peridotite:basalt may vary within the limits 1:1 to 4:1, with 3:1 being the most likely. Column (7) is based on a 4:1 mixture of average anhydrous dunites and an average of tholeiite and alkali basalt. Other model compositions included a primitive Hawaiian olivine tholeiite composition and a hypothetical peridotite with composition calculated from analyzed minerals separated from alpine-type peridotites. Column (5) in Table 6-6 is based on the 3:1 mixture.

Nicholls (1967) approached the problem of upper mantle composition by assuming that it was composed of (a) a volatile fraction, consisting of compounds that have been liberated from volcanoes to the atmosphere and hydrosphere, (b) a basaltic lava fraction derived by partial melting of the mantle, and (c) a residual fraction remaining after depletion in volatiles and basalt. Nicholls examined the possible compositions of each of these fractions to define their limits and then evaluated the proportions in which they might be combined in undepleted mantle material. He concluded that the preferred composition for the volatile fraction is 91% H_2O, 7% CO_2, 2% Cl and traces of other components. A basaltic fraction was calculated, and an estimate of the residual fraction was based on oceanic serpentinites. Column (8) in Table 6-5 is Nicholl's estimate of undepleted upper mantle material, still rich in volatile components, which he suggests may exist beneath the midoceanic ridges: notice the high water content. Column (6) in Table 6-6 represents mantle material from which

TABLE 6-6   Estimates of Primitive Mantle Compositions

Weight percent	Extraterrestrial		Terrestrial		Hypothetical	
	(1)	(2)	(3)	(4)	(5)	(6)
SiO_2	48.09	43.25	44.5	44.2	43.95	45.1
MgO	31.15	38.10	41.7	41.3	39.00	36.7
FeO	12.71	9.25	7.3	7.3	7.50	7.9
Fe_2O_3	—	—	1.5	1.1	0.75	2.0
Al_2O_3	3.02	3.90	2.55	2.7	3.88	4.1
CaO	2.32	3.72	2.25	2.4	2.60	2.3
Na_2O	1.13	1.78	0.25	0.25	0.60	0.6
K_2O	0.13	—	0.015	0.015	0.22	0.0(2)
Cr_2O_3	0.55	—	—	0.30	0.41	0.3
NiO	—	—	—	0.20	0.39	0.2
MnO	0.43	—	0.14	0.15	0.13	0.2
P_2O_5	0.34	—	—	—	—	0.1
TiO_2	0.13	—	0.15	0.1	0.57	0.5
Total	100.00	100.00	100.36	100.02	100.00	100.0

(1) Mason (1966), Table 6-3.
(2) Ringwood (1966), Table 6-3.
(3) White (1967). Upper mantle composition estimated from frequency histograms of 168 ultramafic rocks. Total iron is divided arbitrarily between FeO and Fe_2O_3. Na_2O and K_2O are from Stueber and Murthy (1966) and Stueber and Goles (1967).
(4) Harris et al. (1967). Estimated upper mantle based on analysis (3), and on analysis (5) for nodules in Table 6-5.
(5) Green and Ringwood (1967). Synthetic peridotite used in experimental phase studies; designated pyrolite II. Ratio of dunite: basalt is 3:1.
(6) Nicholls (1967). Estimated mantle material from which volatile components have been removed but not the basaltic components (Nicholls, analysis 1, table 9).

the volatile fraction has been removed, but which still retains the basaltic fraction; this is the analysis plotted in Figure 6-8 along with the residual peridotite.

Mantle Composition

Table 6-6 compares estimates of primitive, undepleted mantle compositions based on extraterrestrial evidence, terrestrial rocks, and theoretical models. Columns (1) and (2) are the estimates of Mason and Ringwood, determined as illustrated in Tables 6-2 and 6-3. The hypothetical compositions were determined as discussed in the preceding section. Column (3) is White's (1967) estimate of mantle peridotite based on frequency histograms for the chemistry of 84 peridotites,

53 serpentinites, and 31 peridotite nodules. White argued that the mode or most abundant class of exposed ultrabasic rock should represent original mantle peridotite. He plotted separately the analyses for rocks with between 1 and 5% CaO plus Al_2O_3 to eliminate the residual dunitic and more basaltic analyses, so that the mode for unfractionated mantle peridotite would not be masked. Harris, Reay, and White compared the dominant ultramafic rock in column (3) with the average compositions of peridotite nodules (Table 6-5) and presented a preferred composition for original or undepleted mantle based on these analyses and other considerations. This is column (4) in Table 6-6.

Element abundances in suites of ultramafic rocks have been studied by activation analysis and spectrochemical analysis, which make possible the accurate determination of elements previously difficult to measure in ultramafic rocks. Fisher, Joensuu, and Bostrom (1969) analyzed 58 ultramafic rocks from diverse sources. Among the rocks they could find no chemically coherent type that might represent the upper mantle. Their average values agree well with estimates for upper mantle compositions (as in Tables 6-5 and 6-6), but the wide spread observed casts doubt on the validity of averages based on only a few samples. Average values of some elements with their ranges are as follows: Al = 1.6 ± 1.6; Cr = 0.3 ± 0.3; Ni = 0.16 ± 0.16. They concluded that either the

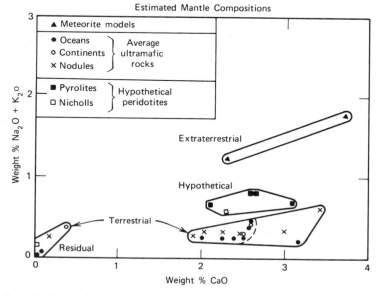

Figure 6-8. Chemical variation of estimated mantle conditions. See Figure 6-9 (after Wyllie, 1970, with permission, Amer. Mineral. Society).

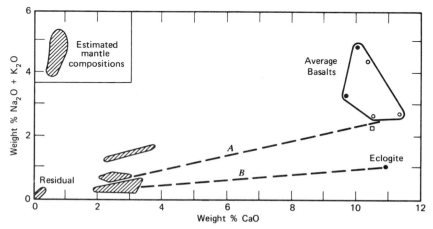

Figure 6-9. Expanded version of Figure 6-8, comparing mantle estimates with average basalts and a picritic eclogite (after Wyllie, 1970, with permission, Amer. Mineral. Society).

mantle is heterogeneous or that none of the rock types included in their study is representative of primitive mantle material.

Despite the different models adopted by Mason and Ringwood for their extraterrestrial estimates and the different approaches used for the other estimates in Table 6-6, these proposed mantle compositions agree with respect to the following:

1. More than 90% by weight of the mantle is represented by the system FeO-MgO-SiO_2 and no other oxide exceeds 4%;

2. Together the components Na_2O-CaO-Al_2O_3 lie within the range 5 to 10%, and more than 98% of the mantle is represented by the six components (with Fe_2O_3 calculated as FeO);

3. No other oxide reaches a concentration of 0.6% of the mantle. FeO substitutes for MgO in most magnesian minerals, and Na_2O substitutes for CaO in plagioclase at low pressures, and in jadeitic pyroxene at high pressures. Therefore the mineralogy of mantle material can be represented in simplified form in the quaternary system CaO-MgO-Al_2O_3-SiO_2.

Chemical Variation Diagrams

The estimated mantle compositions listed in Tables 6-5 and 6-6, together with other estimates that I have reviewed elsewhere (Wyllie, 1970), are compared in Figure 6-8, a standard diagram showing variation in alkali content of rocks with variation in lime content. Distinct areas with no overlap are occupied by each of the four groups of analyses, (a) extraterrestrial, (b) terrestrial undepleted, (c) terrestrial depleted, and (d) hypothetical. This is also true in two other chemical variation diagrams not reproduced here. The first shows the variation in Al_2O_3 using a triangular diagram for CaO-Al_2O_3-alkalis; the CaO/Al_2O_3 ratios for the mantle estimates occupy a limited range with Al_2O_3 being stored relative to lime plus alkalis in the residual types. The second shows the mantle

estimates plotted against two standard differentiation indices, $100(Na_2O + K_2O)/(CaO + Na_2O + K_2O)$, and $100(FeO + Fe_2O_3)/(MgO + FeO + Fe_2O_3)$.

In Figure 6-8 the terrestrial estimates are identified as rocks from oceanic or continental environments or as nodules. The estimates of undepleted mantle based on ultramafic rocks are grouped together closely; only three points depart from the main group. The mantle as a whole, according to the meteorite models, contains significantly more alkalis than the upper mantle according to estimates based on observed terrestrial rocks. Also the observed ultramafic rocks contain less alkalis than the hypothetical models.

Figure 6-9 is an expanded version of Figure 6-8 comparing the mantle estimates with average basalts taken from various sources. The line *A* is drawn through the origin, which corresponds closely to residual dunite, in a position separating the terrestrial mantle estimates from the hypothetical peridotites. The average basalts lie above this line, on the same side as the hypothetical peridotites, confirming that they can be derived by subtraction from the hypothetical peridotite leaving residual dunite. The position of the terrestrial mantle estimates below line *A*, however, demonstrates that the average basalts cannot be derived by direct subtraction from these known ultramafic rocks if the residual product is to be equivalent to those peridotites and dunites normally regarded as residual types.

The hypothetical models assume that the average basalts are primary magmas derived from the mantle by partial melting, uprise, and eruption. If this is correct then the terrestrial ultramafic rocks are too low in alkalis to provide estimates for the upper mantle. In two stimulating reviews, however, O'Hara (1965, 1968) pointed out that the concept of primary magmas is only one limiting case for magma generation (Chapter 8). He argued that basaltic liquids normally regarded as primitive types could not have been in equilibrium with mantle material at the depth where partial melting occurred. The low pressure basaltic compositions erupted at the surface are the end products of a sequence of events acting upon the original high pressure liquid formed at depth. Experimental studies indicate that this liquid is picritic rather than basaltic, and recent petrological studies suggest that deep-seated liquid compositions cluster along the line *B* in Figure 6-9.

Carswell (1968b) examined in detail the garnet peridotites at Kalskaret in Norway and discovered a linear compositional trend between two end members, one the eclogite in Figure 6-9 and the other a dunite plotted with the residual rocks in Figure 6-8. The line *B* connecting these two end members passes through the group of undepleted mantle estimates based on ultramafic rocks. Carswell interpreted the field and textural relationships to indicate tectonic emplacement of the garnet peridotites as relatively cold intrusions from the mantle in the solid state. He concluded that within the upper mantle a picritic partial melt fraction was not completely liberated from a dunitic residual fraction of the original mantle material, but became trapped and mixed with the dunite; there it crystallized as eclogite. The mixed rocks which plot close to the line *B* were subsequently emplaced into the crust.

The high temperature orogenic peridotite masses of Beni Bouchera, Morocco, the Sierrania de la Ronda, Spain, and Etang de Lers, France, have been interpreted by Kornprobst (1969) and Dickey (1970) in terms of upward movement of mantle peridotite. This caused partial fusion (central diagram in Figure 6-6), and the small proportion of liquid later crystallized in magmatic layers forming various pyroxenites with or without garnet. Crystallization occurred within the mantle at high pressure, and tectonic movements emplaced the solid masses into the orogenic root zones. The peridotites in the Beni Bouchera massif plot in the area of terrestrial mantle estimates of Figures 6-8 and 6-9 or very close to it, and

the magmatic pyroxenite layers plot in a narrow zone between 10.5 and 15.5% CaO, which forms a continuation of the line *B*.

Additional evidence for the existence of picritic liquids is provided by a series of minor intrusions in Skye, Scotland, and Ubekendt Ejland in west Greenland studied by Drever and Johnston (1967). These intrusions are termed "picritic" owing to their richness in olivine, but their conventional explanation as simple mixtures of basaltic liquid plus suspended olivine crystals is inadequate to account for the data presented. Chemical analyses of the picritic and related calcic rocks when plotted on Figure 6-9 lie quite close to the line *B* and its extension, between 5.5 and 13.9% CaO. Drever and Johnston suggested that these materials may represent liquids formed at high pressures by partial fusion of an ultrabasic source rock. This is consistent with the interpretations for the other rocks with compositions close to the line *B*.

In view of the evidence that liquid fractions produced by partial melting of mantle peridotite at depth have compositions near the line *B* in Figure 6-9, it seems that this is the material that should be combined with residual peridotite or dunite in order to obtain a hypothetical mantle peridotite and not an average basaltic composition as employed in the models of Ringwood and Nicholls. This procedure does yield a composition corresponding to the terrestrial ultramafic rocks of Figures 6-8 and 6-9 at least with respect to alkali content. Possibly, therefore, the estimated mantle compositions based on selected ultramafic rocks may provide reasonable indications of some peridotite compositions occurring within the upper mantle, whereas the hypothetical peridotites in Figure 6-8 may be too high in alkali content.

Possible explanations for the difference in alkali content between the extraterrestrial estimates and the terrestrial estimates in Figure 6-8 include: (a) the lower mantle retains more alkalis than the upper mantle, (b) the accreting Earth lost more alkalis and other volatile elements than the meteorite parent bodies, (c) the chondrite models for the origin of the Earth are invalid.

Trace Elements and Volatile Components

The concentrations of major elements in the mantle listed in Table 6-6 appear to be reasonably well established, but large uncertainties remain about the concentrations and distribution of trace elements and volatile components. Related to the trace elements is the question of whether the composition of the mantle and crust is approximated more closely by chondritic meteorites or by Ca-rich achondrites. Arguments about the relative merits of chondritic and achondritic models have been concerned with concentrations in the crust, mantle, and meteorites, of the elements K, U, Th, Rb, Cs, Ba, Sr; and with the ratios Rb/Sr, $^{87}Sr/^{86}Sr$, K/Rb, and K/U. The arguments have not been resolved.

It is difficult to measure accurately the abundances of trace elements in mantle-derived ultramafic rocks, and there may be marked variation in trace element concentrations from one specimen to another in the same ultramafic body. Trace element distribution may be controlled to a large extent by the mineralogy. The concentrations of Rb, Sr, and K, for example, are very sensitive to the distribution of small proportions of amphibole. Secondary alteration which may appear quite minor in a petrographic sense can cause serious contamination with respect to whole rock values for elements such as K, Rb, Cs, Sr, and the rare earth group. In some mantle-derived ultramafic rocks U is distributed homogeneously in clinopyroxenes, but in other similar rocks most of the U occurs as contaminants along cracks, grain boundaries, or in microinclusions. There have been many proposals that the upper mantle has trace element concentrations different from the whole mantle arising from differentiation of the mantle at an early stage in Earth history and from the continued depletion of the

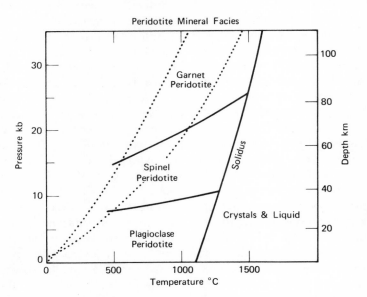

Figure 6-10. Peridotite mineral facies. Generalized phase diagram for an estimated mantle peridotite using data from various sources (after Wyllie, 1970). The lines indicate the solidus of the peridotite and the positions of divariant reaction intervals separating facies for spinel-peridotite from plagioclase-peridotite and from garnet-peridotite. Dotted lines represent geotherms for oceanic and shield regions (Figure 3-12). (With permission, Amer. Mineral Society.)

upper mantle in basaltic components. This makes a difficult problem even more complex.

The regime of volatile components in the Earth is of fundamental significance for many problems. It is generally held that the crust, hydrosphere, and atmosphere have been produced by defluidization of the Earth. The petrographic study of mantle-derived ultramafic nodules and rocks provides evidence that phlogopite and amphibole are primary constituents of the mantle, at least locally, suggesting that some H_2O remains in the mantle. The presence of CO_2 is inferred from the existence of CO_2-filled inclusions in olivines derived from upper mantle sources. The gases given off during volcanic eruptions provide data on volatiles in the mantle and H_2O is the dominant juvenile component. Interpretation is complicated by contamination with meteoritic solutions and by chemical

reactions during uprise and eruption of the magma. The proportion dissolved in most magmas is small but uncertain.

Nicholls (1967) suggested that in certain environments, such as beneath the mid-oceanic ridges, the mantle peridotite might contain as much as 9.7% H_2O and 0.8% CO_2 (analysis (8) in Table 6-5). The peridotite mylonites of St. Peter and St. Paul rocks apparently contained an abundant fluid phase during intrusion from the mantle. The proportion of H_2O and other volatile components in the mantle is probably variable both laterally and vertically and probably no more than 0.1% except in special environments. Even a small proportion of volatiles, however, may have a very pronounced effect on the behavior of mantle rocks with respect to (a) physical strength and (b) temperature for beginning of melting.

PHASE TRANSITIONS IN MANTLE PERIDOTITE AND ECLOGITE

Evidence from extraterrestrial sources, petrology, and geophysics indicates that the mantle is composed of peridotitic material, and that in the upper mantle this is mixed with dunite and eclogite. Experimental studies on these compositions at appropriate pressures and temperatures delineate the mineral facies occurring within the upper mantle. The transition of gabbro to eclogite was reviewed in detail in Chapter 5, and the behavior of eclogite at depths greater than 100 km is probably similar to that of garnet peridotite.

Experiments on Peridotite Mineral Facies

There has been much speculation about the phase diagram for peridotite, and Figure 6-10 shows the general pattern of the phase changes for lherzolite, a peridotite consisting of olivine, clinopyroxene, and orthopyroxene together with an aluminous phase. The aluminous phase is plagioclase feldspar in the low-pressure plagioclase-peridotite facies.

With increasing pressure the plagioclase reacts with olivine to yield aluminous pyroxenes and spinel producing a spinel peridotite. At higher pressures garnet-peridotite is produced by reaction of the spinel with orthopyroxene yielding pyrope-rich garnet and olivine. The boundaries between the subsolidus facies are plotted as univariant lines but in fact they are reaction intervals.

Experimental approaches and problems involved in locating subsolidus phase boundaries for rocks were discussed in Chapter 5, and several of the mineralogical reactions related to peridotite in the system CaO-MgO-Al_2O_3-SiO_2 were reviewed and illustrated in Figures 5-11, 5-12, and 5-13. Figure 6-10 is based on these results and on experimental studies above 1100°C by Green and Ringwood using the synthetic pyrolite II listed in Table 6-6. A different synthetic peridotite, with lower MgO/SiO_2 ratio, introduced an additional phase field just below the solidus; this is pyroxene-peridotite without spinel. With increasing temperature within the spinel-peridotite field the spinel dissolves in the aluminous pyroxenes and solution is completed at about 1250°C. The generation of garnet from the spinel-free peridotite then occurs at a higher pressure about 30 kb on the solidus instead of 25 kb.

The general pattern of phase relationships illustrated is correct, although the positions of the boundaries may change considerably with variation in the content of CaO, and minor trivalent oxides such as Al_2O_3, Fe_2O_3, and Cr_2O_3. The geotherms for continental and oceanic regions indicate that at depths greater than 55 or 70 km the mantle consists of garnet peridotite (possibly with dunite and eclogite). Plagioclase peridotite can have only very limited distribution in the mantle.

Estimates of the mineralogy of mantle spinel-peridotites and garnet-peridotites are compared in Table 6-7. J. L. Carter examined the modal abundances of 150 nodules from five basaltic host rocks in terms of the fayalite content of their olivine. He found continuity except for a small gap between 12–14% fayalite, and he argued that this gap represents primitive upper mantle material. The corresponding mode is listed. H. Fujisawa considered the compositions of olivines in mantle-derived nodules and ultramafic rocks and concluded in 1968 that the olivine in the upper mantle contains 10% fayalite. The other assemblages listed in Table 6-7 are estimated upper mantle compositions recast into norms appropriate for the two facies, using analysis 4 in Table 6-6 and pyrolite III similar to analysis 5 in Table 6-6.

Effect of Water

Addition of water to peridotite causes depression of the melting temperature and introduces three hydrous minerals; serpentine, amphibole, and phlogopite. The melting curve and curves for the breakdown of the hydrous minerals are shown in Figure 6-11. The curve for phlogopite gives its maximum temperature stability, with the phlogopite melting incongruently to a vapor-absent assemblage, forsterite plus liquid. In the presence of excess vapor phlogopite melts together with forsterite at a lower temperature about 1200°C. The curve for serpentinite gives its maximum temperature of stability in the presence of excess water vapor. It would dissociate at lower temperatures if vapor were absent. The curve for hornblende is based on results in the presence of a liquid phase with excess water vapor. With water pressure less than load pressure, as in a vapor-absent system, the amphibole would be stable to higher temperatures and pressures. Notice the opposite effects of decreasing water pressure relative to load pressure for (a) a subsolidus dehydration reaction as for serpentine, and (b) a dissociation reaction involving a liquid phase as for phlogopite and amphibole.

TABLE 6-7   Estimates of Mantle Mineralogy

	Spinel Peridotite		Garnet Peridotite	
	Carter 1966	Harris et al., 1967	Harris et al., 1967	Ringwood, 1969
Olivine	55 ± 10	65.3	67	57
Orthopyroxene	27 ± 5	21.8	12	17
Clinopyroxene	14.5 ± 3.5	11.3	11	12
Spinel	3 ± 1	1.5	—	—
Garnet	—	—	10	14
Total	99.5	99.9	100	100

Given a peridotite containing just a trace of water we see that this could be stored in phlogopite, with no free vapor phase, until the temperature exceeded the phlogopite curve. Then a small amount of liquid would be produced by the incongruent melting of phlogopite.

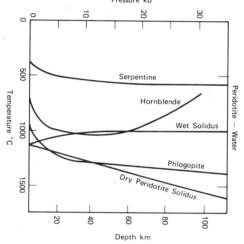

Figure 6-11. Experimentally determined reactions involving peridotite and water, based on various experimental studies (after Wyllie, 1970). Curves are plotted for the solidus of peridotite dry (Ito and Kennedy, 1967), and in the presence of excess water (Kushiro et al., 1968); other curves give the upper temperature stability limits for serpentine (Kitahara and Kennedy, 1966; Scarfe and Wyllie, 1967), hornblende (Kushiro, 1970), and phlogopite (Yoder and Kushiro, 1969). (With permission, Amer. Mineral. Society.)

Addition of water to gabbro causes depression of the melting temperature and introduces one hydrous mineral, an amphibole. Figure 6-12 compares the melting intervals for a dry gabbro, with the position of the gabbro-eclogite transition interval indicated below the solidus (Figures 5-14 and 5-15), and for the gabbro in the presence of excess water. The slope of the solidus (dP/dT) reverses at 15 kb, where plagioclase feldspar is replaced by jadeitic pyroxene. The curve for the breakdown of hornblende is indicated, and the gabbro-eclogite transition is masked by the hornblende. At pressures above 25 kb, the subsolidus assemblage is quartz eclogite. For a gabbro undersaturated in silica to such an extent that it yielded olivine eclogite, or a simple garnet-pyroxene eclogite, we can expect that the solidus would increase in temperature rather sharply within a short pressure interval above 15 kb where plagioclase disappears; the mineralogy would then be similar to that of the high pressure peridotite (Figure 6-11).

Given a gabbro containing only 0.1% of water we can deduce the general pattern of phase relationships from Figure 6-12 and the resulting isopleth is shown in Figure 6-13. The liquidus temperature is depressed only slightly compared to the dry liquidus. Within the stability field of hornblende all of the water is fixed in the hornblende and no melting occurs. There are two vapor-absent fields for hornblende-gabbro and hornblende-eclogite with a transition interval indicated between them. The breakdown curve for hornblende in Figure 6-13 is located at somewhat higher temperatures than in Figure 6-12 because of the vapor-absent conditions.

At pressures above the stability of hornblende in the subsolidus region the water exists as an interstitial film in eclogite. The curve for the beginning of melting is the same as that for melting with excess water as in Figure 6-12, but for an eclogite without quartz this curve would be at a higher temperature. Because of the very small amount of water present only a trace of liquid develops.

Below about 27 kb pressure the solidus is the breakdown curve for hornblende. The subsolidus vapor-absent assemblages are stable until hornblende releases water, which then passes directly into a vapor-absent liquid phase. The light-shaded area represents gabbro or eclogite with a trace of interstitial liquid, undersaturated with water; its composition is intermediate rather than basic. The liquid content increases within the dark-shaded band, which corresponds closely to the melting interval in the dry system (Figure 6-12).

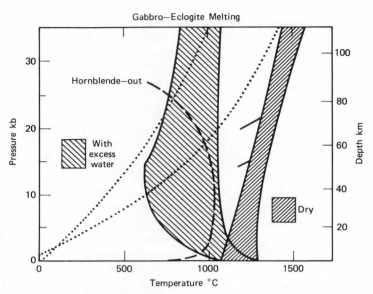

Figure 6-12. Generalized diagram for the melting interval of gabbro, dry, and in the presence of excess water (after Lambert and Wyllie, 1968, 1970). Note the gabbro-eclogite transition interval indicated just below the solidus for dry gabbro. The dashed line gives the upper temperature stability limit of hornblende with excess vapor. The dotted curves are geotherms for oceanic and shield regions. (With permission, Amer. Mineral. Society, Wyllie, 1970).

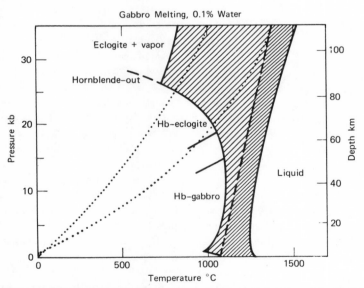

Figure 6-13. Estimated isopleth for the phase fields intersected by material of gabbroic composition in the presence of 0.1% water, based on Figure 6-12 (after Wyllie, 1970; Lambert and Wyllie, 1970). The solidus corresponds to the hornblende-out curve in Figure 6-12, but it is at a higher temperature because of water deficiency. Note the transition interval between gabbro and eclogite just below the solidus. Only a trace of liquid, probably of intermediate composition, is present through the light-shaded area; basic liquid is produced only within the dark shaded interval, which corresponds closely to the dry melting interval in Figure 6-12. (With permission, Amer. Mineral. Society.)

THE OLIVINE-SPINEL TRANSITION

Estimated compositions of the upper mantle (Table 6-6) imply that olivine is the dominant mineral (Table 6-7). When H. Jeffreys first presented evidence for the second order discontinuity in 1936 J. D. Bernal suggested, in the ensuing discussion, that the discontinuity might be accounted for by transition of the orthorhombic mineral olivine into a denser cubic form, spinel. He cited the earlier observation of V. M. Goldschmidt that the analogous compound, Mg_2GeO_4, was known to exist in two forms, one isomorphous with olivine and the other cubic. It has since been claimed several times that the spinel polymorph of Mg_2GeO_4 could not be synthesized.

In 1952 F. Birch published a detailed paper on the elasticity and constitution of the Earth's interior, in which he proposed that within the transition zone of the mantle there occurred a series of phase changes leading to closer packing, and that at depths greater than 1000 km the elastic properties of the mantle were consistent with those of close-packed oxide phases. This was the time when apparatus for the simultaneous development of high pressures and temperatures was being designed, and it was not long before many investigators were attempting to locate phase transitions which might occur within the Earth's interior. An early success was the reproducible formation of the spinel polymorph of Mg_2GeO_4 in the presence of water vapor under pressure by D. M. Roy and R. Roy in 1954; the phase transition curve is shown in Figure 6-14. F. Dachille and R. Roy then approached the problem of locating a corresponding transition in Mg_2SiO_4 by studying the solid solution and transition in the system Mg_2GeO_4-Mg_2SiO_4 with results published in 1956 and 1960.

The greatest contribution to this problem has been made by A. E. Ringwood. In a series of three papers on the constitution of the upper mantle in 1958 he published the results of thermodynamic studies of the olivine-spinel transition based on experimental studies in various model systems at 1 bar and at high pressures. Evidence for the chemical composition of the mantle was reviewed. Relating the composition to the phase transitions anticipated at high pressures and temperatures within the Earth, he set up a specific model for the mantle, in which the transition zone of the upper mantle between 400 and 1000 km (Figure 3-3) was explained by the olivine-spinel transition modified by solid solution effects. In these three papers and following papers on the same topic Ringwood made many predictions about the upper mantle. Ten years later, from his review of the problem at the International Conference on Phase Transformations and the Earth's Interior, held in Canberra in February 1969, it is clear that his percentage of successful predictions is remarkably high.

Between 1956 and 1966 the existence of an olivine-spinel transition was confirmed in many substances, including Fe_2SiO_4, and there were several estimates made for the conditions of transformation of Mg_2SiO_4 based on rather long extrapolations. The development of new types of high pressure, high temperature apparatus in 1966 led to rapid advances in study of the transition in the system Fe_2SiO_4-Mg_2SiO_4. In 1970 Ringwood and Major demonstrated that for compositions near Mg_2SiO_4 the transition is more complicated than previously supposed. Some of the results published between 1956 and 1969 are reviewed below and summarized in Figures 6-14 and 6-15.

Phase Diagram for Fe_2SiO_4

The upper mantle olivine is generally believed to have a composition near 90% forsterite, 10% fayalite and attention was directed early to the transitions in Fe_2SiO_4. The phase diagram for Fe_2SiO_4, published by S. Akimoto and E. Komada in 1967, is

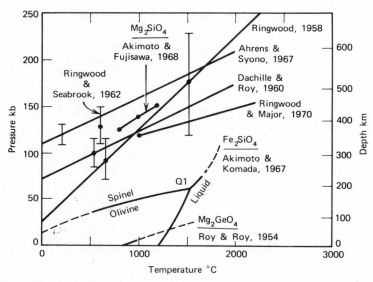

Figure 6-14. Experimental and theoretical results related to the olivine–spinel transition in the upper mantle. Mg_2GeO_4: olivine-spinel transition. Fe_2SiO_4: olivine-spinel transition and melting curves meeting at triple point Q_1. Mg_2SiO_4: olivine-spinel transition, see text and Figure 6-15.

shown in Figure 6-14. The olivine-spinel transition has a slope similar to that illustrated for the same transition in Mg_2GeO_4. The melting curve for fayalite meets the olivine-spinel transition curve at a triple point, Q_1, at 62 kb and the fusion curve for spinel extends upward in pressure after a sharp inflection at Q_1. This shows the pattern of phase relationships sought for the composition Mg_2SiO_4.

The Transition for Mg_2SiO_4

Figure 6-15 summarizes the attempts to determine the conditions for the breakdown of forsterite by extrapolation from within binary systems. Curves through individual points estimated in this way are given in Figure 6-14.

Mg_2GeO_4-Mg_2SiO_4. The univariant curve for the olivine-spinel transition in Mg_2GeO_4 is shown in Figure 6-14. The effect of adding Mg_2SiO_4 as a component was determined by Dachille and Roy in isobaric and isothermal sections through the *PTX* model

required to represent the binary system completely. In 1956 they reported the synthesis of single spinel phases at 20 kb with about 18% Mg_2SiO_4 in solution. Detailed results were published in 1960. Figure 6-15a shows the two-phase field for coexisting olivine and spinel solid solutions determined up to 65 kb pressure at 542°C. Extrapolation of the measured solid solution loop to a closure point (the dashed lines) for the composition Mg_2SiO_4 provided an estimate of 100 kb for the olivine-spinel transition in forsterite at this temperature. This corresponds to one point on the univariant curve for the transition. Dachille and Roy estimated that the slope of the curve passing through this point would be between 13°C/kb and 25°C/kb and an average value is given in Figure 6-14.

In the second of his three papers on the constitution of the mantle Ringwood in 1958 reported data on the extent of solid solution at 660°C and 30 kb between Mg_2GeO_4 and Mg_2SiO_4. From this he calculated the free energy of the transition in forsterite and used this to calculate the pressure required to

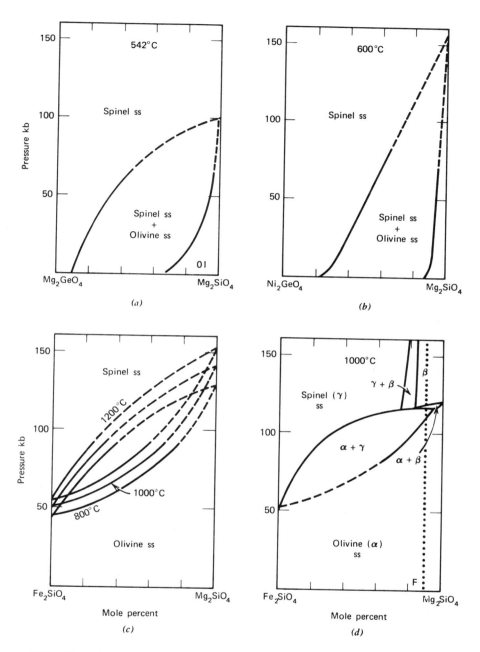

Figure 6-15. Experimental studies on the olivine-spinel transition for Mg_2SiO_4. (a) Isothermal extrapolation from Mg_2GeO_4–Mg_2SiO_4 by Dachille and Roy (1960). (b) Isothermal extrapolation from Ni_2GeO_4–Mg_2SiO_4 by Ringwood and Seabrook (1962). (c) Isothermal extrapolation from Fe_2SiO_4–Mg_2SiO_4 by Akimoto and Fujisawa (1968). (d) Direct determination of phase relationships by Ringwood and Major (1970). Olivine with composition near Mg_2SiO_4 transforms not to spinel but to a spinel-like beta-phase. Composition F is used in Figures 6-16 and 6-17. (With permission, American Jour. Science, Jour. Geophys. Research, Physics Earth and Planet. Interiors.)

cause the transition at 660°C. The most probable pressure was 88 kb, the maximum 116 kb, and the minimum 72 kb. This result is shown in Figure 6-14.

Ni₂GeO₄-Mg₂SiO₄. In the first paper of the 1958 set Ringwood presented a phase diagram for the solvus at 1 bar pressure between forsterite and Ni_2GeO_4 which has a spinel structure. From the solid solution relationships he calculated that the pressure for the transition of forsterite to the spinel structure at 1500°C was 175 ± 55 kb. In his second paper Ringwood drew the univariant curve through this point and the 660°C point estimated from the Mg_2GeO_4 system as shown in Figure 6-14. In an appendix to this paper he reported reconnaissance results up to 70 kb pressure, at temperatures between 560°C and 660°C, on the effect of pressure on the solid solubility of Mg_2SiO_4 in the spinel Ni_2GeO_4. A linear extrapolation indicated that at 600°C the spinel form of Mg_2SiO_4 should be stable at 125 kb (a maximum value). This result is not plotted in Figure 6-14, but the revised value of 130 ± 20 kb at 600°C determined in 1962 by Ringwood and Seabrook is shown. This point was located by the extrapolation shown in the isotherm for the binary system in Figure 6-15b. The extrapolated pressure shown is 155 kb, but this was corrected downward to 130 kb because of a revision in the high pressure scale.

Fe₂SiO₄-Mg₂SiO₄. There have been several studies of the solid solution of Mg_2SiO_4 in the spinel form of Fe_2SiO_4 and Figure 6-15c shows three isotherms located between 43 and 96 kb at 800, 1000, and 1200°C, by S. Akimoto and H. Fujisawa in 1968. Extrapolations of the two-phase fields to closures, as shown by the dashed lines in the figure, give for points on the forsterite-spinel transition the values 127 kb at 800°C, 140 kb at 1000°C, and 151 kb at 1200°C. These three points and the line connecting them are plotted in Figure 6-14. An earlier preliminary report gave an estimate of 150 kb at 800°C.

Figure 6-14 shows that the results described up to 1968 indicate a rather wide range of uncertainty about the position of the transition in pure forsterite. Calculated curves for the transition suffer from the uncertainties in the experimental results. For example T. J. Ahrens and Y. Syono calculated the slope of the curve shown in Figure 6-14, with its range of error, from the thermochemical and compressibility data available in 1967. Having calculated the slope they then passed the curve through a preliminary experimental point at 150 kb and 800°C, which was the following year moved downward through 33 kb by Akimoto and Fujisawa. They gave as the equation for the transition curve: $P = 110 ± 11$ kb $+ 0.050T$°C. In the same year, using essentially the same data, D. L. Anderson calculated an almost identical curve, but noted that the revised pressure scale would reduce the pressure estimates by 20–30% and the curve that he used therefore had the equation: $P = 74$ kb $± 0.053T$°C. This is not plotted in Figure 6-14, but it would be quite close to the curve of Dachille and Roy.

Ringwood and Major reviewed in detail the experimental and theoretical data in this system prior to 1969 and presented the 1000°C isotherm shown in Figure 6-15d. This shows that the previous extrapolations were based on an incorrect assumption. Pure forsterite does not transform to a spinel but to a spinel-like material which they called the beta-phase. Previous claims that spinel was synthesized from Mg_2SiO_4 probably arose from misidentification of the beta-phase or of a hydrated magnesium silicate. No more than 80% of Mg_2SiO_4 dissolves in the spinel (γ) phase at 1000°C below 150 kb.

Figure 6-15d shows the first direct experimental location of the breakdown of pure forsterite with the beta-phase produced at 120 kb and 1000°C. This point is shown on Figure 6-14 and Ringwood and Major assumed that the slope of the curve for this transition was similar to that previously determined for the transition of fayalite to spinel.

The Transition for Olivine Fo₉₀Fa₁₀

The Transition for Olivine Fo$_{90}$Fa$_{10}$

It is generally assumed that the olivine in the upper mantle has a composition close to 90% forsterite. Figure 6-15d shows the changes (under equilibrium conditions) that occur in an olivine of this composition (F) with increasing pressure at 1000°C. At about 108 kb the olivine begins to transform to spinel (γ) with only 48% Mg$_2$SiO$_4$ in solution. With increased pressure both the spinel and the remaining olivine become enriched in magnesium, and at 116 kb their compositions become olivine with 94% Mg$_2$SiO$_4$ and spinel with 75% Mg$_2$SiO$_4$. These phases react at constant pressure to yield the beta-phase with 81% Mg$_2$SiO$_4$ in solution until all of the spinel is used up. With pressure increasing the olivine is transformed into the beta-phase, and the beta-phase composition reaches that of the original olivine at 119 kb just as the last trace of olivine disappears. The olivine is thus transformed to a more dense phase through a divariant transition interval of 11 kb with a univariant (isothermal invariant) reaction occurring within the interval.

The position of this transition interval as a function of pressure and temperature can be estimated from the schematic PTX model for the system Mg$_2$SiO$_4$-Fe$_2$SiO$_4$ which is given in Figure 6-16. The front face of the model is the PT diagram for the system Fe$_2$SiO$_4$ with three unary curves meeting at the invariant point Q_1 (Figure 6-14). Extension of the liquid volume and adjacent phase fields into the binary system are shown for completeness, but they need not concern us here. The back face of the model shows only the transition of forsterite to spinel as anticipated in Figures 6-14 and 6-15c. The phase fields for the beta-phase required by Figure 6-15d have been omitted for geometrical clarity.

The transition interval in which olivine and spinel coexist is given by the intersection of the constant composition plane for 90% Mg$_2$SiO$_4$ with the two-phase space in Figure 6-16. The isopleth showing the estimated transition band is given in Figure 6-17. The width of the interval is based on the experimental value of 11 kb at 1000°C in Figure 6-15d. According to Figure 6-16 the transition interval is a simple two-phase field between olivine and spinel but Figure 6-15d shows that the interval is divided into two parts by the univariant reaction:

$$\text{forsterite}_{ss} + \text{spinel}_{ss} \rightleftharpoons \text{beta-phase}_{ss} \quad (6\text{-}1)$$

This boundary is added as a dashed line in Figure 6-17, and the high pressure field is labelled beta-phase according to Figure 6-15d instead of spinel as indicated by Figure 6-16. Ringwood and Major pointed out that magnesian spinel and the beta-phase had almost identical densities at run conditions.

The geotherm in Figure 6-17 is Ringwood's estimate for shield regions extrapolated from below 250 km (Figure 3-12b) along a curve 250°C higher than the geotherm calculated earlier by Clark and Ringwood (Figure 3-12a). If the upper mantle were composed only of olivine with the composition represented in this diagram, Fo$_{90}$, it would be progressively converted to a spinel phase between A and B; at B the spinel phase would be replaced by the beta-phase; and between B and C, the remaining olivine would be progressively converted to the beta-phase. In the depth interval 385 to 430 km the olivine would be converted completely to the beta-phase with a density increase of 10.6%.

The uncertainties about estimating the temperature at depth within the Earth were discussed in Chapter 3, and the position of the transition interval in Figure 6-17 is only schematic. Figure 6-14 illustrates the uncertainties about the experimental data on which the position and slope of the transition interval is based. As all investigators have concluded, however, the seismic discontinuity at a depth of 400 km does correspond remarkably well with the transition of mantle olivine into a more dense phase.

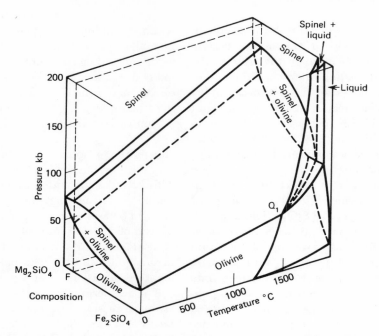

Figure 6-16. Schematic phase diagram in PTX space for the system Mg_2SiO_4–Fe_2SiO_4 using model corresponding to Figure 6-15c, omitting the beta-phase (6-16d) for clarity. The vertical plane through composition F (Figure 6-15d) intersects the transition interval as shown in Figure 6-17.

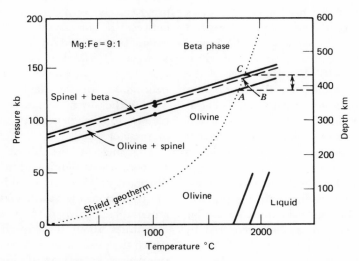

Figure 6-17. Schematic isopleth for the composition F (Figures 6-15d and 6-16) in the system Mg_2 SiO_4–Fe_2SiO_4, showing detail through the transformation interval of olivine to beta-phase (Figure 6-15d). This is related to transitions within the mantle by the depth scale and the shield geotherm from Figure 3-12 (Ringwood 1966). Mantle material of this composition would undergo the transition in the depth interval *A–C*.

The Transition Zone of the Mantle 400 to 1000 km Depth

When experimental and theoretical investigations of the olivine-spinel transition began in the 1950's the aim of the exercise was to explain the seismic wave velocity distribution within the transition zone of the mantle, between 400 and 1000 km, as shown in Figure 3-3. More recent interpretations of the seismic velocity distribution (Figure 3-5) show that the transition zone includes three seismic discontinuities at depths near 400, 650, and 1050 km.

The estimated position of the transformation interval for mantle olivine, as shown in Figure 6-17, is a satisfactory explanation for the sharp discontinuity near 400 km but before this had been detected, the problem was to find some way to extend the effect of the olivine-spinel transformation in order to explain the whole transition zone between 400 and 1000 km.

In 1958 Ringwood suggested that the olivine-spinel transformation in the mantle could occupy a wide depth range if (a) the transformation interval had a much steeper slope than that illustrated in Figure 6-17, and (b) there were a marked increase in the Earth's temperature gradient between the two discontinuities. In papers between 1958 and 1966 Ringwood developed models in which the smoothed velocity distributions from 400 to 1000 km depth (Figure 3-3) were explained by the smearing-out effects associated with solid solution. According to these models the transition zone in garnet-peridotite is dominated by the olivine-spinel transformation at the beginning, but with an additional series of reactions, including the breakdown of orthopyroxene, leading to the progressive increase in density of the material. Solid solution effects spread the individual transitions over substantial depth ranges. In early models the heterogeneous upper mantle garnet peridotite passed through a wide transition interval until at a depth of 1000 km it consisted of a homogeneous rock composed of disordered, defect spinel in which all the other minerals had dissolved. An alternative explanation is that the minerals break down into close-packed oxides. In later models Ringwood suggested that spinels would transform to denser phases with silicon in octahedral coordination.

In his 1969 review Ringwood revised the model taking into account new experimental data and the changed picture of seismic velocities in the mantle (Figure 3-5), and he noted that the effects of solid solution did not have the result previously attributed to them. The apparent success of his previous models emphasizes a point made by Ringwood: however plausible a model may appear the acquisition of data usually requires that it be modified. It is certainly true that limited data can be extrapolated to fit observations far more readily than detailed data.

Figure 6-17 shows that some explanation other than the olivine-spinel transition is required to explain the changes in the seismic velocity profile near 650 and 1000 km (Figure 3-5). Details of Ringwood's 1969 model are shown in Table 6-8.

In 1967 D. L. Anderson had assumed that the spinel broke down to a "postspinel" phase composed of MgO, FeO, and SiO_2 with silicon in octahedral coordination and with the same thermochemical properties as the oxides. He calculated the univariant curves and transformation loops for the spinel-postspinel reaction in the system Mg_2SiO_4-Fe_2SiO_4, and used them in conjunction with similar olivine-spinel intervals based on extrapolations of the type in Figure 6-15c. The depth of a transformation interval such as AC in Figure 6-17 changes as a function of composition (Figures 6-15 and 6-16). By varying the composition Anderson fitted the intervals for both the olivine-spinel and spinel-postspinel transformations to the depths of the seismic transition zones near 400 and 650 km. The transition regions are consistent with temperatures near 1500°C at 365 km, and 1900°C at 620 km (compare Figure 3-12). He concluded that the seismic

TABLE 6-8   Mineralogy of the Mantle Peridotite as a Function of Depth: 1969 Model of A. E. Ringwood[a]. Compare Figure 3-5b

Depth km	Mineral assemblages and transformations	Weight percent mineral	Coordination of elements	Zero pressure density
To 80	*Plagioclase peridotite* *Spinel peridotite*	Table 6-7		
80–350	*Garnet peridotite* olivine orthopyroxene clinopyroxene garnet	57 17 12 14	Si-4 Mg, Fe, Ca, -6, 8	3.38
350–450	Pyroxene → garnet Olivine → spinel[b]			
450–600	Spinel[b] Garnet Jadeite	57 39 4	Si-4, 6	3.66
600–700	Spinel[b] → Sr_2PBO_4 structure or MgO + ilmenite structure Garnet → ilmenite + perovskite structures Jadeite → calcium ferrite structure			
700–1050	Ilmenite solid solution Strontium plumbate structure Perovskite, $CaSiO_3$ Calcium ferrite, $NaAlSiO_4$	36 55 6.5 2.5	Si, Mg, Fe -6	3.99 -4.03
1050–1150	Transformations into phases denser than isochemical oxide mixture			
1150–2900	Speculative. Depends on germanate analog systems and shock wave studies		Si, -6 Mg, Fe 6	7% higher than mixed oxides

[a] See also Ringwood, 1970.   [b] Includes beta-phase, Figure 6-15d.

velocities and the depths to the transition zones are consistent with a mantle composition between about 80 and 60 mole-percent Mg_2SiO_4; the Fe_2SiO_4 content appears to increase with depth from 500 to 700 km. Estimating compositions and temperatures in the mantle by correlating geophysical data with laboratory measurements points the way to future advances. The reliability of the available experimental and theoretical data on phase transitions in mantle material, however, remains rather uncertain as shown by Figures 6-14 and 6-15 for the simple olivine composition itself. According to Table 6-7 the upper mantle may contain only 55–67% olivine. The effects of solid solution have not yet been evaluated.

MINERALOGY AND PETROLOGY OF THE MANTLE

High pressure, high temperature apparatus provides, in effect, an indirect experimental probe into the Earth's interior for comparison with the properties measured by geophysical methods. Before 1966 the experimental probe was limited to less than 200 km in depth. Since 1966 the development of new apparatus has permitted the laboratory reproduction of conditions corresponding to 600 km. Many new phase transformations in silicates have been discovered in particular by A. E. Ringwood and associates. The study of germanate systems, which serve as analogs for silicate systems at higher pressures, has revealed additional phase transformations that may be experienced by the silicates at considerably greater depths. Shock wave techniques are capable of developing transient pressures corresponding to conditions in the lower mantle and phase transformations in silicates have been discovered in this way. Shock wave studies provide information for the formulation of an equation of state for mantle materials.

Ringwood's 1969 correlation of phase transformations in mantle peridotite (pyrolite model) with the seismic properties of the mantle is summarized in Table 6-8. He assumed that the molecular ratio of $Fe/(Fe + Mg)$ remained constant at 0.11 throughout the mantle.

The Upper Mantle, the Transition Zone, and the Low-Velocity Zone

The mineralogy of the mantle at any depth can be estimated by following geotherms through experimentally determined mineral facies in material of appropriate composition. Estimates of mantle composition and the experimentally determined phase diagrams have been reviewed. Combining and extrapolating these data to pressures corresponding to depths of 600 km provides Figures 6-18 and 6-19 which illustrate schematically the phase relationships in probable mantle

materials dry or in the presence of water traces. A guide to the petrology of the mantle is given by the geotherm which is the oceanic geotherm from Figure 3-12, and a corresponding sequence of rock types with depth through the mantle is shown in Figure 6-20. Estimates of temperature distribution in the Earth depend critically upon assumptions regarding the relative significance for heat transfer of conduction, radiation, and convection as discussed in Chapter 3. The combination of extrapolated experimental data and arbitrarily selected geotherms to provide estimated mantle cross sections is intended only to show patterns for the petrology of the mantle; specific temperatures and depths can be modified as required by the acquisition of extended and improved experimental data.

The upper mantle is composed of peridotite and eclogite (Figure 5-18). The density distributions determined by F. Press using Monte Carlo methods (Figures 3-9b and 3-10) suggest that the upper mantle is chemically and mineralogically zoned, laterally as well as vertically, and that there may be as much as 50% eclogite in the suboceanic mantle between depths of 80 and 150 km. In addition to these rocks there is residual peridotite or dunite and precipitated peridotite resulting from the extraction of picritic or basaltic magma and its fractional crystallization during uprise. Superimposed on this heterogeneous mantle is a concentric succession of mineral facies.

Consider first a dry mantle. Experimentally determined mineral facies for peridotite and eclogite down to about 100 km are shown in Figures 5-15 and 6-10, and these are reproduced in Figures 6-18a and 6-19a. The general pattern of transitions for peridotite or eclogite in the mantle according to Figures 6-18a and 6-19a can be seen in Figure 6-20. Figure 6-10 shows that plagioclase peridotite can exist only in oceanic regions or in high-temperature regions with a thin crust. At

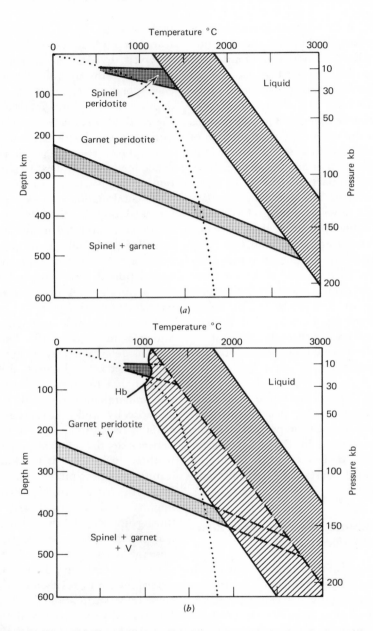

Figure 6-18. Schematic phase diagrams for peridotite and peridotite-water extrapolated to pressures corresponding to 600 km depth in the mantle (after Wyllie, 1971). (*a*) Dry peridotite, based on Figures 8-14*a*, 6-10, and 6-17. (*b*) Isopleth for peridotite with 0.1% water, based on Figures 6-18*a*, 6-11, and the procedure used in construction of Figure 6-13. The melting interval consists of two parts: the light-shaded band represents incipient melting, and the dark-shaded band is almost equivalent to dry melting. The depth scale and the geotherm (dotted) relate the phase diagrams to upper mantle of specific compositions. (With permission, Jour. Geophys. Research.)

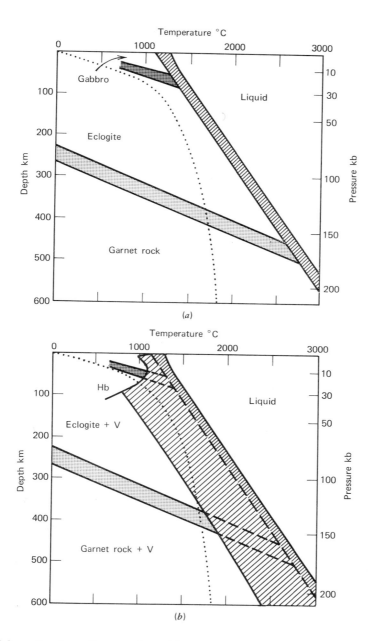

Figure 6-19. Schematic phase diagrams for gabbro and gabbro-water compositions extrapolated to pressures corresponding to 600 km depth in the mantle (after Wyllie, 1971). (*a*) Dry gabbro, based on Figures 8-14, 5-15, and analogy with 6-18*a*. (*b*) Isopleth for gabbroic material with 0.1% water, based on Figures 6-18*b* and extrapolation of Figure 6-13. The melting interval consists of two parts: the light-shaded band represents incipient melting, and the dark-shaded band is almost equivalent to dry melting. The depth scale and the geotherm (dotted) relate the phase diagrams to upper mantle material of specific compositions. (With permission, Jour. Geophys. Research.)

depths of about 30 km this undergoes a transition to spinel-peridotite, and this in turn transforms to garnet peridotite at a depth of about 70 km. In regions where the geotherm is high spinel-peridotite can exist to a maximum depth of about 85 km. Beneath the continental shields a thin layer of spinel-peridotite would be transformed into garnet-peridotite at depths of 55 to 60 km. Figure 5-15 shows that crystallization of basaltic magma in the upper mantle can yield gabbro, but in cooling to the temperature of the normal geotherm the gabbro should transform to eclogite.

According to Table 6-8 the transition zone begins with a seismic discontinuity produced by the breakdown of pyroxene and olivine in the garnet peridotite of the upper mantle. The pyroxene dissolves in the garnet already present producing a complex solid solution. The formation of the β-phase from olivine is illustrated in Figures 6-17 and 6-18 and the formation of garnet in Figure 6-19. Because of the similarity in physical properties and transition parameters Ringwood found it convenient to treat the β-phase and spinel together as spinel. This reaction zone is shown near 400 km depth in Figure 6-20. Accounting for the seismic discontinuity near 650 km is the breakdown of all three minerals, spinel, garnet, and jadeite producing new phases with silica in six-fold coordination. The transition zone ends near 1050 km with a series of small discontinuities; these are interpreted in terms of transformations into phases which are denser than an isochemical oxide mixture and characterized by Mg coordinations higher than six.

Partial melting occurs if the temperature exceeds the solidus curves in Figures 6-18a and 6-19a. These curves have been determined experimentally down to depths of 100 to 150 km (Figures 5-15, 6-11 and 6-12), but results from different laboratories for different compositions vary, and there is a large uncertainty in extrapolation of the slopes of the melting intervals from 100 to 600 km. None of the calculated geotherms shown in Figure

3-12 exceeds the solidus curves of Figures 6-18a and 6-19a which indicates that magma generation requires unusual conditions of temperature distribution (Chapter 8). It also leads us to consider the effect of water on melting in the mantle.

Traces of water in the uppermost mantle will be stabilized in amphibole and phlogopite where potassium abundances are high enough (Figures 6-11, 6-12, and 6-13), and possibly in titanoclinohumite. The hornblende breakdown curves in Figures 6-18b and 6-19b are taken from Figures 6-11 and 6-13. The interval for incipient melting of eclogite in the presence of 0.1% water shown in Figure 6-19b is extrapolated directly from Figure 6-13. The melting relations for peridotite in the presence of 0.1% water in Figure 6-18b were deduced in the same way from the data in Figure 6-11 and the same extrapolation was made. It is experimentally established that the hydrothermal solidus curves change slope (dP/dT) from negative, to positive at pressures between 15 and 30 kb, and it appears from the eclogite-water system that the slope becomes similar to that for the dry rock (Figures 6-12 and 6-13). There is a large uncertainty in the assumption adopted in Figures 6-18b and 6-19b that the solidus curves in the presence of water become parallel with the dry solidus curves, but the general pattern of phase relationships is not likely to differ significantly from that shown. The oceanic geotherm passes through the interval of incipient melting in Figures 6-18b and 6-19b. Therefore, given this temperature distribution and a trace of water in a mantle composed of peridotite and eclogite, a trace of interstitial silicate liquid will exist in the depth zone where the geotherm exceeds the solidus temperature as indicated in Figure 6-20 (see discussion of Figure 6-13).

The existence of a low-velocity zone in the upper mantle for most tectonic environments is now established for both S-waves and P-waves (Figures 3-4 and 3-5b). There have been many attempts to explain this:

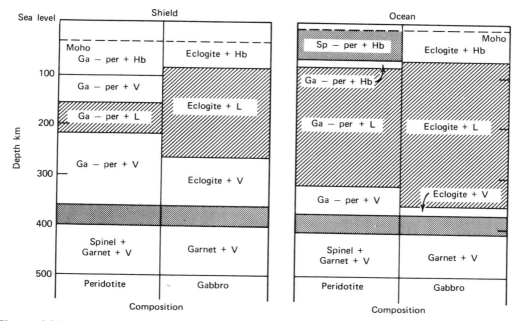

Figure 6-20. Schematic sections through the upper mantle in two different tectonic environments, for mantle material composed of either peridotite or eclogite, in the presence of traces of water (after Wyllie, 1971). These were determined by following geotherms (Figure 3-12) through the mineral facies in Figures 6-18*b* and 6-19*b*. The thickness of the zone of incipient melting depends upon both environment (geotherm) and composition. (See Anderson, 1970). (With permission, Jour. Geophys. Research.)

1. By departure of the geotherm from the critical temperature gradient for which the seismic velocity remains constant,

2. In terms of chemical or mineralogical zoning.

Although these factors may contribute to the low velocities they have not proved adequate to account for the properties of the zone. Geophysicists now conclude that partial melting is required to explain their observations, and the assumption has been usually implicit that melting occurred because the geotherm rose above the dry solidus of peridotite or eclogite. This introduces a number of problems, and it requires that the geotherm be rather sensitively controlled and constrained close to the solidus lest too much liquid is produced. A more likely explanation is that incipient melting occurs in the presence of traces of water as suggested by D. L. Anderson and C. G. Sammis (1969).

The pattern of phase relationships for material of gabbroic composition, containing only 0.1% water, is shown in Figures 6-13 and 6-19*b*. I. B. Lambert and I suggested in 1968 that the low-velocity zone corresponds to the level where amphibole becomes unstable because at greater depths traces of water in eclogite would cause the formation of interstitial magma. At the same time I. Kushiro published the solidus curve for peridotite-water in Figure 6-11 and suggested that water in mantle peridotite could be responsible for partial melting in the low-velocity zone. The geotherms in Figures 6-13 and 6-19*b* show that variation in temperature at depth would cause the boundary between vapor-absent hornblende eclogite and partially melted eclogite to migrate up or down, and

Figure 6-18*b* shows that hornblende may play a similar role in peridotite mantle. The amount of interstitial liquid produced is very small and almost a direct function of the water content; it does not vary significantly for wide variations in temperature and depth. For a given water content the amount of liquid produced in peridotite is much less than in eclogite.

If the low-velocity zone is caused by incipient melting in the presence of water its downward termination could be due to several effects. There would be no liquid below the zone:

1. If the water content became vanishingly small.

2. If the water was stored in high-pressure hydrous crystalline phases.

3. If the bottom of the zone coincided with the lower limit of eclogite; for the same amount of water there would be very much less liquid in the underlying peridotite than in the eclogite-bearing peridotite.

4. If the geotherm passed through the solidus into the subsolidus region. This is the condition illustrated in Figures 6-18*b*, 6-19*b*, and 6-20.

Figure 6-20 shows the patterns for the petrology of hypothetical mantle sections composed of either peridotite or gabbro with a trace of water present, in two different tectonic environments. These are derived by following the continental shield and oceanic geotherms from Figure 3-12 through the schematic phase diagrams of Figures 6-18*b* and 6-19*b*. Note that the water is contained either in (a) hydrous minerals in vapor-absent zones, (b) hydrous silicate liquid in vapor-absent zones, or (c) aqueous vapor phase. Note that changing the position of the geotherm in Figure 6-18*b* could introduce a zone containing vapor between the hornblende peridotite and the peridotite with liquid. The presence of carbon dioxide and other volatile components in the mantle would require modification of this scheme in detail, but the general pattern would prob-

ably remain unchanged (Wyllie, 1970; Lambert and Wyllie, 1970). The thickness of the shaded zone with interstitial liquid in Figure 6-20 varies with the geotherm and therefore with tectonic environment, and the distribution of peridotite and eclogite within the mantle also affects its position. The versatility of this model for the low-velocity zone is one of its attractive features.

The Lower Mantle

The densities of model mantle materials at lower mantle pressures, derived principally from shock wave data, indicate that the lower mantle is denser than the hypothetical peridotite with six-fold coordination of Si, Mg, and Fe, and with $Fe/(Fe + Mg) = 0.11$. This higher density could be caused by (a) increased Fe/Mg ratio in the lower mantle, (b) phase transformations to denser structures with Mg and Fe in coordination higher than six, or (c) a combination of these factors. Ringwood presented evidence suggesting that the required density increase could be caused by phase transformations without significant change in composition. F. Press and D. L. Anderson concluded that a change in composition was required to explain the geophysical data.

The successful Earth models selected by the Monte Carlo procedure (Figure 3-9*b*) were compared by Press with laboratory measurements on rocks and minerals that are candidate constituents for the mantle. He concluded that the transition zone is a compositional boundary as well as a region of phase changes. He estimated that the inhomogeneous mantle is consistent with an increase in the molar $FeO/(FeO + MgO)$ ratio by a factor of two. The increase corresponds to a change from $(Mg_{0.8}Fe_{0.2})_2SiO_4$ to $(Mg_{0.6}Fe_{0.4})_2SiO_4$. Using a completely different procedure reviewed a few pages back Anderson concluded that the Fe_2SiO_4 content of the mantle appeared to increase with depth from 500 to 700 km. He reexamined the problem in more detail the following year, 1968, using a combination of several different types of geophysical evidence, and concluded that if the FeO/MgO ratio in the upper mantle is 0.1 then it is 0.27 in the lower mantle.

If the lower mantle is enriched in iron compared with the upper mantle this has significant geochemical and geophysical implications. It has a bearing on hypotheses of differentiation of the Earth into core and mantle and on mantle convection hypotheses. Concentration of FeO relative to MgO at depth is stabilizing with respect to convection and makes the case for mantle-wide convection more difficult. Convection would then be restricted to the upper mantle.

7. The Structure, Petrology, and Composition of the Earth's Crust

INTRODUCTION

The structure of the Earth's crust can be determined from seismic, gravity, and magnetic measurements. Seismic studies provide data about the physical properties of rocks in the crust and the location of discontinuities between rock units with different properties. The seismic velocities can be compared with velocities measured for rocks in the laboratory at known pressures and temperatures (Figure 5-3). Gravity surveys cannot provide unique crustal sections because the anomalies are capable of explanation by more than one model, but with seismic measurements used as a control in interpretation gravity surveys provide significant data about the density distribution in crustal sections. Magnetic anomalies recorded near the Earth's surface are strongly influenced by local rock masses, but aeromagnetic surveys make it possible to distinguish major crustal units such as shields and folded belts of various ages. During the 1960's there were remarkable advances in technique and interpretation with national and international programs bringing all three methods to bear on specific regions.

The rock types composing the structural units of the crust can be estimated from the distribution of rocks at the surface and in drill cores of limited depth, from deductions about the geological history and petrogenesis of exposed rocks, from geochemical considerations, and from the physical properties of rocks at depth determined by geophysical methods. The composition of the crust can then be calculated from the known average compositions of rocks assigned to the various crustal units and layers.

STRUCTURE OF THE CRUST

Figure 5-1 outlines the main features of the crust in the major tectonic environments: the continents, ocean basins, and continental margins including island arcs and trenches. It also depicts the change in crustal thickness associated with the active belts: the oceanic ridges and the folded mountain chains. Selected velocities for P-waves indicate a sharply defined boundary between crust and mantle and a poorly defined boundary between upper and lower crust. The boundary between the crust and mantle was reviewed in detail in Chapter 5 and, in this context, we also examined the variation and distribution of seismic velocities as a function of depth in various tectonic environments. Figures 5-2, 5-3, and 5-4 show average P-wave velocities in different depth intervals.

Conventionally seismic data have been interpreted in terms of a two-layered crust divided into a "granitic" layer and a "basaltic" layer by the Conrad discontinuity and this is the general pattern depicted in Figure 5-1; the profiles of average seismic velocities in Figures 5-2, 5-3, and 5-4 are consistent with a layered crust. Soviet seismologists have largely abandoned this model in favor of more complex models that are more realistic geologically. A velocity profile for a layered crust is shown in Figure 7-1a, and this differs from the conventional model only by the presence of two velocity reversals, or low-velocity layers, within the crust. Figure 7-1b shows another layered model with the conventional downward sequence of increasing velocities interrupted by layers with high seismic velocities; this fits some observed travel-time curves. A crust composed of blocks rather than layers would give velocity profiles like those shown in Figure 7-1c.

The distribution of the major physiographic features of the Earth's surface was reviewed in Chapter 2, and these were described as tectonically stable or unstable. In Chapter 3 we noted a good correlation of heat flow with the tectonic environment, and

in Chapter 5 we correlated seismic velocities for the crust and upper mantle in these same environments. These geophysical parameters, along with gravity anomalies (Chapter 3) lead to the division of the Earth's crust into the types listed in Table 7-1.

The crustal structures for these types suggest a classification based on both crust and upper mantle. There are two crustal types, continental and oceanic, and two mantle types, stable and unstable. Areas of the Earth's surface can then be identified as one of the four types in Table 7-1. The unstable types include the active tectonic belts reviewed in Chapter 2: the folded mountain chains, the midoceanic ridges, and the island arc systems of the continental margins.

Cross Sections Through Continents

Figure 7-2 shows the results of a transcontinental geophysical and geological survey across the North American continent along a great circle passing approximately through Washington, D.C., Denver, and San Francisco. The schematic geological cross section

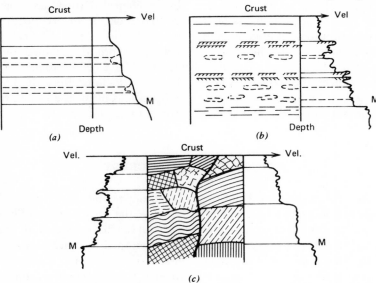

Figure 7-1. Crustal models, after Kosminskaya et al., (1969). (a) Continuous boundaries. (b) Piecewise continuous boundaries corresponding to piecewise continuous travel-time curves. (c) Assumed block-layered crust corresponding to the geophysical and geological points of view. (From Geophysical Monograph **13**, 195, 1969, with permission of Amer. Geophys. Union.)

TABLE 7-1 Tectonic Classification of Earth's Surface (Brune, 1969)

Crustal type	Tectonic characteristic	Crustal thickness (km)	P_n (km/sec)	Heat flow HFU	Bouguer Anomaly (mgal)	Geologic features
Continental crust overlying stable mantle						
Shield	very stable	35	8.3	0.7–0.9	−10 to −30	Little or no sediment, exposed batholithic rocks of Precambrian age. Moderate thicknesses of post-Precambrian sediments.
Midcontinent	stable	38	8.2	0.8–1.2	−10 to −40	
Continental crust overlying unstable mantle						
Basin-range	very unstable	30	7.8	1.7–2.5	−200 to −250	Recent normal faulting, volcanism, and intrusion; high mean elevation.
Alpine	very unstable	55	8.0	variable 0.7–2.0	−200 to −300	Rapid recent uplift, relatively recent intrusion; high mean elevation.
Island arc	very unstable	30	7.4–7.8	variable 0.7–4.0	−50 to +100	High volcanism, intense folding and faulting.
Plateau	Not adequately studied					
Oceanic crust overlying stable mantle						
Ocean basin	very stable	11	8.1–8.2	1.3	+250 to +350	Very thin sediments overlying basalts, linear magnetic anomalies, no thick Paleozoic sediments.
Oceanic crust overlying unstable mantle						
Ocean ridge	unstable	10	7.4–7.6	high and variable 1.0–8.0	+200 to +250	Active basaltic volcanism, litle or no sediment.
Ocean trench	Not adequately studied					

in Figure 7-2a was deduced from seismic measurements. Values for the velocities in the upper mantle and lower crust can be read from Figure 3-7 which also shows the approximate line of the traverse. Contrast the position of the Moho in Figure 7-2 with its position in Figure 5-1a, which was based on data available as recently as 1962.

We noted in connection with Figure 3-7 that the Rocky Mountain system appears to divide the United States into two super-provinces involving both crust and upper mantle. This is confirmed in Figure 7-2 by the contrast in magnetic character west and east of the Rocky Mountains; the amplitudes and wavelengths of the anomalies differ. The

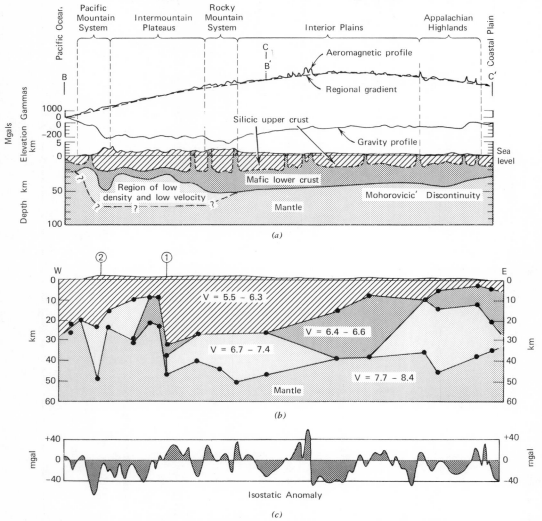

Figure 7-2. (*a*) Transcontinental geological and geophysical cross section of the earth's crust and upper mantle. Indicated distribution of mafic material in the silicic upper crust is schematic only. Sedimentary rocks overlying crystalline basement are not shown. See Figure 3-7b for location of sections (after Pakiser and Zietz, 1965). (*b*) Composite seismic cross section through the crust not far removed from section in *a*, with *P*-wave velocities. (*c*) Isostatic anomalies for the section in *b*. (*b* and *c* after Woollard, 1968.) (From Rev. Geophysics, **3**, 505, 1965; Geophysical Monograph **12**, 312, 1968; with permission of Amer. Geophys. Union.)

magnetic profile may be correlated with heat flow. In regions of low heat flow the geotherms are depressed and there is a greater thickness of rocks below the Curie temperature which produces more intense anomalies. The gravity profile in Figure 7-2a also shows the contrast between the eastern and western provinces of the continent.

Figure 7-2b is a composite seismic cross section through the crust along a profile from Chesapeake Bay, Maryland, to central California; this is not far removed from the great circle profile of Figure 7-2a. Units of the crust with similar ranges of seismic velocities are shown (compare Figure 3-7b), and isostatic anomaly values are given. The striking features of this profile at the Moho are the abrupt transition in structure of the Wasatch Mountain front (location 1) and the local root and negative isostatic anomaly beneath the Sierra Nevada (location 2). The seismic velocities indicate that the predominantly granitic crust of the west is replaced by a predominantly basic crust beneath the Appalachians and eastern seaboard, and that a thick intermediate layer beneath the eastern plains is replaced by granitic material beneath the western plains.

Figure 7-3 shows a crustal cross section through the Russian platform including the Ukrainian shield and the Dnieper graben. The crust is a layered structure with many seismic interfaces. In some areas the seismic velocities indicate that the Moho may be a layer up to 5 km thick between the crust and

mantle. The main difference between this section and those shown in Figure 7-2 is the existence of many deep fractures which extend right through the crust and into the mantle breaking the continuity of the Moho. These fractures break up the continental crust into separate tectonic blocks.

Cross Sections Through Oceanic Crust

The properties of the oceanic crust are often more uniform than those of the continental crust. The median values for normal oceanic crust are shown in Figure 7-8. Three layers can be distinguished: Layer 1 (the sedimentary layer), Layer 2, and Layer 3 (the oceanic layer). The median values for these layers are: (1) Layer 1, 0.3 km thick, with seismic velocity varying between 1.5 and 3.4 km/sec; (2) Layer 2, 1.4 km thick, with velocity between 3.4 and 6.0 km/sec, median near 5.1 km/sec; (3) Layer 3, 4.7 km thick, median velocity 6.8 km/sec.

These features are illustrated by Figure 7-4a, a seismic refraction profile across the Atlantic Ocean between Sierra Leone and Brazil (Figure 9-10). The distribution of seismic velocities as a function of depth is shown in Figure 5-3a. The thickness of the oceanic crust remains quite constant both in the Atlantic and other oceans (Figure 5-3b). Consistent regional variations are associated with the midoceanic ridges (Figure 7-5). Approaching the continental margins abrupt changes in crustal thickness occur (Figures

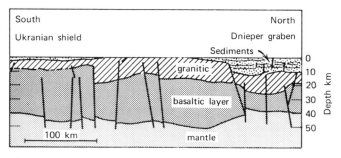

Figure 7-3. Crustal cross section through the Russian continental platform, showing in particular the deep fractures separating main regions and individual blocks (after Sollogub, 1969. From Geophysical Monograph **13**, 189, with permission of Amer. Geophys. Union).

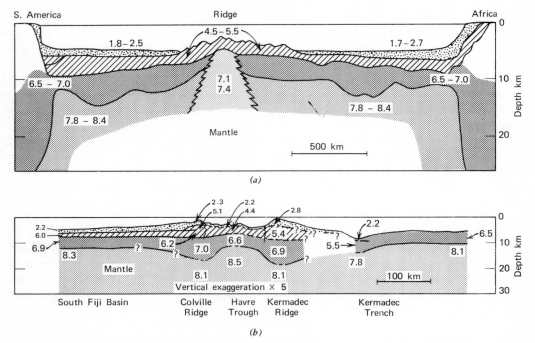

Figure 7-4. (*a*) Schematic cross section across the equatorial Atlantic Ocean, showing *P*-wave velocities (from various sources). (*b*) Crustal section across the Kermadec arc system based on seismic refraction profiles, with *P*-wave velocities (after D. E. Karig, 1970, Jour. Geophysical Res., **75**, 239, with permission).

5-4, 7-4*a*, and 7-6); the sedimentary layer is thicker in the continental rises near the continent.

Figure 7-4*b* shows the oceanic crust in the Pacific Ocean north of New Zealand with a section across the Kermadec-Tonga island arc system. This system includes an oceanic trench and a series of ridges and basins far removed from a continent. The interarc basin, consisting of a series of ridges and troughs with relief of 1000 m, is a region of high heat flow underlain by oceanic crust. The crust thickens to 15 km beneath the major ridges bordering the interarc basin and becomes more complex in structure.

The oceanic ridges form one of the tectonically active regions with anomalous upper mantle where both mantle velocities and mantle depths are less than normal in oceanic regions (Figures 5-4 and 7-8). Figure 7-4*a* shows Layer 2 directly overlying the anomalous mantle material of the mid-Atlantic ridge which extends through the oceanic Layer 3. In a few locations normal mantle with seismic velocity 8.3 km/sec instead of the usual 7.3 km/sec material has been identified.

Figure 7-5*a* shows the Bouguer gravity anomaly across the mid-Atlantic ridge. The minimum in the anomaly shows that the ridge is compensated isostatically by density variations at depth. If it were not compensated then the Bouguer anomalies would have remained constant at 350 mgal across the ridge. The three structure sections shown in Figure 7-5 satisfy both the seismic and the gravity data. The steep slope of the Bouguer anomaly requires that most of the compensation must lie at shallow depths and it is difficult to provide compensation for the ridge flanks except by the density reversal shown. The layer of anomalous mantle de-

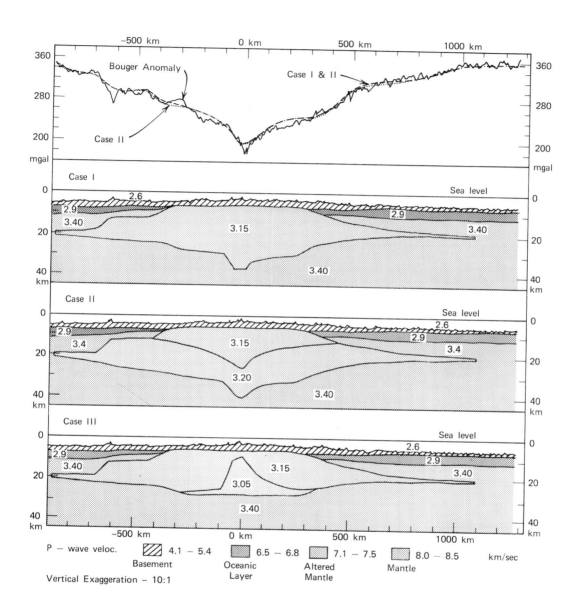

Figure 7-5. Three possible crustal models across the north mid-Atlantic ridge which satisfy gravity anomalies and are in accord with seismic refraction data. In all three models the anomalous mantle found seismically under the crest of the ridge is assumed to underlie the normal mantle under the flanks of the ridge. In Case I the anomalous mantle is assumed to have a uniform density; in Case II its density is assumed to increase downward, and in Case III the material constituting the anomalous mantle is assumed to be lighter near the axis of the ridge (after Talwani et al., 1965). See Figure 14-11. (From Jour. Geophys. Res., **70**, 341, 1965, with permission.)

picted in Figure 7-5 would not be detected by the usual seismic refraction techniques, because it underlies a higher-velocity layer of normal mantle. The configuration of this layer suggests that it may have been converted from normal mantle by a phase transition. Note that the density distributions in the anomalous layer for the three models provide considerable variety for interpretation in geological and petrological terms.

The structure of the East Pacific rise is similar to that of the mid-Atlantic ridge

(Figure 7-8) in that it is underlain by anomalous mantle and it is isostatically compensated. It differs from the mid-Atlantic ridge because the oceanic Layer 3 is continuous across the rise, with the anomalous mantle beneath the crest rising to somewhat higher levels than that beneath the flanks. The compensating layer of anomalous mantle may extend deeper than beneath the mid-Atlantic ridge.

The oceanic crust has not yet yielded all of its secrets. In January 1969 improved refraction measurements in the deep-ocean

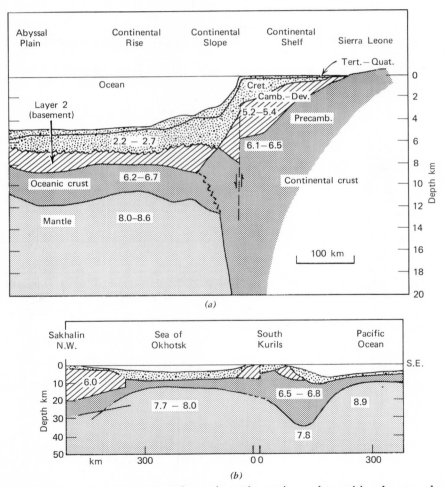

Figure 7-6. Seismic cross sections through continental margins and transitional zones, showing *P*-wave velocities. (*a*) Sierra Leone continental margin, vertical exaggeration 20 : 1 (after R. E. Sheridan et al., 1969, Jour. Geophys. Res., **74**, 2512, with permission). (*b*) Continental margin adjacent to the Sea of Okhotsk, and the South Kurils island arc (from I. P. Kosminskaya and S. M. Zverev, Geophysical Monograph **12**, 122, 1968, with permission Amer. Geophys. Union).

region between the Marshall and the Hawaiian Islands revealed a previously unrecorded basal layer between the mantle and the normal Layer 3. The layer has seismic velocities between 6.9 and 7.6 km/sec with an average of 7.3 km/sec (contrast Figures 7-4, 7-6, and 7-8). G. L. Maynard reported in 1970 that this basal layer may be widely present in the deep Pacific Ocean, and he outlined its implications for interpretation of gravity anomalies and the dispersion of seismic surface waves.

Cross Sections Through Continental Margins

The abrupt changes in the elevation of the Earth's solid surface and in the depth to the Moho in the transitional zones between continents and ocean basins are illustrated in Figure 5-1. Variations in seismic velocities for the continental margins in general have been reviewed in Figures 5-2, 5-3, and 5-4.

Figures 7-4a and 9-10 show crustal sections through the continental margins bordering the Atlantic Ocean, and Figure 7-6a shows a more detailed cross section through one margin. Note the thick accumulations of sediment on the continental shelf and the continental rise. This represents the Atlantic type of continental margin (Chapter 2). The structure of the crust beneath the sediments in the continental margins is complex, and a significant feature is the existence of buried ridges which act as dams for the sediments. Figure 7-7 summarizes several ways in which such ridges may be formed based on interpretations of many crustal sections. Major fault systems appear to parallel many continental edges.

Figure 7-6b represents the Island arc type of continental margin (Chapter 2). The Kurils island arc is separated from the Asian continent by the Sea of Okhotsk which is a small ocean basin. This is comparable with the section through Asia and Japan in Figure 5-1 and fairly typical of continental margins in the north and west Pacific Ocean. The

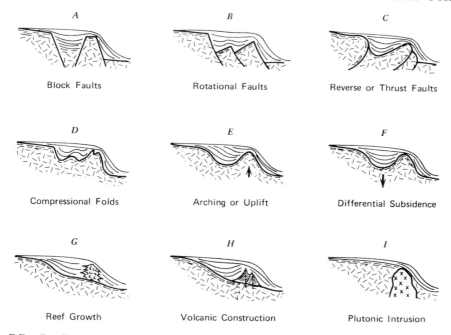

Figure 7-7. Possible origins of buried ridges within continental margins. Other causes can also be imagined, and any particular ridge may have resulted from a combination of processes (after Burk, 1968. With permission of The New York Academy of Sciences.)

depth to the Moho is similar beneath both the Pacific Ocean and the Sea of Okhotsk, but the latter has a thick accumulation of sediments over Layer 3. Sediments of considerable thickness are also shown in the trench bordering the island arc. The crust is greatly thickened beneath the arc-trench system, and the upper mantle here has anomalous seismic velocities. The crust is thicker than beneath the Kermadec-Tonga island arc system (Figure 7-4*b*).

Schematic Structural Subdivisions of the Whole Crust

Figure 7-8 shows a selection of individual crustal sections from various tectonic environments many of which were included in the cross sections in Figures 7-2, 7-3, 7-4, 7-5, and 7-6. Of particular interest are the sections indicating tectonic environments with large thicknesses of sediments. Standard sections for the average oceanic and continental crusts are given for comparison. Figure 7-8 may be compared with Figure 5-4 which provides similar information averaged for many seismic profiles.

Individual crustal sections such as those in Figure 7-8, and crustal cross sections such as those in Figures 7-2 to 7-6, provide the data for schematic representations of the whole Earth's crust as depicted in Figure 7-9. These diagrams were prepared as a basis for computation of the average composition of the crust. Figure 7-9*a* was presented by A. Poldervaart in 1955, and in 1969 with much additional seismic data available A. B. Ronov and A. A. Yaroshevsky presented an improved version Figure 7-9*b*. There are significant differences between the two models in the areas assigned to oceanic crust, continental crust, and the continental margins (transitional, suboceanic, or subcontinental) and in the average thicknesses of the crust in the various regions.

Tables 2-3 and 7-2 give for Figure 7-9*b* the estimates of areas of the structural units, average thicknesses and volumes of the layers

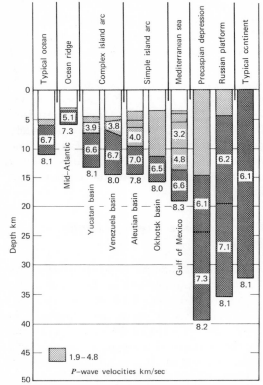

Figure 7-8. Crustal sections for different tectonic environments compared with typical oceanic and continental sections (after Menard, 1967, and Kosminskaya et al., 1969. With permission Amer. Geophys. Union.)

within the crust, and the mass of the layers in each unit. The volumes of the major structural units are (a) the total continental crust, 6,500 km³; (b) the total subcontinental crust, 1,540 km³; and (c) the total oceanic crust, 2,170 km³. The volumes of the three major layers of the crust are (a) the sedimentary layer including volcanic rocks, 985 km³, (b) the "granitic" layer, 3,590 km³; and (c) the "basaltic" layer, 5,635 km³. The composition of the crust can be calculated from Figure 7-9 if the compositions of the different crustal layers and units can be estimated (Tables 7-4 to 7-7).

The detailed seismic data now available for the United States permits a more precise calculation of the volumes of different layers

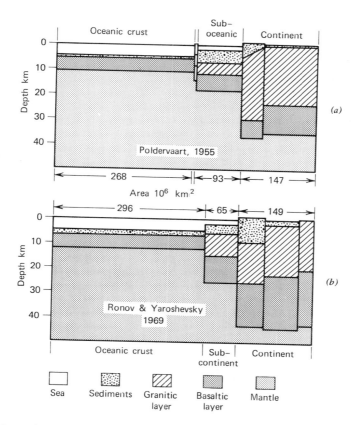

Figure 7-9. Schematic subdivisions of the Earth's crust used as basis for calculating the composition of the crust. Main subdivisions are into (1) sedimentary layer, (2) "granitic" layer, and (3) "basaltic" layer. See text for details. (a) Scheme of Poldervaart (1955, with permission The Geol. Soc. Amer.). (b) Scheme of Ronov and Yaroshevsky (1969. With permission of Amer. Geophys. Union, from Geophysical Monograph **13**, 37).

and their variations from one part of the continent to another. Figure 7-2 shows that there are differences from one tectonic environment to another. L. C. Pakiser and R. Robinson in 1966 estimated the volumes of the upper low-velocity unit (composition assumed to be silicic) and the deeper high-velocity unit (composition assumed to be mafic) within each of ten regions covering the conterminous United States. The percentages of mafic rocks range from a low of 25% in the California coastal region to a high of 77.7% in the Columbia plateaus region with an average of 53.7% for the total crust considered. The regions were then grouped into an eastern superprovince and a western superprovince divided by the Rocky Mountains following the conclusions arising from evaluation of Figure 7-2. The average volume percentages of mafic crust calculated for these superprovinces are 57.4% for the east and 43.6% for the west. Average compositions calculated according to these estimates are listed in Tables 7-5 and 7-7.

PETROLOGY AND MINERALOGY OF THE CRUST

Poldervaart was the first to calculate the average composition of the crust by relating the petrology of the crust to a specific crustal model, Figure 7-9a, and essentially the same approach was followed later by Ronov and Yaroshevsky. Table 7-2 gives their estimates of the percentages of different rock types occurring in the various layers shown in Figure 7-9b.

Volumes of the rock types in the sedimentary layers in different structural units were determined from surface areas of outcrops measured on lithologic maps and on measured or estimated thicknesses. Note that 25.3% of the layer in folded belts consists of volcanic rocks. The abundances of rock types in the granitic layer are based on measured outcrop area in shield areas. Granites, granodiorites, and their gneissic equivalents are dominant. The boundary between the granitic and basaltic layers was arbitrarily chosen to make them of equal thickness. The

TABLE 7-2   Distribution of Rock Types in Large Structural Units of the Crust and the Crustal Layers. (Data from Ronov and Yaroshevsky, 1969)

Crustal layers in different crustal units	Average thickness (km)	Volume (km³)	Mass (10^{24}g)	Types of rocks and abundances. Percent volume of layer, except percent area for oceanic Layer 1
Continental Platform				
Sedimentary	1.8	135	0.35	Sands, 23.6. Clays, 49.5. Carbonates, 21.0. Evaporites, 2.0. Basalts, 3.9
Continental Geosynclinal Folded Belts				
Sedimentary	10.0	365	0.94	Sands, 18.7. Clays and Shales, 39.4. Carbonates, 16.3. Evaporites, 0.3. Basalts, 12.6. Andesites, 10.2. Rhyolites, 2.5
Subcontinental, Shelf and Slope				
Sedimentary	2.9	190	0.48	Similar to above groups
Continental and Subcontinental				
Granitic		3590	9.81	Granites, 18.1. Granodiorites, 19.9.
continental	20.1			Syenites, 0.3. Gabbro, 3.7.
subcontinental	9.1			Peridotites, 0.1.
				Gneisses, 37.6. Schists, 9.0.
				Marbles, 1.5. Amphibolites, 9.8
Basaltic		3760	10.91	Acid igneous and metamorphic
continental	20.1			rocks, 50.0. Basic igneous and
subcontinental	11.7			metamorphic rocks, 50.0
Oceanic				
Layer 1	0.4	120	0.19	Terrigenous, 7.3. Calcareous, 41.5.
Sedimentary				Siliceous, 17.0. Red Clays, 31.2
Layer 2	0.6	175	0.44	Sediments, 50.0.
	0.6	175	0.52	Basalts, 50.0
Layer 3	5.7	1700	4.92	Oceanic tholeiitic basalts, 99.0. Alkaline differentiates, 1.0

constitution of the "basaltic" layer is not known. There are two extreme possibilities; (a) it is petrologically similar to the granitic shell or (b) it is composed essentially of basalt or gabbro. Ronov and Yaroshevsky assumed that in continental and sub-continental regions the layer changed gradually from the granitic rock assemblage to gabbroic material near the Moho. Table 7-2 indicates a shell composed of 50% acid rocks and 50% basic rocks, including plutonic rocks, their metamorphic equivalents, and also paragneisses. In the oceanic crust Layer 3 is assumed to be oceanic tholeiite with minor differentiates. The constitution of Layer 2 is not known, and it is assumed to consist of 50% consolidated sediments equivalent to those in Layer 1, and 50% basic rocks equivalent to those in Layer 3.

The information about distribution of rock types in Table 7-2 was reorganized in order to show the abundances of the main rock types in the crust as a whole with the results shown in Table 7-3. This confirms that the most abundant rock types in the crust are (a) granites, granodiorites, and diorites, (b) basalts, gabbros, and amphibolites, and (c) their metamorphic equivalents. The clays and shales are dominant in the generally thin veneer of sediments at the surface of the Earth.

From the abundance of rock types, and the known mineralogy of the rocks, Ronov and Yaroshevsky worked out the abundance of minerals in the crust. Table 7-3 shows that quartz and feldspars together constitute 63% of the crust. The mafic anhydrous minerals amount to 14% and the hydrated silicates, mica, amphibole, clays and chlorite, amount

TABLE 7-3   Abundances of Main Rock Types and Minerals in the Crust (Data from Ronov and Yaroshevsky, 1969)

Rocks	% Volume of Crust	Minerals	% Volume of Crust
Sedimentary		Quartz	12
Sands	1.7	Alkali feldspar	12
Clays and shales	4.2	Plagioclase	39
Carbonates (including salt-bearing		Micas	5
deposits)	2.0	Amphiboles	5
Igneous		Pyroxenes	11
Granites	10.4	Olivines	3
Granodiorites, diorites	11.2	Clay minerals (+ chlorites)	4.6
Syenites	0.4	Calcite (+ aragonite)	1.5
Basalts, gabbros, amphibolites,		Dolomite	0.5
eclogites	42.5	Magnetite (+ titanomagnetite)	1.5
Dunites, peridotites	0.2	Others (garnets, kyanite, andalu-	
Metamorphic		site, sillimanite, apatite, etc.)	4.9
Gneisses	21.4		
Schists	5.1		
Marbles	0.9		
Totals		*Totals*	
Sedimentary	7.9	Quartz + feldspar	63
Igneous	64.7	Pyroxene + olivine	14
Metamorphic	27.4	Hydrated silicates	14.6
		Carbonates	2.0
		Others	6.4

to 14.6%. Carbonates amount to only 2% and other minerals are even less abundant. This shows the dominance of quartz and feldspars in crustal rocks in sharp contrast with the mineralogy of the underlying mantle (Table 6-7).

The assumption that the "basic" layer contains at least 50% of basic rocks is rooted in the historical development of petrological theory, and recently it has been supported by laboratory studies which confirm that basic rocks have seismic velocities corresponding to those measured in the lower crust. Other rocks have similar seismic velocities, however, and there are now good reasons to conclude that the lower continental crust is not composed largely of basic rocks.

According to Figure 5-15 the pressure-temperature conditions appropriate for the lower crust are inappropriate for the existence of gabbro. Material of gabbroic composition, if present, should exist as eclogite, but the density of eclogite is too high to fit the measured seismic properties of the lower crust. For this reason A. E. Ringwood and

D. H. Green proposed in 1966 that we should abandon the notion of a basic lower crust and consider instead an assemblage of acid-intermediate rocks in the eclogite facies.

Ringwood and Green suggested that as an alternative to this dry assemblage basic rocks could exist in the lower crust as amphibolites in environments where the tectonic evolution had been simple enough that the water was not driven off. Amphibolites are stable in the lower crust and they have appropriate seismic velocities.

The composition of Layer 3 of the oceanic crust is also a matter for debate. The conventional interpretation of a basaltic composition (Table 7-2) is challenged by the hypothesis of sea-floor spreading (Chapter 12). According to this Layer 3 is composed of serpentinite, or hydrated mantle, as shown in Figure 5-6 (model 5) and in Figure 12-23a. The discovery of metamorphosed basalts in dredge hauls from the mid-Atlantic ridge provides some support for an alternative proposal that Layer 3 is composed of greenstone or amphibolite.

COMPOSITION OF THE CRUST

The many estimates of the composition of the crust follow one of five approaches, involving:

1. Averages of available analyses of rocks, such as the average of all igneous rocks or averages of igneous and sedimentary rocks in proportion to their abundances;

2. Averages of rocks weighted in proportion to their occurrence, using geological maps as a basis;

3. Indirect methods such as the analysis of glacial clays which represent a sample from a large area covered by a continental ice sheet;

4. Indirect computations based on various combinations of average granitic and basaltic rocks, with combination ratios being selected to best explain the compositions of sediments

derived by weathering of the crust, or the rare earth abundance patterns in sedimentary rocks;

5. Averages of rocks assigned to crustal models such as those in Figure 7-9. This is the only approach which has taken into consideration the composition and volume of the oceanic crust.

Despite the diversity of approaches most averages are similar in composition to an intermediate igneous rock, and the early estimates have not been greatly revised by later estimates. The uncertainty about the petrology of the lower continental crust, however, and the possibility that much of the oceanic crust might be composed of serpentinite rather than basalt (Chapters 5 and 12) leaves us with uncertainty about the

TABLE 7-4 Average Chemical Composition of Major Layers of the Crust in Weight Percent
(After Ronov and Yaroshevsky, 1969)

Type of Crust	Continental and Subcontinental				Oceanic			
Layer	Sediment	Granitic	Basaltic	Total continental[a]	Layer 1	Layer 2	Basaltic	Total oceanic
SiO_2	50.0	63.9	58.2	60.2	40.6	45.5	49.6	48.7
TiO_2	0.7	0.6	0.9	0.7	0.6	1.1	1.5	1.4
Al_2O_3	13.0	15.2	15.5	15.2	11.3	14.5	17.1	16.5
Fe_2O_3	3.0	2.0	2.9	2.5	4.6	3.2	2.0	2.3
FeO	2.8	2.9	4.8	3.8	1.0	4.2	6.8	6.2
MnO	0.1	0.1	0.2	0.1	0.3	0.3	0.2	0.2
MgO	3.1	2.2	3.9	3.1	3.0	5.3	7.2	6.8
CaO	11.7	4.0	6.1	5.5	16.7	14.0	11.8	12.3
Na_2O	1.6	3.1	3.1	3.0	1.1	2.0	2.8	2.6
K_2O	2.0	3.3	2.6	2.9	2.0	1.0	0.2	0.4
P_2O_5	0.2	0.2	0.3	0.2	0.2	0.2	0.2	0.2
C	0.5	0.2	0.1	0.2	0.3	0.1	0.0	0.0
CO_2	8.3	0.8	0.5	1.2	13.3	6.1	—	1.4
S	0.2	0.0	0.0	0.0	—	—	0.0	0.0
Cl	0.2	0.1	0.0	0.1	—	—	0.0	0.0
H_2O +	2.9	1.5	1.0	1.4	5.0	2.7	0.7	1.1

[a] Total subcontinental is very similar.

average composition of the whole crust. The abundances and distribution of most minor elements within the crust are not well known.

Composition of Crustal Layers

From the abundances and distribution of rocks listed in Table 7-3, and from the known average compositions of these rock types, Ronov and Yaroshevsky calculated the average chemical compositions of the major layers depicted in Figure 7-9*b* for continental and oceanic environments. The results are given in Table 7-4. The composition of the sedimentary layer includes a significant proportion of volcanic rocks as indicated in Table 7-2. Despite this there exist distinct differences between the composition of the sedimentary layer and the granitic and basaltic layers of the continents. The sedimentary layer is higher in the volatile components H_2O and CO_2 and considerably higher in CaO. Poldervaart pointed out that the average composition of sediments does not correspond at all to the average igneous rock of intermediate composition.

Composition of Crustal Units and of the Whole Crust

According to Table 7-4 the average composition of the continental and subcontinental crust (79% of the volume) differs from that of the oceanic crust (21% of the volume). Tables 7-5 and 7-6 show that there are considerable differences among various estimates for the same structural units; these depend largely upon the petrology and composition assumed for the lower crust and upon the values used for areas and thicknesses of layers and units. The differences are carried through to the average compositions of the whole crust above the Moho which are compared in Table 7-7.

TABLE 7-5   Estimates for the Chemical Composition of Structural Units of the Continental Crust (Weight Percent)

Type of Crust	Shield		Young Folded Belts	United States			Total Subaerial Continental
				West	East	Average	
Author[a]	P	R & Y	P	Pakiser and Robinson			R & Y
SiO_2	59.8	66.0	58.4	60.0	57.1	57.9	60.2
TiO_2	1.2	0.6	1.1	1.1	1.3	1.2	0.7
Al_2O_3	15.5	15.3	15.6	15.1	15.2	15.2	15.2
Fe_2O_3	2.1	1.9	2.8	2.3	2.3	2.3	2.5
FeO	5.1	3.1	4.8	4.9	5.7	5.5	3.8
MnO	0.1	0.1	0.2	0.1	0.2	0.2	0.1
MgO	4.1	2.4	4.3	4.5	5.6	5.3	3.1
CaO	6.4	3.7	7.2	6.3	7.5	7.1	7.1
Na_2O	3.1	3.2	3.1	3.2	3.0	3.0	3.0
K_2O	2.4	3.5	2.2	2.4	2.0	2.1	2.9
P_2O_5	0.2	0.2	0.3	0.2	0.3	0.3	0.2
CO_2	—	—	—	—	—	—	1.2
H_2O +	—	—	—	—	—	—	1.4

[a] P, Poldervaart (1955); R & Y, Ronov and Yaroshevsky (1969); Pakiser and Robinson (1966).

TABLE 7-6   Estimates for the Chemical Composition of Oceanic and Suboceanic (Subcontinental) Crust (Weight Percent)

Type of crust	Oceanic		Suboceanic or Subcontinental	
Author[a]	P	R & Y	P	R & Y
SiO_2	46.6	48.7	49.5	59.5
TiO_2	2.9	1.4	1.9	0.7
Al_2O_3	15.0	16.5	15.1	15.1
Fe_2O_3	3.8	2.3	3.4	2.5
FeO	8.0	6.2	6.4	3.9
MnO	0.2	0.2	0.2	0.2
MgO	7.8	6.8	6.2	3.2
CaO	11.9	12.3	13.2	5.9
Na_2O	2.9	2.6	2.5	2.9
K_2O	1.0	0.4	1.3	2.8
P_2O_5	0.3	0.2	0.3	0.3
CO_2	—	1.4	—	1.5
H_2O+	—	1.1	—	1.4

[a] P, Poldervaart (1955); R & Y, Ronov and Yaroshevsky (1969).

Table 7-5 shows that the shield composition estimated by Ronov and Yaroshevsky is more silicic than that estimated by Poldervaart, but the difference is reduced for their estimates of the total continental crust. Poldervaart's estimate of suboceanic crust (Table 7-6) is more mafic than that of Ronov and Yaroshevsky for the subcontinental crust which refers to similar major structural units (Figure 7-9). The estimate of continental crust composition by Pakiser and Robinson, which is based on detailed structural sections similar to those in Figure 7-2, is more mafic than those based on the crustal models of Figure 7-9 and more mafic than estimates obtained by other methods. In contrast Ringwood and Green concluded that the average chemical composition of the crust is more acidic than the estimates reviewed here because of the instability of gabbro in the lower crust. The new seismic results used by Pakiser and Robinson reverse the conclusion of Poldervaart that the folded belts are less silicic than the shields. Table 7-5 shows that the tectonically active western superprovince of the United States is more silicic than Poldervaart's young folded belts (his analogous crust), and the eastern superprovince is more mafic than Poldervaart's shields (his analogous crust).

The differences in estimated compositions of oceanic crust in Table 7-6 result from the composition adopted for the basalt of Layer 3. Poldervaart used a value for average basaltic rocks. Since 1955 more attention has been paid to the study of basalts dredged from the ocean floors and the estimate of Ronov and Yaroshevsky is based on the compositions of oceanic basalts.

Table 7-7 compares the two estimates for the average composition of the whole Earth's crust above the Moho based on the crustal models of Figure 7-9. Ronov and Yaroshevsky included in their contribution a minor revision which led to insignificant changes in the analysis listed; they reported estimates for CO_2 and H_2O+ of 1.40% and 1.37%, respectively.

The average composition of the Earth's crust is very similar to the average composition of andesites as shown by McBirney's averages

listed in Table 7-7. McBirney distinguished between andesites of island arcs and continental margins.

Markhinin studied the distribution and composition of volcanic rocks erupted from the Kuril Islands, an island arc system extending northward from Japan. A cross section through the Kurils is shown in Figure 7-6b. The volcanic rocks included basalts, andesites, and rhyolites, but the average composition listed in Table 7-7 is very similar to an average andesite. Markhinin pointed out that the entire observable geological sequence of the Kuril Islands consists either of volcanic material or of the products of reworking of the volcanic material. He estimated that since the beginning of the Cretaceous, about 6.5×10^6 km^3 of volcanic material, mostly pyroclastic, had been erupted from the volcanoes of the Kuril Islands, and he concluded that this volume of material was sufficient to convert an original oceanic crust into a crust of continental type. He reviewed the thesis

TABLE 7-7   Estimates for the Average Chemical Composition of the Earth's Crust Compared with Average lavas (Weight Percent)

	Earth's Crust		Andesites		Kuril Islands[c]
			Island arcs[a]	Continental Margins[b]	
Author	Poldervaart 1955	Ronov and Yaroshevsky 1969	McBirney 1969		Markhinin 1968
SiO_2	55.2	59.3	58.7	58.7	58.1
TiO_2	1.6	0.9	0.8	0.8	0.7
Al_2O_3	15.3	15.9	17.3	17.4	17.1
Fe_2O_3	2.8	2.5	3.0	3.2	3.4
FeO	5.8	4.5	4.0	3.5	4.1
MnO	0.2	0.1	0.1	0.1	0.1
MgO	5.2	4.0	3.1	3.3	3.4
CaO	8.8	7.2	7.1	6.3	7.1
Na_2O	2.9	3.0	3.2	3.8	2.8
K_2O	1.9	2.4	1.3	2.0	1.2
P_2O_5	0.3	0.2	0.2	0.2	—

[a] Average of 89 calcic andesites..
[b] Average of 29 calc-alkaline andesites.
[c] Average of 427 analyses of volcanic rocks erupted in Kuril Islands.

that the Earth's continental crust was formed progressively through geological history by the eruption and reworking of volcanic material on island arcs and related structures (see Figure 4-1 for an outline of the Rock Cycle).

Vertical Distribution of K, U, and Th in the Continental Crust

The average chemical composition of the crust changes with depth from the sedimentary layer, through the granitic layer, and through the basaltic layer. According to Table 7-4 an increase in Fe, Mg, and Al is accompanied by a decrease in the amount of combined H_2O and CO_2. There is an increase of Si, Na, and K from sediment to granitic shell and then Si and K decrease with depth. The ratios K/Na, Ca/Mg, and Fe_2O_3/FeO decrease, and Al/Si increases with depth.

The abundances and distributions of trace elements are less well known than those of the major elements. I. B. Lambert and K. S. Heier have detected trends in the abundances of some trace elements with depth in the continental crust. Of particular significance for problems related to heat generation, heat flow, and orogenic processes are the distributions of K, U, and Th. In 1967 Lambert and Heier compared the abundances of these elements in rocks of similar bulk composition in three groups:

1. Igneous and sedimentary rocks of the Paleozoic folded belts of eastern Australia;
2. Intrusive granites and metamorphic greenstones and gneisses of the shield regions in western Australia;
3. Medium to high pressure pyroxene granulite subfacies rocks of the shield regions.

This represents a sequence corresponding to greater depths in the continental crust. The relatively rare outcrops of pyroxene granulite provide samples from the deepest levels of the crust that are available in any areal extent at the surface.

In all three groups there is a wide and similar range of K. The overall average of Th is similar in groups 1. and 2., although the metamorphic shield rocks of group 2. are somewhat depleted in U. The pyroxene granulites of group 3. are depleted in U and Th by factors of five and nine respectively, compared with the concentrations in the lower grade shield rocks of group 2., and the average K content is also lower. The decrease in U does not strictly parallel that of Th, but major depletion of both U and Th occurs between the shield rocks of group 2. and the pyroxene granulites of group 3. Values of U, Th, Th/K, and U/K in the mafic rocks studied showed no significant variation with metamorphic grade.

Lambert and Heier concluded that metamorphic processes involving movement of a vapor phase and partial melting contribute towards upward migration of granitophile elements including K, Th, and U. They suggested that the pyroxene granulite facies rocks may represent the residue remaining after partial melting and upward migration of an interstitial magma, transporting Th and U, which leaves the deep-seated rocks depleted in the heat-producing radioactive elements.

Heat Production in the Crust

If radioactive heat production in the crust is concentrated into upper levels, then erosion through a long period may be capable of removing enough of the upper layer that heat production in the crust and heat flow at the surface could be strongly affected. The results of Lambert and Heier confirmed that the overall average contents of the radioactive elements in the surface rocks of the western Australian basement shield are similar to those in the Paleozoic rocks in eastern Australia; yet the heat flow from the shield is lower than that from the younger orogenic region in the east. Lambert and Heier suggested that the higher heat flow in the east is accounted for because the upper layer with concentrated radioactive elements is thicker in this region than in the shield region. In the shield areas the radioactivity of the surface rocks is representative of only a thin layer resting on crust that has been depleted in Th and U.

The concentrations of K, Th, and U in large granitic plutons at 38 localities in the United States were measured by R. F. Roy, D. D. Blackwell, and F. Birch in 1968, and these were converted into heat productivity values expressed in units of HGU (1 HGU = 10^{-13} cal/cm³sec). Heat flow measurements at each locality were recorded in HFU (1 HFU = 10^{-6} cal/cm²sec). The variations in heat flow, when plotted against the heat generation of the surface rocks at each site, lie close to three lines of the form $Q = a + bA$, where Q is the surface heat flux in HFU and A is the heat generation in HGU. The three lines, shown in Figure 7-10, define three heat flow provinces; the eastern United States, the Sierra Nevada, and a zone of high heat flow in the western United States which includes the Basin and Range province. Figure 7-10 shows that there is lateral variation in heat flow within each province and that the ranges of heat flow overlap from one province to another. The distinctive character of each province becomes clear only when the heat flow is plotted against the heat productivity of the surface rocks.

Figure 7-10 shows that the lateral variation in heat flow within a province is a linear function of the heat productivity of the surface rocks. The linear relation is consistent with only a few classes of radioelement distributions in the crust. The limiting cases are (a) nearly constant heat production to the lower limit of the plutons, and (b) an exponential decrease in heat production with depth. In both cases the variability of heat flow over large areas is caused by variations in a near-surface layer.

The linear relationship is most simply interpreted if there is a constant heat flux from the upper mantle and lower crust, corresponding to the intercept a on the ordinate,

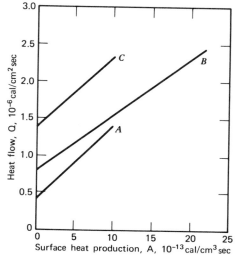

Figure 7-10. Summary of relationship between heat flow and surface heat production in the Sierra Nevada (A), eastern United States (B), and Basin and Range province (C). (After Roy, et al., 1968.) See also Lachenbruch (1970) and Tilling et al. (1970). (With permission of North-Holland Publishing Co.)

to which is added the radioactive heat generation within a surface layer of depth b. It is assumed that the radioactivity measured at the surface is constant from the surface to the depth b; the variations from place to place within the province produce the range of heat flow values. The values of a and b determined from the lines plotted in Figure 7-10 are as follows:

A. Sierra Nevada, $a = 0.40$ HFU, $b = 10.1$ km;

B. Eastern United States, $a = 0.79$ HFU, $b = 7.5$ km;

C. Basin and Range province, $a = 1.4$ HFU, $b = 9.4$ km.

These results are consistent with the conclusion of Lambert and Heier based on the geochemistry of metamorphic rocks, but whereas Lambert and Heier relate them to variation in thickness of the surface layer, the United States measurements are interpreted in terms of variations in heat productivity of the layer from place to place. The thickness (b) appears to remain fairly constant within the range 7 to 11 km.

The heat flow from beneath the plutons consists of a component from the upper mantle and a component from the lower crust. These must be insignificant over large areas of the United States, which confirms the conclusion of Lambert and Heier that there is an abrupt decrease in heat production a few kilometers below the surface of the crust. The similar slopes for the lines in Figure 7-10 suggest that the thickness of the heat producing layer varies little, and therefore the different regional values must arise largely from variations in the heat flux from the lower crust and upper mantle.

The curve for the eastern United States (the Central Stable Region) appears to apply also to the Australian shield; this is considered to be the reference curve for normal continental heat flow. Heat flow from the upper mantle is estimated as 0.4 HFU. If the lower crust in the Basin and Range province has a heat productivity similar to that inferred for

the Central Stable Region, then it has approximately 1.1 HFU flowing from the mantle. These values are summarized in Figure 7-11 for the three provinces. From the values of heat flow from the mantle and heat productivity in lower and upper crust the temperature-depth curves for the three provinces were calculated assuming steady state conditions (compare Figure 3-12). There are only small temperature differences within each province, the heat flow variations being produced by heat production only in the surface layer, but there are large temperature differences from one province to another because of the variations in heat flow from the mantle.

The heat flow from the normal, stable continental regions has a mode of 1.1–1.2 HFU, and this is apparently controlled by a distribution of static heat sources. Roy, Blackwell, and Birch pointed out that the coincidence of this mode with the average heat flow in ocean basins (Table 3-2) "revives the argument for equality of radioactive sources beneath the continents but with different vertical distributions."

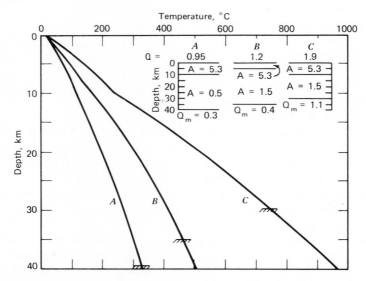

Figure 7-11. Temperature-depth curves for the three heat-flow provinces for the models shown (assumes steady state). The thermal conductivity is 6.5×10^{-3} cal/cm sec °C for the upper layer and 5.0×10^{-3} cal/cm sec °C for the lower layer of the crust. (After Roy et al., 1968.) Compare Hyndman et al. (1968). (With permission of North-Holland Publishing Company.)

DEEP STRUCTURE OF CONTINENTS

In a series of papers in the first half of the 1960's G. J. F. MacDonald reviewed the evidence that vertical segregation had been the dominant feature in the process of continent formation. This implies that the observed differences between continental and oceanic crust (Figures 5-1, 7-3, 7-4, and Tables 7-5 and 7-6) must extend into the mantle to depths of several hundreds of kilometers. MacDonald's most detailed review on the deep structures of continents was published in 1963.

According to MacDonald's arguments, large scale lateral convective movements have not been involved in the formation of continents. If the continental structure extends downwards to depths of 500 km or so, then severe restrictions are placed on theories of continental drift (Chapter 11). Lateral movements would have to involve blocks of crust and mantle with a thickness of 500 km, which implies that any convection in the mantle must occur at greater depths than this. Since about 1966 other arguments supporting sea-floor spreading and plate tectonics (Chapters 13 and 14) appear to have received more wide-spread credence than those of Mac-Donald. We saw in the preceding section, however, that the relationship between heat flow and heat production in surface rocks revives the arguments for different vertical distributions of radioactive elements beneath the continents and the oceans, which is one of the major arguments for the deep structure of continents.

The evidence considered by MacDonald to indicate that continental structure extends to depths of about 500 km involves observations on (a) heat flow, (b) gravity, (c) surface waves in the upper mantle, and (d) concentration of deep-focus earthquakes along continental borders. The rocks composing the continental crust contain greater concentrations of radioactive elements than those composing the oceanic crust, and the continental crust is much thicker than the oceanic

crust. The continental crust therefore contributes much more thermal flux to the measured value at the surface than the oceanic crust. In fact it appears that more than half the heat flow from the continents is generated within the crust and probably within a rather thin layer of the crust at the surface (Figures 7-10 and 7-11). We saw in Chapter 3 that the average values for heat flow through the continents and ocean floors are approximately the same. This observation requires that there is a greater contribution from the mantle beneath the oceanic crust than from beneath the continental crust. MacDonald concluded that this could be explained only if the abundance of radioactive elements in the suboceanic mantle was greater than that beneath the continents, which implies considerable vertical differentiation of elements and major differences in chemical composition between the suboceanic and subcontinental regions of the mantle.

Others have proposed that the thermal flux contribution from the suboceanic mantle is augmented by large-scale convection and that this is sufficient to equalize the heat flow from oceans and continents, but Mac-Donald presented additional arguments against the feasibility of mantle convection based on viscosity estimates. Roy, Blackwell, and Birch showed that the variations in heat flow from the surface within a continental heat flow province appear to be controlled by a distribution of static heat sources (Figures 7-10 and 7-11). They noted that if the oceanic heat flow were controlled by dynamic processes such as convection it would be surprising if this were equal to the average from static heat sources in the continental crust.

Artificial satellites have improved our knowledge of the Earth's gravity field (Chapter 3) and confirmed that there are regional variations which are of such extent that they cannot be accounted for in terms of crustal structures. This indicates lateral variations in mantle structure which could be explained in terms of temperature variations or compositional variations.

The present equatorial bulge is larger than that calculated for hydrostatic equilibrium. The present bulge corresponds to the equilibrium figure 10^7 years ago, assuming that the current rate of deceleration of the Earth remained constant through that interval. If we accept a model for the Earth with an elastoviscous mantle the interval of 10^7 years can be interpreted as a relaxation time which led MacDonald to estimate an average viscosity for the mantle of 10^{26} poises. This would prohibit convection. Some lower estimates of mantle viscosity are reviewed in Chapter 3.

The study of surface waves has revealed significant differences in the structure of the upper mantle beneath the oceans and continents. This is shown by the velocity profiles in Figure 3-4. Figure 3-5b shows that there are differences in wave velocity profiles in the mantle beneath different tectonic environments of the continents as well.

The association of deep-focus earthquakes with continental margins (Figures 2-5 and 2-6) can be explained in several ways (Chapters 12 and 14). MacDonald interpreted the deep zones of weakness as due to the effects of thermal stress which he considered to be consistent with his conclusions based on heat flow, gravity, and surface wave observations. The equality of heat flow from oceans and continents requires that at a given depth the temperature below the oceans is greater than that below the continents if convection is prohibited by the high viscosity deduced from gravity observations. MacDonald estimated temperature differences of 50 to 150°C. The temperature differences, and the induced zone of thermal stresses, are assumed to disappear at a depth of about 700 km corresponding to the deepest earthquake foci. MacDonald estimated that this pattern of thermal differences beneath oceans and continents would be produced by differ-

ences in radioelement content extending down to about 500 km in the mantle.

The evidence for the deep structure of the continents supports the models for vertical segregation of mantle material in the development of continents, and MacDonald outlined the following scheme. The early protocontinents were localized by restrictive fracture in the cool outer layers of an Earth initially expanding through radioactive heating. The continents were produced by uprise of basaltic and andesitic magmas including concentrations of the radioactive elements. Depletion of the subcontinental mantle in radioactive elements produced different thermal structures beneath the continents and oceans and caused the localization of deep fractures in the zones of thermal stress beneath the protocontinental margins. Further growth of the continents thus tended to be concentrated around the margins.

This is the hypothesis for the growth of continents by marginal accretion. The other major hypothesis for the origin and evolution of the continents is that a thin granitic crust formed very early in the Earth's history (either from the mantle or from extraterrestrial material) and was subsequently broken up into continents, recycled repeatedly, and resorbed by the mantle in a process of oceanization. In order to establish the pattern of continental evolution we need to know the distribution in the crust and mantle of the radioactive elements and their daughter products. Geochronological and isotope tracer studies should permit us to delineate continental age patterns and to determine the incidence of orogenic disturbances.

Concentration of Elements into the Crust

S. R. Taylor has calculated the percentage concentration of the elements in the crust relative to the whole Earth using a model of type I carbonaceous chondrite for the Earth (Chapter 6) and an overall andesitic crustal composition (Table 7-7) derived by vertical

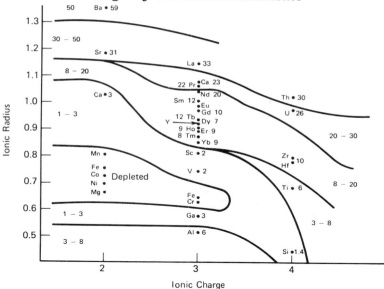

Figure 7-12. Concentration of chemical elements in the continental crust assuming an andesitic crustal composition and a chondritic Earth model. The numbers represent the percentage concentration of the elements in the crust on a vertical segregation model. The effect of ionic radius in fractionating elements of the same valency (e.g. rare earths) is striking (after Taylor, 1968, with permission of Pergamon Press).

segregation. The nonvolatile elements are plotted in Figure 7-12 on the basis of their crystal chemical properties of valency and ionic size. The boundaries between groups of elements with the same ranges of calculated concentrations show the strong control of crystal chemistry on the crustal enrichment.

The elements such as Mn, Fe, Co, Ni, Mg, and Cr are retained in the major minerals of the mantle (Chapter 6) and consequently depleted in the crust. Elements with size or valency different from these elements are strongly concentrated in the upper mantle or crust. These include K, Rb, Cs, Ba, U, and Th. The concentrations of the rare earth elements into the crust show a striking dependence upon ionic size. Figure 7-12 shows that there has been a very strong upward concentration of certain elements within the Earth and this conclusion remains valid whether or not the dominant process is vertical segregation.

AGES OF CONTINENTAL BASEMENTS

The tectonic structure of the North American continent has fostered the concept of the growth of continents by marginal accretion, and a compilation of apparent ages from radiometric data by A. E. J. Engel in 1963 strengthened this concept. Figure 7-13 shows the broad pattern of ages and geological provinces in North America as defined by major granite-forming, mountain-building events. When stripped of its thin veneer of younger sediments the continent has an ancient core, six times older than its margins and a zonal pattern with age increasing from the core to the margins.

Each broad province in Figure 7-13 is composed of a series of overlapping volcanic-sedimentary and granite-forming cycles. The granite-forming events tend to be localized

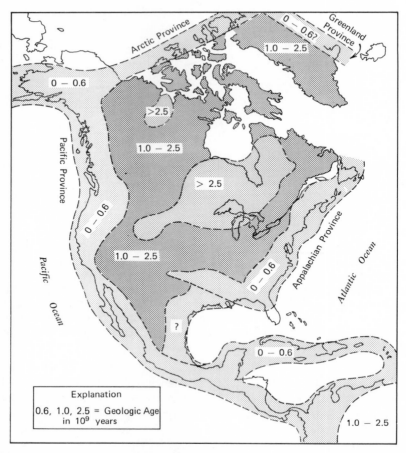

Figure 7-13. Gross patterns and ages of geologic provinces in North America as defined by major granite-forming, mountain building events dated by 1963 (after Engel, 1963). Compare Figure 7-14. (From Science, with permission. Copyright 1963 by the American Association for the Advancement of Science.)

along the sites of maximum crustal instability delineated at the time of intrusion by sinuous mountain belts. Each province overlaps pre-existing provinces by 20 to 60%. It has been estimated that the Appalachian province $(0.2-0.8 \times 10^9$ years) overlaps the Grenville province (about 1×10^9 years) by about 60%, and in western North America there is near obliteration of older provinces by a succession of younger provinces dated 0.1, 0.4, 1.0, and 1.8 billion years ago. Engel illustrated the extent of overlap in another figure which also showed that the $1.0-2.5 \times 10^9$ year zone consists of two rather distinct parts:

basement rocks in the north and west are in the range $1.8-2.5 \times 10^9$ years and those in the south and east are $1.0-1.8 \times 10^9$ years.

The oldest rock complexes in the core of North America are composed of rocks characteristic of present day island arcs and continental margins. The geology of the younger provinces is well known; they form elongate sheaths to the continent built in part on preexisting older provinces but including constituents of fringing island arcs. The successive volcanic-sedimentary belts involving the adjacent continental crust have become folded, partially melted, and stabilized by welding onto the continental crust. The liquid formed by partial melting forms the granitic intrusions which engulf much of the geosynclinal pile. This is the model reviewed by Engel.

Many more measurements were available for a similar study by W. R. Muehlberger, R. E. Denison, and E. G. Lidiak in 1967 and their results are summarized in Figure 7-14. They studied buried basement rocks from more than 3000 drill-holes and scattered outcrops in the continental interior and combined the petrographic data with isotopic ages to outline a series of geological units and their geological histories.

They presented four maps showing the areas known to be underlain by rocks with isotopic ages equal to or greater than 2.5, 1.7, 1.35, and 1.0 $\times$ 10^9 years respectively and on each map they also outlined the area showing

the probable minimum size of the continental crust at the stipulated time. They showed that more than 50% of the present continent was in existence 2.5 $\times$ 10^9 years ago and concluded that lateral continental accretion is less important than usually assumed. They stated that Engel's diagram (Figure 7-13) showed not continental accretion but the percentage of the continent that was stabilized by a particular time *and* was not involved again in a major granite-forming orogenic event. Engel did point out that much of the younger crust was built of reworked older crust, and certainly he implied that the original 2.5 $\times$ 10^9 year old continent was larger than the 16% of the present continent shown in Figure 7-13, but the case was made that the oldest rocks formed a core for the continent as a whole.

Figure 7-14 shows the outcrop areas of

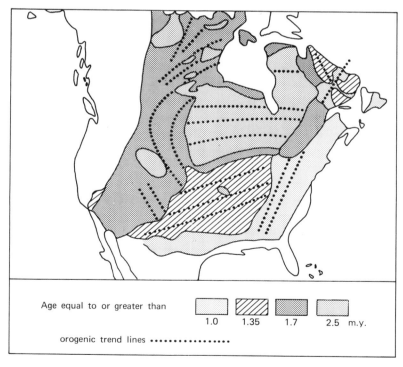

Figure 7-14. Generalized orogenic trend lines and ages of geologic provinces in North America according to Muehlberger et al. (1967). Each region delineated is known to be underlain by rocks giving isotopic ages equal to or greater than that indicated. Compare Figure 7-13.

successively younger continental belts and their generalized orogenic trend lines. This shows that the 1.0–2.5 × 10⁹ years belt is more complex than indicated in Figure 7-13. The older tectonic trends are truncated by the younger trends. Many of these rocks have had a multiple history that isotopic studies may elucidate.

Two years later, in 1969, P. M. Hurley and J. R. Rand surveyed the available age data for the basement rocks of North and South America, Africa, Europe, India, Australia, and Antarctica. The results extended over two-thirds the land area of the Earth. Examination of the natural grouping of the apparent ages led to consideration of three groups of rocks, (a) older than 1.7×10^9 years, (b) in the range 0.8–1.7×10^9 years, and (c) younger than 0.8×10^9 years. There seems to have been a universal quiet interval between 0.7 and 0.9×10^9 years ago, and there is a slight dip in the world-wide histogram of reported ages at 1.7×10^9 years.

For comparison of the data from all continents they were plotted on a geographical reconstruction of the continents based on the hypothesis of continental drift (Chapter 11, Figures 11-1 to 11-4). The result is shown in Figure 7-15. Compare the age provinces on North America in this figure with those in Figures 7-13 and 7-14.

The continental basement complexes older than 1.7×10^9 years occupy two areas as shown in Figures 7-15 and 11-4, and these are transected by younger geologically active belts. Older relict ages are found in many of the younger transcurrent zones. The two areas enclosing the oldest rocks are encircled almost entirely by younger continent. Hurley and Rand proposed that the two areas of old rocks represent former continental nuclei surrounded by successively younger belts. The continental age patterns on the pre-continental drift reconstruction thus appear to be consistent with marginal continental accretion in two supercontinents, neither of which was broken up prior to the onset of drift about 200 m.y. ago. Note, however, that

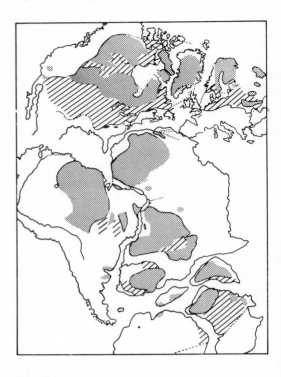

Figure 7-15. Continents reassembled in a pre-drift reconstruction—see Chapter 11. Lighter hatching, regions underlain by rocks having apparent ages in the range 800 to 1700 million years; heavier hatching, regions having apparent ages >1700 million years. It appears that there are two (or one) central regions of older rocks transected and totally surrounded by belts of younger rocks. This suggests that there was no significant fragmentation or scattering of the continental nuclei prior to the last great drift episode. (After Hurley and Rand, 1969. From Science, with permission. Copyright 1969 by the American Association for the Advancement of Science.)

in addition to marginal activity orogenic belts also developed within the continental cores or platforms.

The evidence outlined in this section supports the view that continents have grown through geological time. Estimating the rate of continental growth is difficult because of the problem of deciding how much of an orogenic belt of a given age represents new accreted material and how much represents reworked older continental crust. In the early 1960's Hurley and his associates and Engel showed that for North America, to a first approximation, there has been a linear growth rate. This is indicated by a plot of the areal extent, including known overlap, of each age province against the age interval. Equal surface areas of crust appear to have been generated in equal time intervals. Comparison of the ratio of $^{87}Sr/^{86}Sr$ with the age of granitic rocks led to an estimate of an average growth rate of about $7 \times 10^3 km^2/10^6$ years. In 1969 Hurley and Rand developed a more complex model for estimating the rate of generation of new continental crust using the partition of Rb into the crust relative to Sr as a criterion. K-Ar ages were calibrated against whole-rock Rb-Sr isochron ages and then used to estimate the distribution of area relative to age. Under the terms of their definitions and assumptions it appears that starting about 3.8×10^9 years ago there has been an accelerating generation of new crustal material, with a rate of increase of $20 km^2/10^6$ years/10^6 years, or about $600 km^3/10^6$years/10^6 years.

In order to trace the evolution of the crust and its relationships with the mantle through geological time it is necessary not only to map the apparent ages of crustal rocks but also to work out the strontium isotope and lead isotope evolution trends. Strontium studies have been applied to the problem for more than a decade but the data are still sparse. The available lead isotope data are still insufficient to resolve such fundamental questions as whether:

1. The continental crust has grown continually through geological time by addition of new material from the mantle,

2. The formation of continental crust was largely completed during the period $2.5-3.5 \times 10^9$ years ago with younger crust representing regeneration of the initial crust.

8. *Magma Generation*

INTRODUCTION

Magma is a mobile assemblage of rock matter with the essential ingredient a silicate liquid; it usually contains suspended crystals and sometimes a separate gas phase. Magma is generated whenever conditions become appropriate for partial fusion of rocks within the Earth. Because of its lower density a magma tends to rise and differentiation of the liquid phase occurs in response to the changing conditions. Magmas may break through the crust to produce extrusive volcanoes or lava floods, or they may crystallize at depth within the mantle or crust as intrusive plutonic rocks. Since early Earth history repeated magma generation and uprise has caused progressive defluidization of the mantle.

The Earth's crust is composed essentially of igneous rocks and their metamorphic equivalents (Tables 7-2 and 7-3); it is thus the product of volcanism and plutonism. Its average chemical composition is very similar to that of andesites or the average of the whole suite of volcanic rocks building the island arcs (Table 7-7). It is through the study of igneous rocks and processes that we hope to gain some insight into the question of whether (a) the continents have grown continuously through geological time by addition of new material from the mantle or (b) younger crust represents regeneration of older, initial crust, by magmatic processes. The formation of magma also provides a guide to temperatures within the Earth.

There are many different igneous rock types and rock associations, and classification of these provides the essential basis for petrogenetic theories. The problem is complex as illustrated by our brief review in Chapter 6 of the occurrences, distribution, and petrogenesis of ultramafic rocks; Table 6-4 lists five major associations occurring in different environments. Despite the complexity in detail, if we are concerned with major Earth processes such as the origin of continents and mountain ranges (Chapters 7 and 9), we may concentrate our attention on the most abundant rocks: the extrusive (a) basalts, (b) andesites, and (c) the intrusive diorites, granodiorites and granites of batholiths

IGNEOUS ROCK ASSOCIATIONS

It is in the oceanic tectonic provinces that we seek samples of magmas generated in the mantle and uncontaminated by crustal materials. Here we find two varieties of basalt with minor differentiates and ultramafic rocks. Subalkaline basalt, saturated or nearly saturated with silica and with the rather unfortunate name "tholeiite," is the dominant basalt of the ocean floors and midoceanic ridges and the oceanic volcanic islands. Alkaline basalts, undersaturated in silica, occur in oceanic islands. Ultramafic rocks occur on the midoceanic ridges, and their relationship to the ophiolite complexes and other ultramafic rocks of the orogenic belts was reviewed in Chapter 6 and will be mentioned again in Chapter 14.

Tholeiitic basalts and alkaline basalts are also widely distributed on the continents with minor associates of andesite and rhyolite or trachyte and phonolite. The ultramafic rocks of the large layered intrusions (Table 6-4) are produced by crystal accumulation from a parent subalkaline basaltic magma. The rifted portions of otherwise stable continental platform areas are characterized by the rarer highly alkaline lavas. Plutonic equivalents of these lavas form usually small feldspathoidal intrusions which may be associated with ultramafic rocks (Table 6-4) and carbonatites.

The calc-alkaline andesites with associated basalts and rhyolites occur characteristically in orogenic belts of the continents and in the volcanic island arcs of the continental margins. They are interspersed with sediments in geosynclinal piles, and they have been repeatedly erupted during and after the folding and uplift of the geosynclinal rocks.

Batholiths are emplaced in orogenic belts and in some respects they appear to be the plutonic equivalents of the andesites. Their field relationships and petrology have often been adduced as evidence that the plutonic rocks were derived by reworking of crustal material by metamorphism and metasomatism without the intervention of magmatic activity.

DEVELOPMENT OF PETROGENETIC THEORY IN THE TWENTIETH CENTURY

Figure 4-5 summarizes the main processes in petrogenesis beginning with magma genesis and outlining various sequences of events that occur between melting and emplacement or eruption of the magma. The composition of the liquid which finally crystallizes as a plutonic intrusion, or erupts from a volcano, is the product of a very complex history. Petrogenesis is concerned with unravelling that history. The problem has been to examine the rocks, the end-products of the history, and to construct a petrogenetic scheme by inductive reasoning. During the past twenty years new approaches to the subject have made it easier to devise tests for petrogenetic theories. These include (a) geophysics of the solid Earth, (b) high pressure experimentation, and (c) geochemistry.

Improved techniques in geophysics have advanced our knowledge of the Earth's interior to such an extent (Chapters 3 and 6) that petrogenetic theory must now be based on the geophysical data. The development of apparatus for high pressure, high temperature studies makes it possible to reproduce crustal and mantle conditions in the laboratory, and experimental melting of

minerals and rocks places rather close restrictions on the conditions of magma generation. Improved analytical techniques for measuring isotopes and trace element concentrations in igneous minerals and rocks have led to a wealth of new data, much of which can be interpreted in terms of the composition of a mantle source or the extent of contamination by crustal material. These results provide limits for petrogenetic theories.

The First Half of the Century: Primary and Derivative Magmas

The concept of primary magmas has held a dominant position in petrogenesis. A primary magma is one formed either by partial or complete fusion of crystalline rock at depth or, according to older concepts, by tapping a body or layer of liquid or glassy magma persisting at depth from some early stage of the Earth's development. Basaltic and granitic magmas are the most abundant primary magmas. Figure 8-1 illustrates some of the models proposed for magma generation.

Until the mid-1930's many petrologists accepted a hypothesis, promulgated by R. A. Daly, that a worldwide substratum of basaltic

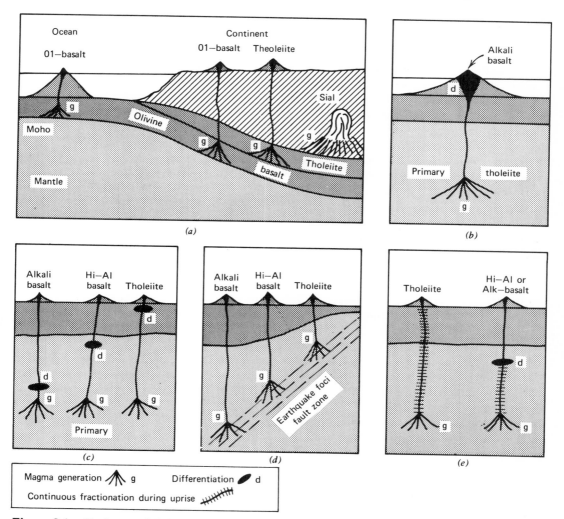

Figure 8-1.   Various models for magna generation and fractionation in the crust and mantle. See text for review.

magma existed beneath the lithosphere. Daly subsequently abandoned the idea when it appeared that neither basaltic magma nor glass had suitable seismic properties, and he then favored fusion of basaltic rocks near the base of the crust. The concept of a worldwide subcrustal magma zone with composition equivalent to an olivine-rich basalt, however, was retained by A. Rittmann in 1962; he maintained that at high pressures magma viscosity would become high enough for the transmission of shear waves because the periods of the seismic transverse waves are shorter than the relaxation time of the magma. In the second edition of his petrology text T. F. W. Barth (1962) cites Rittmann and tentatively reverts to the hypothesis of a glassy basaltic substratum beginning at a depth of 50 to 70 km beneath the surface. Rittmann proposed that this extended to depths of at least 1400 km and perhaps to more than 2000 km. The hypothesis of a liquid substratum has not supplied explanations for either the properties of the mantle

(Chapter 6) or the relationships among the several kinds of basaltic magma.

Given a primary basaltic magma N. L. Bowen argued convincingly, from about 1915, that most igneous magmas could be derived from this parent by fractional crystallization. He based his arguments mainly on study of the crystallization of silicate liquids in the laboratory, but he did not neglect the standard petrological approaches. Other processes are listed in Figure 4-5.

During the 1920's the detailed study of basalts in the Tertiary province of Scotland led E. B. Bailey and others to develop the concept of magma types. Both subalkaline and alkaline basalts occur, and each type is associated with a characteristic group of differentiates representing distinct magma series. In 1933 W. Q. Kennedy proposed that each of these basaltic magma types was an independent primary magma, and he gave them the names tholeiite (subalkaline) and olivine basalt (alkaline). He examined the distribution of similar basalts on a worldwide basis and presented the petrogenetic scheme illustrated in Figure 8-1a. There are two basaltic crustal layers to provide for the independent formation of the two primary magmas with the subalkaline tholeiite layer being present only beneath the continents. Kennedy distinguished between volcanic igneous associations developed by differentiation of the primary basalts and plutonic igneous associations developed by fusion of sialic material of the crust above the basaltic layers. Figure 8-1a shows explicitly the development of batholithic magmas by fusion of the crust. Present interpretations of temperature distribution at depth do not permit the generation of basaltic magmas by fusion of basaltic layers above the Moho (Chapters 3 and 7). As long ago as 1928 before these models were developed Bowen had proposed that basaltic magmas were generated by partial fusion of feldspathic peridotite or eclogite beneath the Earth's crust, and this forms the basis of most recent models for the generation of primary basalts.

Kennedy proposed that these two basaltic types were independent primary magmas, but others have sought differentiation processes to derive one from the other. There has been incredible confusion in successive chemical, petrographical, and mineralogical descriptions and definitions of these two magma types, as well as uncertainty about their tectonic environments, which were reviewed in a delightful paper by F. Chayes in 1966.

Kennedy made a distinction between the olivine basalt and the tholeiite on the presence or absence of olivine. C. E. Tilley in 1950 noted that the olivine basalts of the Pacific had the chemical characteristics of tholeiites despite their olivine content. He therefore defined the subalkaline tholeiitic basalt magma as one relatively silica-saturated in such a way that olivine exhibits a reaction relationship with the liquid during crystallization. The alkaline olivine basalt, on the other hand, is relatively silica-under-saturated and the olivine exhibits no reaction relationship. A chemical distinction is that the tholeiitic rocks have hypersthene in the norm. In his review Chayes recognized the real existence of tholeiitic basalts and alkali olivine basalts but, because these two names had been used in so many different ways, he proposed that they be replaced by the terms subalkaline basalt and alkaline basalt respectively which are unambiguous and well known. Unfortunately it appears that his plea will be ignored and "tholeiites" may be with us for evermore. Tilley also recognized the existence of basalts unusually rich in alumina, and high-alumina basalt is now regarded as a third primary magma.

The celebrated "Granite Controversy" raged bitterly through the 1930's and 1940's. According to Bowen's thesis the diorites, granodiorites, and granites comprising the batholiths, as well as the eruptive andesites, were derived by fractional crystallization of primary basalts. Others suggested that the enormous volumes of granitic magma were augmented by contamination of the parent basalt by assimilation. Many petrologists, especially those who had worked in Precambrian gneissic terranes, were convinced that the batholiths were formed by metasomatism of preexisting rocks in the solid state. The necessary components were introduced and subtracted from the original rock either by diffusion of ions through crystal lattices or by the flow of interstitial pore fluids. "Magmatists" and "transformists" had hardly a civil word for each other.

At the turn of the century J. J. Sederholm had systematically studied metamorphic rocks that had undergone anatexis, or "remelting," with the formation of migmatite. This is a mixed rock part metamorphic and part magmatic. Figure 8-1a shows Kennedy's interpretation of the plutonic igneous associations being produced by fusion of the crust. These ideas eventually were formulated into experimentally based anatectic models, and the bitterness of the debate subsided during the 1960's as many petrologists found at least partial satisfaction in the model that produced magmatic rocks by partial fusion at the culmination of metamorphism.

Influence of High Pressure Experiments Since 1950

When Tilley's presidential address was published at the midpoint of the century this coincided with a turning point in petrology. His review of the primary basalts brought reasonable order into the problem. At the same time O. F. Tuttle designed a simple pressure vessel which permitted the routine study of mineral and melting reactions in the presence of water vapor under pressure. For years petrologists had expressed the need for such experiments. Now here was an apparatus that could be used with ease by students.

The first results were published in 1950 by Bowen and Tuttle on the feldspar system, $KAlSi_3O_8(Or)$-$NaAlSi_3O_8(Ab)$-H_2O. In 1958 Tuttle and Bowen published a monograph on the "Origin of granite in the light of experimental studies in the system

$NaAlSi_3O_8$-$KAlSi_3O_8$-SiO_2-H_2O" in which they developed an anatectic model. They showed how partial melting of crustal rocks in the presence of small amounts of water offers a mechanism for producing large batholithic masses of granite. Similar conclusions were reached by H. G. F. Winkler and associates in a series of melting studies on sedimentary rocks and metamorphic rocks.

The Tuttle vessel and others utilizing a fluid medium to transmit the pressure to the sample are effectively limited in their pressure range to conditions in the Earth's crust. For the reproduction of mantle conditions an apparatus compressing a solid pressure medium is required. In the 1950's L. Coes was the first to synthesize minerals in this type of apparatus. Many solid-pressure designs have been used but most of them are elaborate and rather difficult to operate. There is one design, a single-stage piston-cylinder apparatus, which is now functioning routinely in many laboratories reproducing conditions in the mantle corresponding to depths of 100 km and more. This was described in 1960 by F. R. Boyd and J. L. England and developed also by G. C. Kennedy and associates. Since 1960 experimental data yielded by this apparatus have strongly influenced petrogenetic theory.

In the early 1960's most piston-cylinder experiments involved single minerals or simple assemblages but in 1962 H. S. Yoder and C. E. Tilley published a pioneering study on the effect of water on the melting of various basalts up to 10 kb pressure, together with exploratory work on eclogite under mantle conditions (the latter with the aid of S. P. Clark, F. R. Boyd, and J. L. England). From their study it became clear that the depth of magma generation and fractionation would have a marked effect on magma composition. They introduced the basalt tetrahedron, Figure 8-12d.

In 1965 A. E. J. Engel and associates concluded that in the ocean basins the only primary magma is a low-potassium tholeiitic basalt derived by partial fusion of the peridotite mantle. Various arguments led them to conclude that the alkaline basalts capping the oceanic islands were formed by differentiation processes occurring near the surface. Yoder and Tilley had shown that this process could not be crystallization differentiation because of the low-pressure thermal divide between these two liquids (Figure 8-12d), so Engel invoked other processes such as gas transfer for the conversion of the oceanic tholeiites into alkaline basalts. The model is illustrated in Figure 8-1b.

Yoder and Tilley showed that shifting thermal divides could cause a given basaltic composition to follow quite different fractionation paths depending on the pressure. These observations led them to the model illustrated schematically in Figure 8-1c with tholeiitic basalt being the only primary magma developed in the mantle. If the basalt rises directly to the surface and undergoes low pressure fractionation only a tholeiite and its usual derivatives can be erupted. On the other hand if the primary tholeiite undergoes a period of differentiation at depth within the mantle this causes formation of a derivative alkali olivine basalt. High-alumina basalt is produced by differentiation of the primary tholeiite at a high level within the mantle or possibly deep within the crust.

H. Kuno was the first, in 1960, to explain the different basalts by pressure effects; his model includes three primary magmas. Kuno has shown that the three types of basaltic magma occupy successive belts across island arcs, and he assumed that the position of the Benioff zone of earthquake foci gives the sites of magma generation. Figure 8-3d shows that the depths of magma generation are successively deeper beneath volcanoes erupting tholeiites, high-alumina basalts, and alkali olivine basalts. Hence Kuno's conclusion that the compositions of the three magma types are controlled by depth of generation, and their classification as separate primary magmas.

M. J. O'Hara published a stimulating review in 1965, revised and expanded in 1968, in which he pointed out that the concept of primary magmas was only one extreme model for magma generation. Implicit in the model is the assumption that a magma is generated at depth and then rises rapidly to the surface with little or no change of composition. It is the frequent eruption of tholeiite basalt according to the concept which permits one to recognize magma of this composition as being the primary product of melting at depth; Figure 8-1c shows two variations of the model which involve the derivation of other magmas from a primary tholeiite by a period of high pressure differentiation. O'Hara pointed out that enough experimental information was available on the compositions of liquid fractions in equilibrium with peridotites at high pressures to indicate that these compositions did *not* correspond to the lavas erupted at the surface and interpreted as primary magmas. Therefore he concluded that a new model for magma generation and fractionation is required.

The other extreme model is one where a magma is generated by partial fusion of mantle peridotite, and this experiences continuous fractionation from source to surface eruption. This is illustrated in Figure 8-1e which should be contrasted with the primary magma models of Figures 8-1b and d. O'Hara also discussed models between these two extremes as being somewhat more reliable than either extreme and an example is shown in Figure 8-1e.

In order to develop this type of model in detail it is necessary to work out the effect of bulk composition, pressure, and temperature on the composition of the liquid coexisting with residual peridotite. A rising magma can follow an infinite number of different paths through a system with these variables. Most evidence indicates that basaltic magma contains only traces of juvenile water although additional water may be dissolved during transit through the crust.

Even traces of water have significant effects, however, so we turn next to the general pattern of phase relations in rock-water systems. This provides the framework for more detailed examination of magma generation from materials of the crust and mantle based on high pressure experimental studies published since 1967.

ROCK-WATER SYSTEMS AS GUIDES TO MAGMA GENERATION

Basaltic magmas are produced by melting of the mantle and granitic magmas may be produced by melting of crustal material; andesitic magmas may be produced by differentiation of basaltic magmas or by melting in the mantle or crust. As shown in Figure 4-5 magmas may be generated by partial fusion of three types of assemblages: (a) anhydrous minerals, (b) anhydrous and hydrous minerals, and (c) minerals in the presence of a pore fluid usually considered to be aqueous, although carbon dioxide and other components are expected. For models of magma generation in the mantle we therefore consider the experimental results in the systems peridotite-gabbro, peridotite-water, and gabbro-water and for the crust we consider the systems gabbro-water and granodiorite-water.

Most magmatic processes occur with no vapor present and with water pressure less than total pressure: these are water-deficient conditions which have received little experimental attention. The dry and water-excess experimental results provide limits for the natural conditions, and the general pattern in water-deficient systems can be interpolated between these limits. C. W. Burnham reviewed the conditions for anatexis in the crust and mantle in water-deficient systems in 1967.

Water-Absent and Water-Excess Systems

The availability of piston-cylinder apparatus during the 1960's permitted the measurement of melting curves for dry minerals and rocks at pressures greater than 10 kbars. The effect of pressure is to raise the melting temperature as shown for various minerals

and rocks in Figures 4-3, 4-4, 5-13, 5-15, 6-10, 6-11, 6-12, 6-14, 6-17, 6-18a, and 6-19a.

Chapter 4 included a review of melting reactions in silicate-water systems, Figures 4-3e and 4-3f. Figure 6-11 shows the effect of excess water on the solidus of peridotite, and Figures 6-12 and 8-18 compare the melting interval for gabbro composition dry and with excess water. The initial study of Bowen and Tuttle in 1950 on the system $NaAlSi_3O_8$-$KAlSi_3O_8$-H_2O was followed by many others. By 1967 the curves for the beginning of melting for many silicate assemblages in the presence of excess water had been measured to 10 kbars; the Tuttle pressure vessel and internally-heated hydrostatic vessels of the type used successfully by H. S. Yoder and C. W. Burnham yielded these results. In 1967 A. L. Boettcher and I used the piston-cylinder apparatus to extend the melting curves in feldspar-quartz-water and rock-water systems to pressures greater than 10 kb in order to evaluate the effects of water on magma generation in the mantle. By 1970 melting curves in the presence of excess water for individual feldspars for most feldspar-quartz combinations and for most major rock types had been located through a wide range of pressures. Selected results are shown in Figures 8-2 and 8-3.

Figure 8-2a shows the univariant reaction curves for individual feldspars and for quartz in the presence of excess water. The effect of water under pressure is to lower the melting temperature. The curve for orthoclase continues to pressures greater than 20 kb, but the curves for albite and anorthite terminate at invariant points where the feldspars break down to yield dense minerals such as jadeite, zoisite, and kyanite (Figures

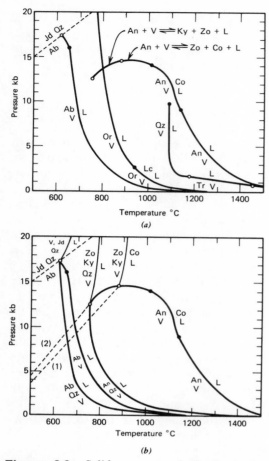

Figure 8-2. Solidus curves (univariant) in feldspar-quartz-water systems with excess water, compiled by Merrill et al. (1970). Heavy lines are solidus curves; light lines are solidus curves at pressures above the stability limit of feldspars; light dashed lines are subsolidus reactions; open circles are invariant points; closed circles are singular points. An = anorthite, Ab = albite, Or = Sanidine, Qz = quartz, Tr = tridymite, Lc = leucite, Co = corundum, Zo = zoisite, Ky = kyanite, Jd = jadeite, L = liquid, V = vapor.

(a) The system SiO_2-H_2O, and reactions involving single feldspars.

(b) Univariant reactions in the system $CaAl_2Si_2O_8$–$NaAlSi_3O_8$–SiO_2–H_2O.
  Reaction (1): An + Co + V $\rightleftharpoons$ Zo + Ky
  Reaction (2): An + V $\rightleftharpoons$ Zo + Ky + Qz

(Sources of data: Qz, Kennedy et al., 1962; Kushiro, 1969. An and An–Qz, Stewart, 1967;

4-3*b*, 5-11, and 5-12). The curve for quartz terminates at a second critical end-point.

The melting curves for plagioclase feldspars and plagioclase feldspar-quartz mixtures in the presence of excess water are shown in Figure 8-2*b*. The light dashed lines show the subsolidus reactions which limit the high pressure stability of the plagioclase feldspars, and the light solid lines indicate the hydrothermal melting curves which replace the feldspar and feldspar-quartz curves at higher pressures. Involvement of the dense mineral phases in the assemblage reverses the slope of the solidus so that with increasing pressure in the presence of excess water the temperature of beginning of melting increases just as in the dry systems.

The mineralogy of the major igneous rocks of the Earth's crust is dominated by feldspars and quartz (Table 7-3), and the similarity of the curves for the beginning of melting of rocks in the presence of excess water to pressures of 15 kb shows this influence (Figure 8-3). The effect of water on the melting temperature of peridotite is much less than on the crustal rocks that include feldspars and quartz. Curves for granite, tonalite, and gabbro, representing acid, intermediate, and basic compositions in the calc-alkaline suite are roughly parallel with each other; they are separated by only 30°C between 8 and 15 kb. At pressures above 15 kb the solidus curves reverse slope and become similar to those for dry silicates and rocks (see also Figures 6-18*b* and 6-19*b*).

Determination of the solidus curves for the beginning of melting of rocks is only the beginning of elucidation of magma generation. In order to examine the products of partial melting it is necessary to know the phase relationships through the melting

Figure 8-2 continued

Boettcher, 1970. Or and Or–Qz, Lambert et al., 1969. Ab and Ab–Qz, Boettcher and Wyllie, 1969; Tuttle and Bowen, 1958). (From Jour. Geology with permission. Copyright 1970 by The University of Chicago Press.)

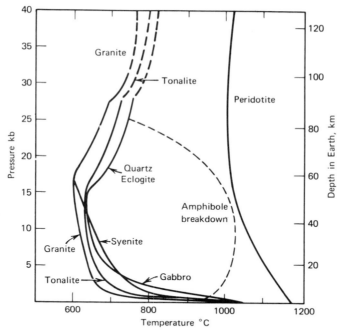

Figure 8-3. Solidus curves for rocks in the presence of excess water, compiled by Merrill et al. (1970). Granite-H_2O (rhyolite-H_2O) after Piwinskii (1968), Boettcher and Wyllie (1968). Tonalite-H_2O (andesite-H_2O) after Piwinskii (1968), Lambert and Wyllie (1970). Gabbro-H_2O (basalt-H_2O) after Lambert and Wyllie (1968, 1970), Hill and Boettcher (1970), Tuthill (1969) at 5 kb, Holloway and Burnham (1969) at 2, 5 and 8 kb. Syenite-H_2O after Merrill et al.(1970). Peridotite-H_2O after Kushiro et al. (1968). (From Jour. Geology, with permission. Copyright 1970 by The University of Chicago Press.)

interval of the source rock. The first detailed experimental work of this kind was the 1962 study of Yoder and Tilley on basalt compositions with excess water to 10 kb and dry at greater pressures. Additional studies of dry basalt compositions and peridotites began to appear in 1967 by A. E. Ringwood, D. H. Green, and associates at Canberra, and by K. Ito, G. C. Kennedy, and L. H. Cohen at Los Angeles. From Chicago A. J. Piwinskii and I in 1968 reported phase relationships at crustal pressures through the melting intervals of granites, granodiorites, and tonalites in the presence of excess water, and in 1968 and 1970 I. B. Lambert and I published similar results for gabbro and tonalite at mantle pressures between 10 and 25 kb.

Figures 6-12 and 8-18*a* for the composition gabbro-water illustrate the main differences in phase relationships produced at pressures up to 10 kb, if excess water is added to dry rock:

1. The solidus and liquidus temperatures are lowered.

2. The temperature interval between liquidus and solidus is increased.

3. Hydrous minerals are stabilized and they become involved in reactions including liquid.

Two other factors are introduced at higher pressures corresponding to mantle conditions:

1. The slope of the solidus is reversed so that with increasing pressure in the presence of excess water the temperature of beginning of melting increases.

2. At high pressures the slope (dP/dT) for the breakdown of hydrous minerals changes

from positive to negative so that with increasing pressure the dehydration temperature becomes lower.

For amphibole this effect becomes marked at about 15 kb, as shown in Figure 8-3, and for the gabbro composition the amphibole breaks down below the solidus at pressures greater than about 25 kb. An amphibole stability curve intersects the peridotite-water solidus between about 8 and 17 kb (Figures 6-11 and 6-18).

Water-Deficient Systems

Terms such as vapor-absent and water-deficient have been widely used without adequate definitions probably because experimental studies on melting relationships in water-deficient systems were not reported until the late 1960's. In 1971 J. K. Robertson and I prepared a consistent set of definitions for four types of subsolidus assemblages in silicate-water systems, providing a basis for models of magma generation. These types also have relevance for processes of magma crystallization:

Type I: Water-absent. An assemblage of anhydrous silicate minerals with no aqueous vapor phase.

Type II: Water-deficient and vapor-absent. An assemblage of silicate minerals including hydrous minerals but with no aqueous vapor phase.

Type III: Water-deficient and vapor-present. An assemblage of silicate minerals with or without hydrous minerals and with an aqueous vapor phase present. There is insufficient water present to saturate the liquid when the crystalline assemblage is completely melted at the existing pressure.

Type IV: Water-excess. An assemblage of silicate minerals with or without hydrous minerals and with more than sufficient water to saturate the liquid when the crystalline assemblage is completely melted at the existing pressure.

A system at a given pressure is water-deficient if it contains less water than necessary to saturate the liquid formed by complete melting (Type III) or less water than that required to fully hydrate the subsolidus assemblage (Type II). Water-deficient systems are characterized by vapor-absent conditions through at least one temperature interval. It is convenient for petrogenetic models to distinguish between the two types of water-deficient systems. Figure 8-4 illustrates the pattern of melting for all four types of systems.

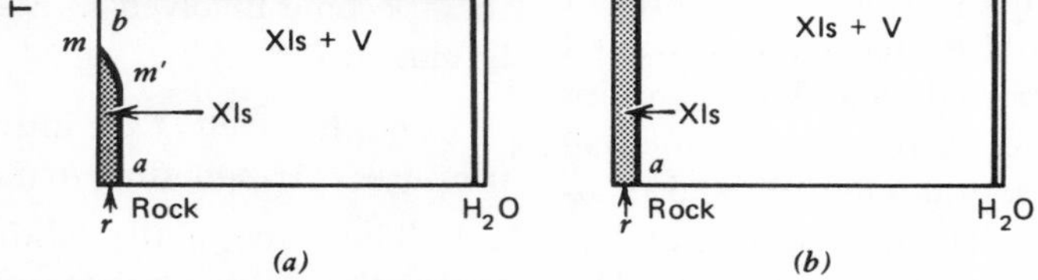

Figure 8-4. Schematic isobaric temperature-composition sections for rock–water systems. Shaded areas are vapor-absent; b–c is saturation boundary; m–m' is dissociation interval of hydrous minerals. See text for significance of lettered points. (a) Hydrous minerals dissociate below solidus temperature. (b) Hydrous minerals remain stable to temperatures above vapor-present solidus (after Robertson and Wyllie, 1971, from Amer. Jour. Sci., with permission).

Figure 8-4 shows a rock composition, *r*. If this rock is completely dehydrated it becomes a water-absent system. The anhydrous rock melts in the temperature interval *n′–n*. If the rock is held at some temperature below the solidus in the presence of excess water vapor under pressure, additional hydrous minerals may be formed changing the rock composition to point *a*. The shaded area within the saturation boundary *a–b–c–d* is vapor-absent. The line above *a* marks the limit of water-deficient systems of Type II which are shown by the shaded area for "crystals." Subsolidus assemblages between *a* and the H_2O axis include systems of Types III and IV. The point *m′* gives the temperature where dissociation of the hydrous minerals begins.

In Figure 8-4a dissociation is completed in the temperature interval *m′–m*, before the solidus temperature is reached, and the Type II assemblages therefore do not melt; for all compositions, melting begins in the presence of excess vapor. The point *c* gives the amount of water required to saturate the liquid when the rock is completely melted. Therefore for temperatures above point *m* the compositions between *b* and *c* are water-deficient Type III and the others are water-excess Type IV.

In Figure 8-4b, compositions in the shaded area to the left of line *a–b* are water-deficient Type II, compositions between points *b* and *c* are water-deficient systems of Type III, and those on the H_2O side of point *c* are water-excess Type IV. Dissociation of the hydrous minerals does not begin until the solidus temperature for Types III and IV has been exceeded, and Type II compositions therefore do melt. As soon as dissociation begins, at *m′*, a small amount of H_2O-undersaturated liquid is produced.

The shape of the saturation boundary between *b–c* is different for every rock. The boundary gives the amount of water required to saturate the assemblage of crystals + liquid. Its position at each temperature is therefore given by adding (a) the water content of the crystals to (b) the product of the amount of liquid present with the solubility of water in the liquid. $P_{H_2O} = P_{total}$ for all assemblages with vapor and $P_{H_2O} < P_{total}$ in vapor-absent regions.

Figure 8-4 permits comparison of the melting relationships in systems of Types I, II, III, and IV. Note the high temperature and narrow temperature interval, *n′–n*, for xls + L in Type I. The solidus temperature for Type II, if this assemblage does melt as in Figure 8-4b, coincides with the temperature of beginning of dissociation of the hydrous minerals and is thus quite independent of Types I, III, and IV; the liquid produced is undersaturated with water and $P_{H_2O} < P_{total}$. The solidus temperatures for the vapor-present Types III and IV are identical and much lower than for Type I; the melting temperature is independent of the amount of water and the initial liquid is saturated with water, $P_{H_2O} = P_{total}$. The melting pattern of water-deficient Type III assemblages is initially the same as for water-excess Type IV assemblages. Melting of Type IV assemblages is completed with excess vapor present throughout, but for Type III assemblages the vapor is dissolved where the temperature reaches the saturation boundary, *b–c*; further increase in temperature yields a vapor-absent assemblage with the liquid becoming progressively more undersaturated with water, and with P_{H_2O} decreasing. The temperature of the saturation boundary *b–c* for Type III assemblages depends sensitively upon the water content and the mineralogy of the rock; with increasing water content it migrates from solidus to liquidus, *b* to *c*, providing the transition from Type II to Type IV. The liquidus temperature in the vapor-absent region is successively lowered with increasing water content from Type I to Type IV.

Figure 8-4b is the normal model for magma generation in the crust. The pattern in Figure 8-4a would prevail for crustal rocks such as granodiorite at pressures below a few tenths of a kilobar (Figure 8-10a), for the system gabbro-water at pressures below about 1 kbar (Figure 6-12), and for peridotite-water below about 8 kbars (Figures 6-11 and 6-18b). The same pattern would be repeated for gabbro(eclogite)-water at pressures above 25 kbars (Figures 6-12, 6-19b, and 8-18a) and for peridotite-water at pressures above about 16 kbars (Figures 6-11 and 6-13b). Thus Figures 8-4a and 8-4b are the two basic models for magma generation in the mantle, the appropriate model being governed by composition and depth.

For details of the composition and amount of liquid produced during progressive fusion we need detailed phase relationships within the melting interval in the water-deficient assemblages for specific rocks. Experimental data are not yet available but interpolation between experimental results for dry Type I systems and for water-excess Type IV systems locates the positions of phase boundaries in the water-deficient region with little variation possible.

The Conditions $P_{H_2O} = P_{total}$ and $P_{H_2O} < P_{total}$

P_{total} refers to the applied pressure in experiments or to the lithostatic pressure resulting from the superincumbent rocks. P_{H_2O} is the partial pressure of water or the pressure that water alone would have if it occupied the volume of the system. Most experimental results using water have been in water-excess systems, and they have often been presented as a function of P_{H_2O}; there has been a tendency in the literature to apply the term P_{H_2O} only to vapor-present systems. Therefore it may be worthwhile to emphasize that the use of P_{H_2O} need not imply the existence of a vapor phase. We have assumed that $P_{H_2O} = P_{total}$ in all vapor-present assemblages (Types III and IV), and noted that $P_{H_2O} < P_{total}$ in vapor-absent assemblages including hydrous minerals or undersaturated liquids (Types II and III).

Solution of solids in the vapor phase reduces P_{H_2O} compared with P_{total}; the effect is small in most silicate-water systems at crustal pressures but at mantle pressures this becomes a factor to consider. For most magmatic processes $P_{H_2O} < P_{total}$, first, because a vapor phase is not invariably present. Second, when a vapor phase is present it contains components additional to water. Therefore even if pore fluid pressure equals P_{total} the P_{H_2O} is less. Third, rocks may exist with pore fluid pressure less than P_{total}, with the differential pressure being supported by grain-to-grain contacts; under these conditions P_{H_2O} in the multicomponent pore fluid is even lower.

In vapor-absent assemblages P_{H_2O} at a given pressure and temperature is a function of water content: it is a dependent variable. In vapor-present assemblages at a given pressure and temperature P_{H_2O} is a function of the water content of the vapor phase. In experiments and in the Earth vapor phase components can be transferred in or out of the system so that P_{H_2O} can be varied independent of P_{total}. If there are two independent pressure variables a system gains an extra degree of freedom. A univariant melting reaction such as Ab + Qz + V $\rightleftharpoons$ L in Figure 8-5 becomes divariant. From the univariant melting curve for the condition $P_{H_2O} = P_{total}$ there extends a divariant four-phase solidus surface referred to three orthogonal axes, P_{total}, P_{H_2O}, and T. Figure 8-5a shows contours for the divariant surface projected onto the P_{total} T plane.

P_{eH_2O} is used in Figure 8-5a instead of P_{H_2O} because it is a term more amenable to rigorous definition. It was introduced and defined by H. G. Greenwood in 1961 specifically in connection with subsolidus reactions. The equilibrium pressure of water in a system at a given temperature, P_{eH_2O}, is the pressure of pure water that would be in equilibrium with the system through a membrane permeable only to water. The contours for constant P_{eH_2O} in Figure 8-5a are based on the first quantitative experimental results for $P_{eH_2O} < P_{total}$ with a liquid phase

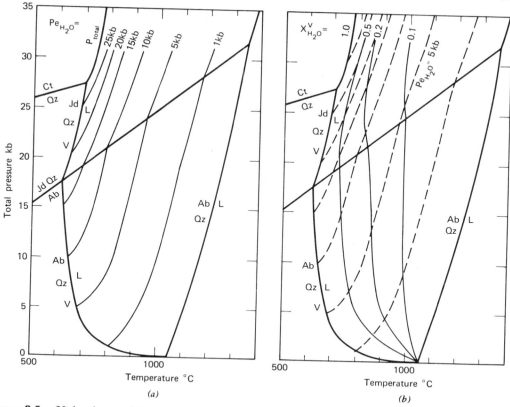

Figure 8-5. Univariant melting reactions for mixtures of albite and quartz both dry and in the presence of excess water, with the jadeite formation curve and coesite transition curve. Estimated values for divariant surfaces are long extrapolations from experimental data from below 6 kb. (a) Each light line (schematic) gives the locus of a melting reaction in the presence of a vapor phase with P_{eH_2O} maintained constant at the specified value. These curves are contours for the divariant solidus surfaces represented in terms of three independent variables, temperature, P_{total}, and P_{eH_2O}; the surfaces connect the dry curves and the water-saturated solidus reaction curves. (b) Each light line (schematic) gives the possible locus of a melting reaction in the presence of a vapor phase with $P_{vapor} = P_{total}$, and with the water proportion in the vapor phase fixed at the specified value. These curves are contours on the divariant solidus surface of (a). The dashed lines, transferred from figure (a), show the general relationship of the composition contours to the water pressure contours.

present. These were obtained by G. Mill-hollen from experiments on the effect of H_2O-CO_2 mixtures on the solidus curve for a nepheline syenite up to 6 kb. The slope of the projected contours shows that with P_{eH_2O} constant the melting temperature increases with increasing pressure. The limiting curve is the dry melting curve with P_{eH_2O} constant at zero. At a given total pressure the melting temperature increases with decreasing P_{eH_2O}.

The solidus curve changes slope at 17.5 kb where albite breaks down to yield jadeite and quartz. The family of contours is constrained to remain between the dry melting curve and the solidus with excess water. The P_{eH_2O} contours cross the jadeite reaction curve with a slight change of slope. At pressures above the jadeite curve the geometrical constraints indicate that the P_{eH_2O} contours become more closely spaced near the solidus than they were at lower pressures. This means that a larger decrease in P_{eH_2O} at constant P_{total} should cause a smaller increase in melting temperature than it did in the low pressure range.

Figure 8-5*b* shows another way of contouring the divariant melting surface again based on numerical results obtained by Millhollen up to 6 kb and extrapolated up to the jadeite reaction curve. The contours show the effect of mixed volatile components of fixed composition; the curves shown are for mixtures with $X_{H_2O}^V$ = 0.1, 0.2, and 0.5, the balance being composed of CO_2 in Millhollen's experiments. These contours also must change slope at the jadeite reaction curve and, because of the crowding together of the P_{eH_2O} contours, it seems likely that the $X_{H_2O}^V$ = 0.5 contour and possibly the 0.2 contour would migrate toward the solidus. These curves need experimental determination, but the restrictions imposed by the known geometry of the phase relationships suggest that in the high pressure range the aqueous vapor phase can be diluted considerably without producing much increase in the melting temperature. This kind of effect was reported by R. E. T. Hill and A. L. Boettcher in 1970 in experiments with gabbro-H_2O-CO_2, but they attributed it to increased solubility of CO_2 in the basaltic liquid at high pressures. Figure 8-5*b* offers an alternative or a contributory explanation.

We can conclude from Figure 8-5 that although additional components such as CO_2 may have a marked effect on the temperatures of melting and crystallization under crustal conditions this effect may be reduced under mantle conditions.

The effect of lowering P_{H_2O} at constant P_{total} by diluting the vapor with another component is to increase the solidus temperature. Similarly the dissociation temperature of a hydrous mineral is lowered if P_{H_2O} is reduced relative to P_{total} (Figure 8-6*a*). The effect of multicomponent pore fluids on conditions of melting in systems including Types II and III assemblages is therefore complicated. This is readily illustrated by the system muscovite (Ms)-quartz (Qz)-water.

Figure 8-6*a* is a schematic *PT* projection for reactions involving muscovite, quartz, and H_2O vapor showing the relevant univariant curves around an invariant point, Q. Other curves extending from the univariant point have been omitted. Assemblages of Ms + Qz (Type II systems) dissociate at pressures below Q and melt at pressures above Q. The dissociation curve is unchanged if vapor is added. The melting curve for the vapor-absent Ms + Qz assemblage is at a higher temperature than that for the vapor-present assemblage (Type III). The high temperature melting curve for Ms + Qz is a Type II reaction equivalent to m' in Figure 8-4*b*. The effect of reducing P_{H_2O} below P_{total} on the two reactions below Q is shown by the dashed lines, contours for constant P_{H_2O}. These contours are considered here as if they are identical with P_{eH_2O} contours (see Figure 8-5).

Figure 8-6*b* shows the phase fields for a specific assemblage of Ms + Qz + V where P_{H_2O} = P_{total} up to 3 kb, but with P_{H_2O} remaining constant at 3 kb as P_{total} increases further. This is not entirely an arbitrary choice because D. S. Korzhinskii believes that during regional metamorphism, P_{fluid} = P_{total}, but P_{H_2O} reaches a maximum value of only 1.5 or 2 kb through the Earth's crust. The Ms + Qz + V assemblage dissociates along the contour for P_{eH_2O} = 3 kb which is situated at a temperature below Q. Therefore muscovite dissociates below the solidus; melting begins at the solidus for the Qz-Or-Si-V assemblage. At a depth corresponding to the invariant point, Q, the estimated temperature difference between dissociation and beginning of melting is about 150°C; with P_{H_2O} = P_{total} Ms + Qz + V melts at Q without dissociation.

Figure 8-6*b* also shows the curve for Ms + Qz transferred as a dashed line from Figure 8-6*a*. The vapor-absent assemblage remains stable right up to the dashed line. This is at a

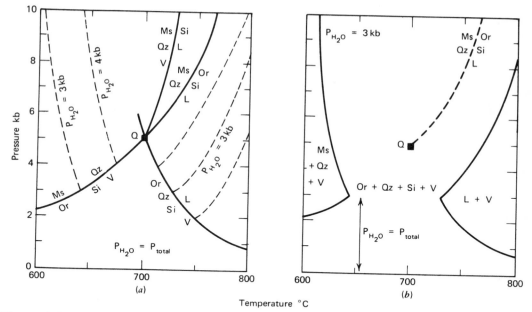

Temperature °C

Figure 8-6. Dissociation and melting reactions involving muscovite, quartz, and aqueous vapor. (*a*) Condition $P_{H_2O} = P_{total}$, in part schematic (based on Lambert et al; 1969). Dashed lines (schematic) contour divariant surfaces for $P_{H_2O} < P_{total}$ (compare Figure 8-5). (*b*) Condition for excess vapor, with $P_{H_2O} = P_{total}$ up to 3 kb, and P_{H_2O} constant at 3 kb for higher values of P_{total}. Dissociation temperatures are lowered and melting temperatures are raised. The dashed line, transferred from figure (*a*), shows the melting temperature for muscovite-quartz mixtures with no vapor phase present. New mineral abbreviations: Ms = muscovite, Si = sillimanite.

lower temperature than the melting curve for the vapor-present assemblages with $P_{H_2O} < P_{total}$. We are accustomed to thinking that the effect of aqueous pore fluids is to lower the melting temperatures of rocks, but if $P_{H_2O} < P_{total}$ because of a multicomponent pore fluid Figure 8-6*b* shows that a pore fluid can cause melting to occur at a temperature higher than it would if there were no fluid phase present at all.

MAGMA GENERATION IN CRUSTAL ROCKS

The variables affecting the temperature of melting of crustal rocks and the composition of the liquid produced are (a) mineralogy especially the Ab/Or ratio in alkali feldspar, the An/Ab ratio in plagioclase, and the feldspar/quartz ratio, (b) total pressure, (c) P_{H_2O}, and (d) the amount of water. If no pore fluid is present the temperature of melting depends on the dissociation temperature of the hydrous minerals (Type II system).

Mineral Variation in Crustal Rocks

The greater part of the Earth's crust is composed of the calc-alkaline rock series or their metamorphosed equivalents (Tables 7-2 and 7-3). The initial liquids produced by their partial fusion approach compositions in the system $NaAlSiO_4(Ne)$-$KAlSiO_4(Kp)$-$SiO_2(Qz)$ which is illustrated in Figure 8-7a. This was termed the Residua System by Bowen because fractional crystallization of

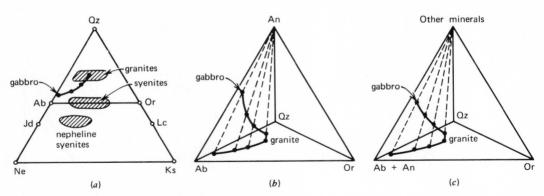

Figure 8-7. Mineral variation (normative) in dominant crustal rocks, the calc-alkaline series gabbro-diorite-granodiorite-granite and their metamorphic equivalents. (*a*) Plotted in the Residua System. (*b*) Plotted in the Granodiorite System, and projected down onto the Granite System, a portion of the Residua System. (*c*) Plotted in terms of feldspars, quartz, and all other minerals.

magmas produces residual liquids approaching compositions in this system. The alkali feldspar join is a thermal divide separating SiO_2-saturated liquids of granitic composition from SiO_2-undersaturated liquids of the nepheline syenite clan. The compositions of syenites and trachytes, expressed in terms of their normative components, project close to the alkali feldspar join. The SiO_2-saturated part of the Residua System has been known as the Granite System since the work of Tuttle and Bowen on the system Ab-Or-Qz-H_2O.

In the calc-alkaline series from granite, granodiorite, tonalite (quartz diorite) to gabbro there is a progressive increase in the anorthite component of the plagioclase feldspar. Addition of anorthite (An) to the Granite System produces the Granodiorite System shown in Figure 8-7*b*. The four representative rock types are plotted in this tetrahedron in terms of their normative feldspars and quartz recalculated to 100%. They lie close to a curved line within the tetrahedron, and if they are projected down through the An apex on to the base, as illustrated, the result is a triangular diagram showing the rock series in terms of normative plagioclase-Or-Qz. The rock series is shown projected in just this way on the base of Figure 8-7*c*, where all other normative

components of the rocks (the mafic minerals) are represented by the apex of the tetrahedron. This shows the increasing content of mafic minerals in rocks ranging from granite through to gabbro.

The phase relationships in the system in Figure 8-7 can be determined by three approaches:

1. Experimental study of synthetic systems in the presence of water under pressure (dry reactions are too sluggish).

2. Experimental study of rocks and rock series occupying different positions within the tetrahedron.

3. Petrographic and mineralogic study of plutonic and volcanic rocks in which the compositions of phenocrysts and quenched liquid or groundmass can be determined.

Approaches 1. and 2. are considered in this chapter. I. S. E. Carmichael in 1963 presented a detailed study involving the third approach with respect to the Granodiorite System.

The Granite System

The phase relationships in the Granite System are sufficiently well known that it can be used to illustrate the effects of the variables influencing melting temperatures and liquid compositions.

Figure 8-8a shows a schematic version of the system Ab-Or-Qz-H_2O at 1 kb pressure. There are no hydrous minerals represented and so no Type II assemblages. The approximately triangular surface *efdgc* is the vapor-saturated liquidus surface giving the compositions of silicate liquids saturated with H_2O. Starting assemblages with compositions above *efg* are Type IV and those below *efg* are Type III. For all subsolidus assemblages and for all vapor-present hypersolidus assemblages $P_{H_2O} = P_{total}$, and for all vapor-absent hypersolidus assemblages $P_{H_2O} < P_{total}$. The surface *ijh* gives the compositions of vapors coexisting with liquids on the surface *efg*. The volume between these surfaces is a miscibility gap between silicate liquids and aqueous vapors.

There are temperature minima, *m* and m_1, on the phase boundaries for the system Ab-Or-H_2O. These minima are reflected in the dry system Ab-Or-Qz by the minimum *M* on the field boundary *ab* which separates fields for the primary crystallization of alkali feldspar and a silica phase, and on the vapor-saturated quaternary liquidus surface by the minimum M_1 on the field boundary

cd. The surface *abcd* separates the phase volumes for the primary crystallization of alkali feldspars and a silica phase. There is a temperature trough, or minimum, on this surface connecting the points *M* and M_1 (actually this dotted line need not begin and end at these points but near enough for our purposes).

Effect of mineralogy and water content. The composition of the first liquid depends on the ratio of Ab/Or. For quartz and a feldspar with composition between Ab-Or the first liquid has a composition on the boundary *cd* and approaching M_1 for a wide range of feldspar compositions; it is identical with M_1 only for one particular feldspar composition. For most Type IV assemblages a considerable amount of liquid with composition on the boundary cM_1d forms within a narrow temperature interval. When all of the feldspar or quartz dissolves the liquid composition changes across the vapor-saturated liquidus surface towards the bulk composition of the mixture with a considerable increase in temperature producing much less liquid. The feldspar/quartz ratio determines whether later liquids are in area *cdg* or *cdef*.

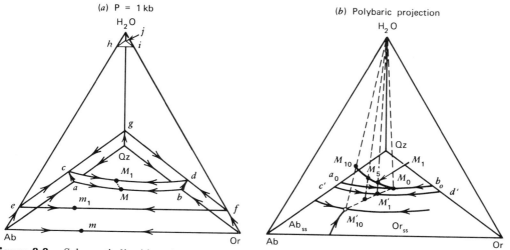

Figure 8-8. Schematic liquidus phase relationships in the Granite System, Ab-Or-Qz-H_2O. (a) Polythermal isobar at 1 kb. (b) Polythermal polybaric projection showing compositional variation of specific points as a function of pressure. *c'-d'* is a projection of *c-d* in figure *a*.

For a water-deficient Type III mixture with bulk composition below the surface *efg* the composition of the first liquid developed is also on the boundary cM_1d, close to M_1 for most bulk compositions. Fusion continues until all of the H_2O available has been used to saturate the liquid. Then with continued fusion the liquid composition migrates downwards across the surface *cdab* in the vapor-absent region with P_{H_2O} decreasing. Further changes in composition within the H_2O-undersaturated volume beneath surface *efg* depend upon the bulk composition of the starting mixture.

Whatever the water content of a mixture the first liquids developed are H_2O-saturated with compositions along *cd* and clustering around M_1. They are independent of the amount of H_2O and dependent only upon the feldspar composition. For the H_2O-deficient systems, however, the later liquids developed are constrained to follow paths on the surface *cdab*.

At any fixed pressure compositions on the boundary *cd*, when expressed in terms of anhydrous components, differ from those of the boundary *ab* for the dry system Ab-Or-Qz. Therefore as liquid compositions migrate down the surface *cdab*, with increasing temperature and decreasing H_2O content, their compositions change with respect to compositions on the boundary *cd*. The compositions of H_2O-undersaturated liquids coexisting with alkali feldspar and quartz are thus dependent not only upon the Ab/Or and feldspar/quartz ratios but also upon the amount of H_2O present in the mixture.

Effect of total pressure and of P_{H_2O}.
With increasing pressure more water dissolves in the silicate liquids, more solids dissolve in the vapor phase, and the surfaces *efg* and *ijh* therefore approach each other. With increasing pressure the boundary *cd* of Figure 8-8a migrates towards the side *ef*, and the position of the quaternary minimum M_1 migrates towards the corner *e*. In the dry system similarly with P_{H_2O} remaining zero

the position of the boundary *ab* changes with changing pressure but to a lesser extent.

Figure 8-8b shows selected composition changes caused by increasing total pressure from 1 bar to 10 kb. For vapor-present assemblages the change is caused in effect by increasing P_{H_2O} from zero to 10 kb. The minimum temperature on the boundary a_0b_0 at 1 bar is at the point M_0. If P_{H_2O} is increased to 1 kb the boundary moves to *cd* in Figure 8-8a with minimum M_1. Figure 8-8b shows both the position of M_1 within the tetrahedron, and that of the 1 kbar boundary in terms of its anhydrous components by its projection $c'M_1'd'$.

The effect of increasing total pressure to 1 kb with P_{H_2O} remaining at zero changes $a_0M_0b_0$ (Figure 8-8b) to *aMb* (Figure 8-8a). The boundaries a_0b_0, *ab*, and *c'd'* (projected) do not coincide. In particular the lines M_0M_1' (projected) and M_0M do not coincide. Therefore the effect of P_{H_2O} on the anhydrous composition of the quaternary temperature minimum, M_0M_1', differs from that of total pressure, M_0M.

Figure 8-8b shows the positions of the quaternary temperature minima for P_{H_2O} zero (M_0), 1 kb (M_1), 5 kb (M_5), and 10 kb (M_{10}), and their projections on to the anhydrous base. At a pressure of about 5kb the solidus temperature is lowered enough that two feldspars are stable with liquids on the vapor-saturated surface, and at higher pressures the minimum is replaced by a quaternary eutectic (e.g. M_5 and M_{10}) but this does not affect the present discussion. The change in pattern of field boundaries is shown by the projections for 10 kb meeting at the projected eutectic M_{10}'.

Consider an assemblage containing only 1% water at 10 kb with feldspar compositions such that the first liquid formed by partial fusion is close to the eutectic M_{10}; this contains about 15% dissolved water. All of the available water is concentrated in this first liquid, and further melting is accomplished by the liquid changing composition with decreasing water content in the vapor-absent region; P_{H_2O} decreases. The liquid path extends downward from the vicinity of M_{10}, following a path corresponding closely to M_1M in Figure 8-8a. The path probably follows the same general trend as $M_{10}M_0$ in Figure 8-8b but it is certainly not identical with it. Thus with progressive fusion of a Type III assemblage an originally H_2O-saturated, albite-rich liquid (near M_{10}) becomes deficient in water with P_{H_2O} decreasing as the liquid becomes more undersaturated and with the liquid composition becoming enriched in quartz/feldspar. The later stages of fusion are controlled by the original quartz/feldspar ratio.

These few examples are sufficient to demonstrate that the compositions of liquids developed by partial fusion in silicate-water systems are dependent not only upon the silicate mineralogy but also upon the total pressure, and independently upon P_{H_2O}, if (a) this becomes less than P_{total} because of low water content, or (b) if the vapor phase is diluted by nonreacting volatile components.

The Granodiorite System: Effect of Mineralogy

The Granite System provides a model of limited applicability because the plagioclase feldspar in a rock has a marked influence not only upon the temperature and composition of the initial liquid, but also on the composition of liquids developed with progressive fusion.

Figure 8-9 shows the phase relationships in the Granodiorite System in the presence of excess water at 1 kb pressure with liquid compositions expressed in terms of anhydrous components. No vapor-absent relationships are represented. The complete tetrahedron corresponds to the vapor-saturated surface *efg* in Figure 8-8a. The phase diagram for Ab-Or-Qz on the base of the tetrahedron is the surface *efg* of Figure 8-8a projected downward radially through the apex H_2O onto the anhydrous base; the compositions of liquids cM_1d in Figure 8-8a are thus

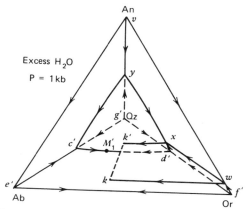

Figure 8-9. Schematic isobaric liquidus phase relationships with excess vapor at 1 kb pressure for the Granodiorite System (Figure 8-7b).

plotted in terms of their anhydrous components as $c'M_1'd'$ from Figure 8-8b. The other triangular sides of the tetrahedron in Figure 8-9 are similar projections. Rising from the field boundary $c'M_1'd'$ to the edge An-Qz is a surface separating the volume for the primary crystallization of quartz (in the presence of excess vapor) from the volumes for the primary crystallization of feldspar. The surface *wxkk'* separates the small alkali feldspar volume from the large plagioclase feldspar volume.

The field boundary xk' gives the compositions of liquids coexisting with vapor and with three crystalline phases: an alkali feldspar, a plagioclase feldspar, and quartz. It extends from x towards the temperature minimum M_1' with a very small content of An. The three surfaces emanating from this field boundary give the compositions of liquids coexisting with vapor and two of the crystalline phases; and the volumes give the range of liquid compositions coexisting with vapor and one of the crystalline phases. At 1 kb the boundary xk' terminates at the critical point k' where the two coexisting feldspars become coincident at the top of the solvus. Similarly the quinary surface *wxkk'* terminates at the critical line kk' situated just above the base of the tetrahedron.

Figure 8-7*b* shows that the compositions of calc-alkaline igneous rocks, represented in terms of normative quartz and feldspars, project into the plagioclase volume of Figure 8-9. Consider a granodiorite composed of plagioclase feldspar, alkali feldspar, quartz, and mafic minerals in the presence of excess water (Type IV system). The mafic minerals can be neglected in the early stages of fusion. The composition of the rock in Figure 8-9 is represented by a point on the triangle produced by joining the composition of the quartz and two feldspars. The composition and temperature of the first liquid developed for this rock, and for any other rock composition represented in the same triangle, is given by the apex of the four-phase tetrahedron plagioclase + alkali feldspar + quartz + liquid; the liquid lies on the field boundary xk'.

With increasing temperature, the liquid composition changes up the boundary towards x, with the plagioclase feldspar and alkali feldspar changing composition simultaneously, until one of the crystalline phases is dissolved; this is probably alkali feldspar for a granodiorite. Further fusion causes the liquid to follow a curved path across the surface $c'k'xy$, with simultaneous solution of quartz and plagioclase until all of the quartz is dissolved. Then the liquid follows a curved path through the plagioclase volume towards the bulk composition of the rock, while the remaining plagioclase becomes further enriched in anorthite.

For a given plagioclase feldspar composition the composition of the coexisting alkali feldspar is fixed; it does not change much with change in plagioclase composition. The first liquid produced does not have a composition at M_1', the minimum liquidus temperature in the system, nor at the composition k', the minimum temperature on the liquidus boundary for the coexistence of three crystalline phases. The first liquid lies on the boundary $k'x$, and it is sensitively dependent upon the composition of the plagioclase. The effect of increasing An content in original plagioclase

is to increase the Or content of the first liquid developed. In the limit for a plagioclase of composition An the first liquid is x. Liquid k' would develop from a three-phase assemblage with a specific albite-rich plagioclase.

A small temperature change while the liquid composition changes along the boundary $k'x$ produces a high proportion of liquid. Less liquid is produced within similar temperature intervals after the liquid leaves this boundary. Thus, the products of partial melting of crustal rocks containing the components of the minerals orthoclase, plagioclase and quartz tend to be concentrated along the field boundary $k'x$, which remains quite close to the Residua System so that the liquids are essentially granitic. A high proportion of granitic liquid is produced within a small temperature interval. The temperature of beginning of melting increases for rocks with plagioclase feldspars richer in anorthite (Figures 8-3 and 8-7*b*).

Alkali feldspar is less widely distributed in crustal rocks than quartz and plagioclase, but there are many rocks not containing free alkali feldspar which nevertheless contain the Or component in other minerals. Plagioclase-quartz-muscovite gneisses, for example, undergo reactions involving the breakdown of muscovite and the release of the component Or at temperatures close to those for the beginning of melting (Figure 8-6). Similarly breakdown of biotite can yield the Or component for melting. The addition of hydrous minerals to assemblages in Figure 8-9 introduces Type II systems as shown for the system granodiorite-H_2O in Figure 8-11.

Now let us consider briefly how the illustrated phase relationships might change under other conditions. With increasing P_{H_2O}, the positions of the surfaces will change, as illustrated by the migration of the boundary $c'd'$ shown in projection in Figure 8-8*b*. If $P_{H_2O} < P_{total}$ with no change in P_{total} all temperatures are increased (Figure 8-5): the positions of the surfaces are changed in the same sense as but not coincident with the changes that are produced by decreasing P_{H_2O} with the condition $P_{H_2O} = P_{total}$. If the amount of water available for a mineral assemblage is less than that required for saturation of the liquid, at a given confining pressure, then the first liquids developed are H_2O-saturated on the line $k'x$ until all the water is used up, and then the H_2O-undersaturated liquids ($P_{H_2O} < P_{total}$) follow paths through narrow volumes close to the surfaces shown in terms of anhydrous components but not coincident with them.

The System Granodiorite-Water

Figure 8-10 shows partly schematic phase relationships for a granodiorite. Figure 8-10*b* shows the Type I system, dry, and Figure 8-10*a* shows the effect of excess water in the Type IV system. The Type IV results show a standard pattern for calc-alkaline igneous rocks of intermediate composition, although in some rocks pyroxene coexists with hornblende down to subsolidus temperatures. The minerals quartz, potash-feldspar, and the sodic portion of the plagioclase yield a granite liquid within a narrow temperature interval above the solidus, and biotite and hornblende persist to temperatures where they dissociate. The hornblende and the more lime-rich plagioclase coexist with granite liquid through a considerable temperature interval. Piwinskii and I reviewed these results in 1970.

The equivalent of Figure 8-4*b* for granodiorite-water provides the basis for models of magma generation. The assemblages of Types II and III up to the saturated liquid at *c* are significant. Figure 8-11 shows an enlarged portion of Figure 8-4*b* constructed from available data for the system granodiorite-water including Figure 8-10. The temperatures of phase boundaries in the vapor-present fields were located from the

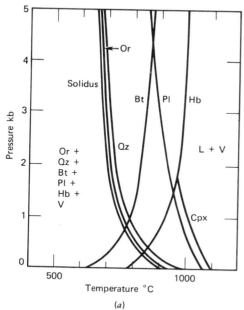
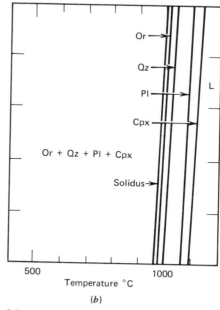

Figure 8-10. *P-T* projections contrasting water-excess and dry melting relationships for a granodiorite, representative of the dominant rocks of the Earth's crust (after Robertson and Wyllie, 1971). (*a*) Water-excess, extrapolated from data by Pinwinskii and Wyllie (1968). (*b*) Inferred melting relationships for the anhydrous equivalent of granodiorite. New mineral abbreviations: Bt = biotite, Hb = hornblende, Pl = plagioclase, Cpx = clinopyroxene. (From Amer. Jour. Sci., with permission.)

boundaries at 2 kb in Figure 8-10*a*. Isothermal boundaries for the solidus, liquidus, and the upper limit of each mineral extend into the water-deficient region as far as the saturation boundary, *b–c*. The shape of the saturation boundary was determined from estimates of the percentage of liquid at each temperature between the solidus and liquidus with excess vapor and the known solubilities of water in similar silicate liquids at each temperature. The dry rock at this pressure melts between the points *n'* and *n* taken from solidus and liquidus in Figure 8-10*b*.

Field boundaries in the vapor-absent region of Figure 8-11 were interpolated from the water-absent points between *n'–n* to the corresponding points on the saturation boundary between *b–c*. There is very little variation possible from the estimated curves shown for feldspars and quartz. More variation is possible for biotite and hornblende, because they have no stability on the water-absent axis, and their dissociation temperatures in the vapor-absent region have not been measured experimentally. There are two alternatives: congruent or incongruent dissociation. Figure 8-11 illustrates the former; each hydrous mineral phase boundary reaches a temperature maximum, *m*, at the boundary between Types II and III compositions or within the Type II range, and with further decrease in water content the stability curve is lowered to a point, *p*, and then to the anhydrous axis.

The major changes in phase relationships in Figure 8-11 caused by passing from water-excess to water-deficient conditions with progressive decrease in water content include the following:

1. The liquidus temperature increases.

2. The temperature interval between solidus and liquidus increases.

3. The temperature stability of hydrous minerals (biotite and hornblende) increases.

4. The temperature interval for the coexistence of two feldspars with liquid increases.

5. Quartz is stable with liquid through a significant temperature interval.

6. The amount of liquid produced within a given temperature interval above the solidus decreases.

Similar schematic diagrams have been constructed for the system gabbro-water at 10 and 20 kb in Figure 8-18.

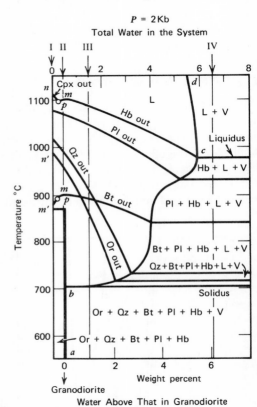

Figure 8-11. Schematic isobaric *T–X* section for granodiorite-water at 2 kb pressure. Composition I is for the dry dehydrated rock (Figure 8-10*b*), and composition IV corresponds to the water-excess rock (Figure 8-10*a*). Phase relationships are vapor absent to left of line *a–b–c–d*, and compositions II and III are thus water-deficient at this pressure. (After Robertson and Wyllie, 1971.) This corresponds to the generalized Figure 8-4*b*. Congruent reaction is indicated for Bt and Hb (*m*); incongruent reactions provide alternative arrangements. (With permission from Jour. Geology. Copyright 1971, University of Chicago Press.)

Anatexis in the Crust

Figure 8-11 provides a guide to the pattern of magma generation from the dominant calc-alkaline rock types of the continental crust. The confining pressure of 2 kb corresponds to a depth of about 8 km; the requisite temperatures for fusion would not be attained at this depth but the essential pattern remains unchanged for greater depths. There are two models corresponding to fusion paths for assemblages III and II.

The Type III composition represents an intermediate rock or its metamorphic equivalent with 1% pore fluid, which is probably a generous estimate for the fluid content of a metamorphic rock in the lower crust. Melting begins with the fusion of quartz and two feldspars and the solution of vapor, producing a water-saturated liquid of granite composition. The solidus temperature is controlled by P_{H_2O} and is independent of the amount of water. The amount of water-saturated liquid produced is controlled by the water content. A small increase in temperature above the solidus produces much additional water-saturated liquid. The pore fluid disappears within a few degrees of the solidus, except for components other than water which are insoluble in the silicate liquid or which have very low solubilities; carbon dioxide is such a component but we need not consider its effects (Figure 8-5). The temperature interval within which the water-saturated liquid exists (the temperature interval between the solidus and the saturation boundary *b–c*) increases with water content but it remains small for any reasonable amount of pore fluid.

With further increase in temperature the liquid becomes progressively more unsaturated with water. Within the vapor-absent region the percentage of liquid developed as a function of temperature is less than that for excess vapor. For a wide temperature interval, about 200°C in Figure 8-11, the undersaturated granite liquid coexists with quartz, alkali feldspar, plagioclase feldspar, and more refractory minerals until the quartz and alkali feldspar dissolve. Then through another wide temperature interval, the liquid composition becomes closer to that of a granodiorite as the granite liquid dissolves the more refractory minerals. The rock is not completely melted until a temperature of nearly 1100°C is attained.

In deep crustal environments where a granitic liquid has formed and migrated upward carrying with it the original pore fluid the remaining rocks are Type II systems. Subsequent anatexis can then occur only at considerably higher temperature, corresponding to *m′* in Figures 8-4*b* and 8-11 where water is released by dissociation of hydrous minerals. The liquid so produced is undersaturated with water, and its composition is different from that produced in Type III models. There may be major differences in the products of anatexis of crustal rocks depending upon whether or not a pore fluid is present.

Water-saturated granite liquids produced by anatexis in the crust can exist only for a few degrees above the solidus. It appears that the normal product of anatexis is a mush composed of crystals and water-undersaturated granite liquid. The amount of pore fluid controls the amount of liquid generated from the granitic minerals. Water pressure for a liquid within the vapor-absent region may be considerably less than the load pressure and therefore upward migration of such a liquid, or liquid-crystal magma, can proceed without excessive crystallization until the load pressure is decreased to a level approaching the water pressure in the undersaturated liquid.

High temperatures are required for complete melting of a granodiorite, about 1000°C with excess water and nearly 1100°C for more reasonable crustal models (Types III and II), even under the most favorable conditions with water being the only volatile component. Such high temperatures make it improbable that liquid magmas of intermediate composition can be generated by anatexis in the crust.

For consideration of the partial fusion of basic rocks in the crust the experimental determination of diagrams such as Figure 8-18*b* is required.

MAGMA GENERATION IN THE MANTLE

The basaltic magmas erupted at the Earth's surface were derived originally by partial fusion of the mantle but, as shown by Figure 4-5, many things may have happened between magma generation and eruption. Figure 6-18*b* shows the general pattern for melting of mantle peridotite in the presence of traces of water. The melting interval consists of two parts: the low temperature band shows where traces of liquid are produced by water, and the upper portion above the dashed line shows where significant melting occurs producing basaltic or picritic liquids. Factors affecting the composition of the liquid at its source include:

1. The mineralogy which varies as a function of pressure and temperature.

2. The depth (pressure).

3. The temperature interval above the solidus which controls the percentage of melting.

4. The amount of water present and its physical state (in pore fluid or combined in hydrous minerals).

Other factors affecting its composition during transit from source to surface include: (a) the speed of movement towards the surface, (b) fractionation during uprise, (c) fractionation at specific depths due to interruption of uprise, (d) changes in P_{H_2O} and f_{O_2} during crystallization.

A major objective of petrology is to determine the composition of liquids developed by partial fusion of mantle peridotite, as a function of pressure (depth), temperature (and percentage melting), and composition including especially the water content; and to determine the changes in composition of these liquids during crystallization at various pressures. This is an enormous experimental task.

Mineralogy of Mantle and of Basalts

We saw in Chapter 6 that the peridotite of the upper mantle consists of olivine (Ol), orthopyroxene (Opx), clinopyroxene (Cpx), and an aluminous mineral, which is plagioclase (Pl), spinel (Sp), or garnet (Ga), depending upon the depth (Table 6-7, Figure 6-10). For some bulk compositions the spinel in spinel peridotite dissolves in the pyroxenes before melting begins, leaving an assemblage of olivine plus aluminous pyroxenes (A-px). The system $CaO-MgO-Al_2O_3-SiO_2$ contains representatives of all these minerals and has been widely used to illustrate mantle reactions.

Figure 8-12*a* shows the compositions of the mantle minerals plotted on the ternary systems bounding the quaternary tetrahedron. Figure 8-12*b* shows the garnet solid solution series, Py-Gr, on the plane Wo-En-Al_2O_3 extending through the tetrahedron. Another geometrical feature to note is that the garnet join intersects the mid point of the An-Fo join at the composition Gr_1Py_2 (Figure 8-12*d*). Partial fusion of the assemblage Fo + En + Di + (An, Sp, or Ga) in the system $CaO-MgO-Al_2O_3-SiO_2$ yields a liquid representing basalts, the isobaric invariant liquid at *A* in Figure 8-13. Figure 8-12*c* shows the relationship of the inner tetrahedron Fo-En-Di-An to the other aluminous minerals, garnet solid solution and calcium-tschermak's molecule (CaTs), and to quartz. This is analogous to the simple basalt tetrahedron of Figure 8-12*d* introduced

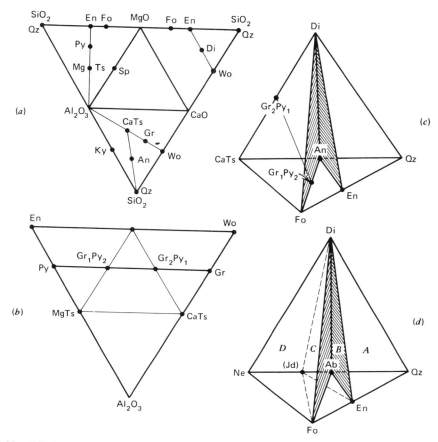

Figure 8-12. Model mantle systems, showing mantle mineralogy. (*a*) System CaO-MgO-Al$_2$O$_3$-SiO$_2$; the tetrahedron has been opened up at apex SiO$_2$ for clarity. (*b*) Garnet and Tschermak's molecule joins on the plane Wo-En-Al$_2$O$_3$ through the tetrahedron represented in (*a*). (*c*) Relationship of garnet join to join Fo-An in the basaltic compositional region of (*a*). (*d*) Simplified basalt tetrahedron (after Yoder and Tilley, 1962); compare (*c*). Major basaltic types occupy volumes *A*, *B*, *C*, and *D*. The planes separating these volumes are replaced by other planes at pressures where jadeite becomes stable at the expense of albite. (From Jour. Petrology, with permission.)

by Yoder and Tilley in 1962 as a means for classifying the major basaltic types in terms of their normative mineralogy.

Figure 8-12*d* includes representatives of the basaltic minerals, olivines (Fo), pyroxenes (En, Di), and plagioclase (Ab); the line nepheline (Ne)-quartz (Qz) shows the degree of silica-saturation. In a generalized basalt tetrahedron these iron-free minerals would be replaced by the normative minerals of the rock. The planes En-Di-Ab (plane of silica saturation) and Fo-Di-Ab (critical plane of silica undersaturation) divide the tetrahedron into three volumes, each of which encloses a specific group of basaltic types. The tholeiites have hypersthene (Hy) in the norm; these are divided into quartz-normative tholeiites (silica-oversaturated in volume *A*) and olivine-normative tholeiites (silica-saturated in volume *B*). The silica-undersaturated alkali olivine basalts lie in volume *C* on the Ne side of the critical plane Fo-Di-Ab. High-alumina basalts plot closer to the plagioclase composition (Ab) than the

normal basalts. Highly alkaline undersaturated rocks (basanites and nephelinites)
contain more normative nepheline and
occupy the volume near *D*.

The plane Fo-Di-Ab is a thermal divide at
low pressures indicating that tholeiites and
alkali olivine basalts cannot be related to
each other by liquid-crystal relationships.
Yoder and Tilley pointed out that at higher
pressures, where plagioclase broke down to
yield jadeite and other minerals (Figure 8-2),
this barrier would be replaced by another as
shown by the dashed lines in Figure 8-12*d*.
M. J. O'Hara in 1968 reviewed the limita-
tions which such divides place upon petrogenetic theories. In the system CaO-MgO-
Al_2O_3-SiO_2 (Figures 8-12*a* and 8-12*c*) he
noted

1. The existence of two low-pressure
divides, the planes Fo-Di-An and Fo-Qz-An.

2. The plane En-Di-An at low and intermediate pressures between about 5 and
15 kb.

3. The eclogite plane garnet (Py-Gr)-
pyroxene(Di-En) at high pressures above
27 kb.

4. Possibly another high pressure divide
Fo-Ga-Px above 40 kb.

Melting Relationships of Peridotites and Basalts

Information about the relationships of
peridotites to basalts may be obtained by
two approaches:

1. By studying the melting relationships of
mineral assemblages in portions of the
synthetic model systems shown in Figure 8-12.

2. By working with a series of natural rocks,
or with glasses of the same compositions.

The first approach is illustrated in Figure
8-13 and the second in Figure 8-14.

Synthetic model systems. Figure 8-13
is a portion of the system CaO-MgO-Al_2O_3-
SiO_2 (Fo-Di-Qz-Al_2O_3 in Figure 8-12*b*)
showing paths of melting for the model
mantle assemblage Fo + Opx + Cpx + Ga.
The range of bulk compositions of mantle
peridotites is represented by the small
volume *P*. The shaded surface *ABCD*
separates the volumes for the primary
crystallization of Ol and Opx; the field
boundary *AC* gives the compositions of
liquids where Ol and Opx coexist also with
Cpx; the field boundary *AB* gives the
compositions of liquids where Ol and Opx

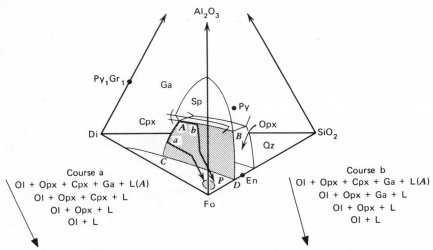

Figure 8-13. Pattern of melting of a simplified garnet peridotite in the system CaO-MgO-Al_2O_3-SiO_2
(after Kushiro, 1969). Compare Figure 8-12*a*. Peridotite compositions occupy shaded volume P; first
liquids form at *A*, and path followed by liquid varies according to the peridotite mineralogy. (From
Tectonophysics, with permission.)

coexist also with Ga solid solution (note the points for Py and Py_1Gr_1 and see Figure 8-12b); the point A is an isobaric eutectic giving the composition of the liquid formed from the four minerals. The liquid does not change composition or temperature until one of the minerals is dissolved, and then with temperature increasing the liquid will follow either a path such as a or b depending upon the compositions and relative proportions of the Ga and Cpx. The compositions of liquids formed by partial melting of similar mineral assemblages may diverge quite significantly, following either AB or AC because of minor variations in the proportions of Cpx and Ga.

O'Hara and Yoder reported in 1967 the results of melting experiments for mixtures on the join Di-Py and adjacent parts of Wo-En-Al_2O_3 (Figures 8-12b and 8-13) at 30 kb; the results were matched by parallel experiments using similar combinations of natural analyzed minerals separated from eclogite and garnet peridotite nodules from kimberlite. They concluded that the initial liquid produced by melting the assemblage Fo + En + Di + Ga at high pressures (liquid A in Figure 8-13) is silica-undersaturated and probably picritic in character; it is analogous to a hy-normative picrite basalt; orthopyroxene has a reaction relationship with this liquid. These results form the basis for the scheme of generation and fractionation of basaltic magmas in Figure 8-15.

In 1968 I. Kushiro published results of experimental studies in sub-systems Mg_2SiO_4-SiO_2-X, where X represents $CaMgSiO_4$, $CaAl_2O_4$, $MgAl_2O_4$ and $NaAlSiO_4$, in the pressure range 7 to 40 kb. The idea was to determine the effect of increasing pressure on the composition of the univariant liquid A in Figure 8-13 and, in the composition join where $X = NaAlSiO_4$, to determine as well the distribution of alkalis between crystals and liquid as a function of pressure. In each composition join the liquid produced at the beginning of melting coexists with Fo, En solid solution, and one other mineral (Di

solid solution, Sp, Ga, Ab, or jadeitic pyroxene). Three effects were recorded as a function of increasing pressure:

1. The two pyroxene fields expand at the expense of the olivine, so that an increasing proportion of olivine dissolves in the first liquid produced.

2. The first liquid shifts in the direction of decreasing silica content.

3. The first liquid with $X = NaAlSiO_4$ changes from Qz-normative below 10 kb (volume A in Figure 8-12d) to Ol-Hy-normative within the interval 10 to 30 kb (volume B) to Ne-normative at 30 kb (volume C).

4. At a given pressure increasing temperature enriches the liquid in dissolved olivine, producing picritic liquids (paths a and b in Figure 8-13).

From these results Kushiro concluded that quartz tholeiites could be generated by partial fusion of the mantle between about 40 and 100 km, olivine tholeiites between 40 and at least 130 km, and the nepheline-normative basaltic liquids could be produced only at depths greater than 100 km. The results support Kuno's model of magma generation depicted in Figure 8-1d.

Natural peridotite and basalt. The general pattern of the melting interval for peridotites and basalts, both dry and in the presence of water, has been illustrated in Figures 5-15, 6-12, 6-13, 6-18 and 6-19. Following the early work of Yoder and Tilley in 1962 it was 1967 before more experimental details became available.

Figure 8-14a shows results for a garnet peridotite nodule published by K. Ito and G. C. Kennedy in 1967, and Figure 8-14b shows results for a tholeiitic basalt glass published in a companion paper by L. H. Cohen, Ito, and Kennedy. At mantle pressures the Cpx, Sp, and Ga of the peridotite pass into the liquid within 50°C above the solidus producing a wide field for the coexistence of Ol + Opx with liquid. This is

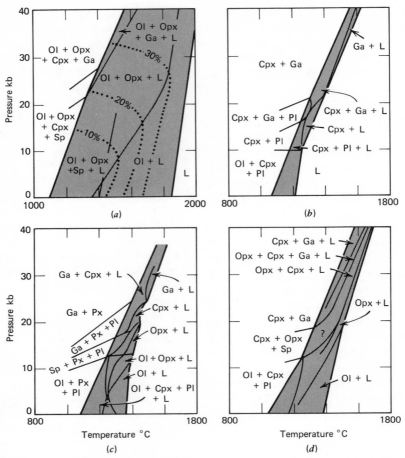

Figure 8-14. Experimentally determined melting relationships for natural peridotite and basalts. (*a*) Peridotite (after Ito and Kennedy, 1967). The dotted lines show percentage of normative olivine in liquid from estimate by O'Hara (1968). (*b*) Tholeiitic basalt (after Cohen et al; 1967). (*c*) Tholeiitic basalt (after Green and Ringwood; 1967; phase boundaries estimated on basis of their data points). (*d*) Olivine-rich tholeiite (after Ito and Kennedy, 1968). Note that there are three different basalts, but the compositions of *b* and *c* do not differ greatly. (With permission from Amer. Jour. Science, and Contr. Mineral. and Petrol. Berlin-Heidelberg-New York: Springer 1967, 1968.)

consistent with results reported by I. Kushiro, Y. Syono, and S. Akimoto in 1968 for a spinel peridotite nodule. Once the first liquid forms at 20 kb there is not much increase in the amount of liquid between 1320 and 1600°C, and presumably little change in composition. In 1968 O'Hara concluded that melting should proceed in step-like form. In contrast Green and Ringwood assume that the percentage of liquid formed will be nearer to a linear function of

temperature above the solidus (Figures 8-17 and 8-19). At 20 and 40 kb the liquid is picritic as shown by the dotted lines. These are estimates by O'Hara for the normative olivine content of the liquid in a partially melted peridotite: with increasing pressure the liquid dissolves an increasing quantity of olivine.

The olivine tholeiite glass of Figure 8-14*b* has 14% normative olivine; subsolidus results have been reviewed in Figures 5-14*b*,

5-15b, and 5-17. The solidus of the basalt and eclogite is coincident with that of the peridotite within the limits of experimental measurement, and the melting interval is almost exactly the same as the interval for the solution of Cpx + Ga in the peridotite. The liquidus phase is Ol to about 9 kb, Cpx between 9 and about 24 kb, and Ga at higher pressures. The olivine tholeiite liquid with composition similar to that of the oceanic tholeiites is therefore not in equilibrium with Ol-bearing peridotite at pressures greater than 9 kb or about 30 km and it cannot be an unmodified partial fusion product of peridotite from depths below 30 km. Ito and Kennedy concluded that this cast serious doubt that basalts are primary magmas from the mantle and favored O'Hara's proposal that they are residual liquids of well advanced crystal fractionation processes. They shared O'Hara's view that the common basalts may be derived from deep-seated picritic liquids which undergo extensive olivine fractionation during uprise (Figure 8-1e).

In 1967 D. H. Green and A. E. Ringwood published results for a glass of olivine tholeiite composition, with 20% normative olivine; this is equivalent to the tholeiite believed to be the parental magma in Hawaii. The subsolidus transitions for this material were reviewed in Figures 5-15a and 5-17. The phase boundaries given in Figure 8-14c are my attempt to correlate the run points plotted between 5 and 27 kb by Green and Ringwood. The liquidus phase is Ol to about 12 kb, Opx to about 20 kb, Cpx between 20 and 24 kb, and Ga at higher pressures. Opx is not recorded in runs with liquid present at pressures above 18 kb. There are two significant differences between results for this basalt and that of Figure 8-14b; the melting interval is wider at low pressures because of the higher content of normative Ol, and Opx occurs as the liquidus phase instead of Cpx between 12 and 20 kb. These results formed the basis for a comprehensive series of experiments designed to trace paths of fractional crystallization of basaltic magmas at different pressures and the scheme is summarized in Figure 8-16; orthopyroxene plays a dominant role in this fractionation scheme. Tilley and Yoder suggested that the orthopyroxene might be metastable, crystallizing in place of stable clinopyroxene. Green and Ringwood repudiated this suggestion reporting runs which indicated that clinopyroxene rather than orthopyroxene may crystallize metastably in short runs.

In 1968 Ito and Kennedy presented results for an olivine tholeiite and a picrite which, together with the results in Figure 8-14a and b, provide a picture of the melting relationships on the composition join basalt-peridotite. They extended this toward nepheline basanite compositions with preliminary data on four other compositions. Results for the crystalline olivine tholeiite composition with 26% normative olivine are shown in Figure 8-14b. The phase boundaries drawn by Ito and Kennedy must be regarded as schematic; they are based on only 11 run points between 10 and 20 kb, with three more at 40 kb. A discontinuous reaction series, Ol → Opx → Cpx, occurs during crystallization at high pressures. Although this material has composition very similar to the olivine tholeiite in Figure 8-14c, there are significant differences in the two sets of results. The liquidus phase is Ol at 20 kb and Opx is given as the liquidus phase at pressures greater than about 21 kb. Ol is replaced by Opx at 12 kb in Figure 8-14c.

Iron capsules were used by Ito and Kennedy and platinum capsules by Green and Ringwood. Ito and Kennedy attributed the different results to errors introduced by the platinum capsules. Green and Ringwood evaluated the loss of iron to the platinum capsule and concluded that the effect is not a major factor in modifying either the major element composition or the crystalline phases of the experimental runs. According to Ito and Kennedy, on the other hand, the platinum capsules probably dissolved appreciable amounts of iron, causing the normative olivine to decrease by "25% or more during the course of most runs, accompanied by an increase in hypersthene."

Ito and Kennedy concluded that their additional results supported their previous conclusion. The discontinuous reaction series gives the prospect of fractional crystallization of the picritic liquid at depth with the formation of SiO_2-poor alkali basalts and other liquids.

Generation and Fractionation of Basaltic Magmas

There have been many schemes developed for basalt petrogenesis and some of these are illustrated in Figure 8-1. The most comprehensive schemes with an experimental basis are those presented by O'Hara and by Green and Ringwood.

Petrogenetic scheme of M. J. O'Hara 1965 and 1968. O'Hara's conclusion that extrusive basalts are residual liquids of advanced crystal fractionation and not primary magmas was discussed in connection with Figure 8-1e, and the experimental basis for his alternative model was outlined in connection with Figures 8-12 and 8-13. His model for a comprehensive scheme for the genesis of basaltic igneous magmas was a stimulating departure from the standard views of the time. A simplified version is shown in Figure 8-15 with an addition on the right from his 1968 paper. O'Hara did not indicate specific pressure intervals; I have added estimates of these to permit comparison with the fractionation scheme of Green and Ringwood in Figure 8-16.

O'Hara distinguished between high, intermediate, and low pressure regimes on the basis of subsolidus mantle mineralogy: garnet, spinel, and plagioclase peridotites. There are three subdivisions, $L1$, $L2$, and $L3$, in the low pressure regime. From available data he estimated the composition of the univariant liquid produced from the four minerals (equivalent to the composition A in Figure 8-13) as a function of pressure. In Figure 8-15 this liquid is shown as hy-normative picrite

in the H regime, Ne-normative picrite in the I regime, and varying from an alkali olivine basalt-like magma, to a high-alumina basalt-like magma, to a tholeiitic magma in the regimes $L1$, $L2$, and $L3$ respectively. This shows the types of primary magmas generated at various depths, and it shows how the liquid composition would change during steady movement upward if it remained just saturated with Ol, Opx, Cpx, and an aluminous mineral. The inevitable eruptive product is a Qz-normative tholeiite. There are many alternative pressure-temperature-time paths by which a deep primary liquid could reach the surface. The effects of interrupting the ascent with fractionation occurring at depth within the specific pressure regimes are illustrated by the horizontal arrows. High-alumina basalt is produced by fractionation in the $L2$ pressure regime, and when this rises to the surface it is split on the low-pressure thermal divide (D) between Ne-normative and Hy-normative basalts. Green and Ringwood disagree with O'Hara's interpretation of the effect of eclogite fractionation in the high pressure regime, H; they conclude that this does not produce silica-poor residual liquids. For partial melting of peridotite in the presence of water in the low pressure regime $L2/L1$ or for a basaltic liquid becoming water-saturated under these conditions O'Hara included the generation of parent andesite liquid but he subsequently concluded that this was not important in the generation of andesite provinces.

In 1968 O'Hara reviewed the available experimental data of the type illustrated in Figures 8-13 and 8-14 with special attention to the compositions of liquids in equilibrium with Ol and Opx at various pressures. He presented various subprojections of the data onto planes within the system CaO-MgO-Al_2O_3-SiO_2 and noted that

"the results obtained by ten workers in three different laboratories are wholly consistent with each other. Moreover these results can be interpreted at each pressure to yield a wholly consistent phase equilibria diagram despite the gross simplification involved in reducing the analyses to a system of four effective components prior to making the subprojections within that system."

D. H. Green has argued that the use of projections in which Mg and Fe are equated and on which basalts, peridotites, and minerals with very different Fe/Mg ratios are plotted should be used only with great caution for the derivation of liquidus phase fields and cotectics.

On the basis of his review O'Hara revised his model for the evolutionary paths for basalt magma. He rejected the concept that tholeiitic magmas *in general* pass through a nepheline-normative stage at depth and introduced explicitly the concept that olivine fractionation has been a continuous factor in the evolution of all volumetrically important basalt magmas. His revised petrogenetic scheme, making provision for advanced stages of partial melting, is more complex than his 1965 model and that of Green and Ringwood (Figure 8-16). He presented a schematic phase diagram for peridotite, like Figure 8-14a, and considered the composition of erupted surface liquids when a magma batch was transported from a specific depth-temperature area at such a speed that it fractionated only olivine during ascent; the source areas of various erupted types are illustrated schematically with respect to the pressure regimes on the right-hand side of Figure 8-15 with temperature above the peridotite solidus indicated by the arrow.

Petrogenetic scheme of D. H. Green and A. E. Ringwood 1967.

Green and Ringwood determined the phase relationships for glass with the composition of an olivine tholeiite from Hawaii believed to represent closely the primary magma existing in the depth range 20–60 km (Figure 8-14c). Partial analyses of Ol, Cpx, Opx, and Ga coexisting with the high pressure liquids were obtained using a microprobe, and these results were used to estimate the directions of isobaric fractionation of the basaltic magma during crystal-

lization at various pressures. Additional basaltic glasses lying on the calculated fractionation trends were prepared. A few experiments on these compositions permitted them to follow the fractionation path into regions where there would have been very little liquid remaining in the original basalt; thus they examined the later stages of fractionation in detail. The program was very well conceived.

Figure 8-16 shows a simplified version of their fractionation scheme related to the subsolidus mineralogy of mantle peridotite. This incorporates later deductions by Green (1969). Note the absence of spinel peridotite in their model and the greater depth for garnet. Green and Ringwood concluded that there are three distinct trends of fractionation characteristic of the pressure intervals corresponding to:

1. Low pressure or shallow crustal fractionation above 15 km (5 kb) to yield Qz-normative residual liquids.

2. Intermediate pressure fractionation at depths of 15 to 35 km (9 kb experiments) to yield high-Al_2O_3 olivine tholeiites with 3–10% normative Ol.

3. High pressure fractionation at depths of 35–70 km (13.5 and 18 kb experiments) to yield Ol-rich alkali basaltic magmas.

They concluded that garnet does not play a significant role in the genesis of magmas by fractional melting at depths less than 100 km. The olivine tholeiite and tholeiitic picrite starting materials in Figure 8-16 are assumed to represent 20 to 40% partial melting of the mantle peridotite, much more than in O'Hara's scheme. Figures 8-16 and 8-15 merit careful comparison. Figure 8-16 includes two sequences added to the 1967 diagram by Green: the high pressure fractionation sequence and the two trends to olivine nephelinites produced by increasing water content and fractionation of aluminous orthopyroxene.

Note that according to Figures 8-16 and 8-14c fractionation at intermediate and high

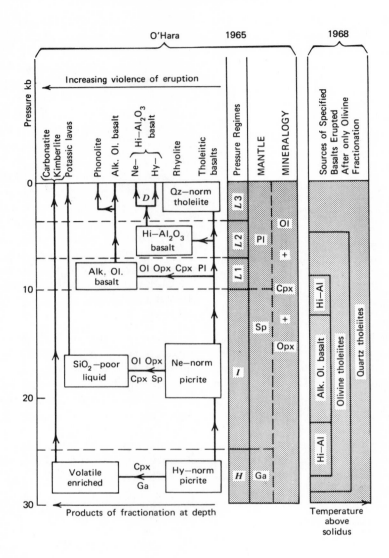

Figure 8-15. Summary of suggested basalt fractionation schemes dependent upon depth of origin and rate of migration to the surface according to O'Hara (1965, 1968). The main portion is based on O'Hara's Table 1 (1965). The main vertical sequence involves steady movement towards the surface through the successive pressure regimes (H = high pressure regime, I = intermediate, L = low, subdivisions 1, 2, and 3. Pressure scale is my estimate to permit comparison with Figure 8-16). Interrupted or delayed ascent at various depths yields the fractionation products shown. Minerals fractionating are listed near arrows. The right hand portion is a revised scheme (1968) showing the character of the erupted liquid when a magma batch formed and fractionated at various levels is taken from a particular pressure and temperature at such a speed that its composition remains near the boundary of, but always just within the olivine primary phase volume; hence it fractionates olivine only during its ascent. This is possible because of the increase in normative olivine content of liquids with pressure, as shown in Figure 8-14a. Olivine nephelinites and melilitites are erupted from fractionated materials at depth, with temperatures below that of the solidus depicted in the figure.

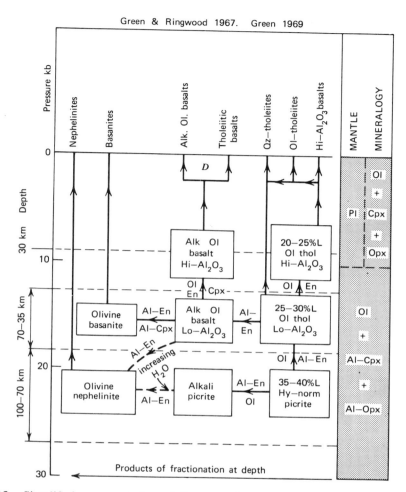

Figure 8-16. Simplified representation of Green and Ringwood's deduced crystal fractionation relationships among various basaltic magmas at moderate to high pressures; "closed system" fractionation. Based on Green and Ringwood (1967) and Green (1969). Contrast the mantle mineralogy between 10 and 30 kb (35 to 100 km depth) in this figure and Figure 8-15. Minerals fractionating are listed near arrows. Compare with Figure 8-17.

pressures is dominated by aluminous ortho-pyroxenes. Both O'Hara and Kushiro (1969) questioned the validity of this fractionation scheme; the olivine tholeiite treated as a parental basalt cannot be the partial melting product of peridotite between 13 and 18 kb because olivine is not a liquidus phase (Figure 8-14c). Green replied in 1969 that he and Ringwood had discussed the absence of olivine in their original paper. He con-

cluded that careful consideration of the experimental data shows that while olivine would be the liquidus phase of a primary olivine tholeiite magma formed at 18 kb cooling results in precipitation of major orthopyroxene and very minor olivine; if so then the dominant role attributed to ortho-pyroxene fractionation is justified. Additional experimental data supporting this conclusion was presented by Green in 1971.

Green and Ringwood applied the fractionation scheme of Figure 8-16 to magma generation through partial melting as in Figure 8-17 by assuming the following:

"In one sense, fractional melting may be regarded as the reverse of fractional crystallization, providing that the nature of the crystalline phases are similar in both cases. This relationship is independent of the actual proportions of phases which may be present." (p. 164).

This is a crucial step in the development of their model for magma generation and this statement needs clarification. The process of fractional melting referred to is actually partial melting under equilibrium conditions (Figure 8-17). The reverse of equilibrium partial melting is equilibrium crystallization not fractional crystallization. The path of fractional crystallization of a liquid is the same as that for equilibrium crystallization only if the minerals involved exhibit

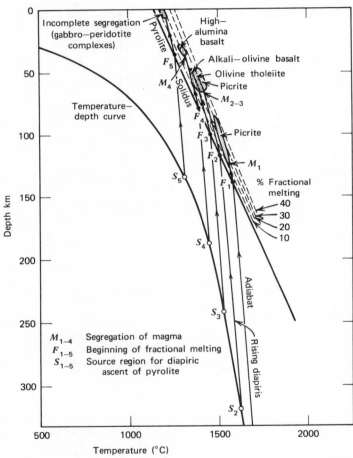

Figure 8-17. A model for magma formation by fractional melting of mantle pyrolite (after Green and Ringwood, 1967). S_1–S_5 represent arbitrary source regions from which there is diapiric ascent of bodies of subsolidus pyrolite to intersect the pyrolite solidus at points F_1–F_5. Partial melting begins at these points and the rising crystal-liquid mushes follow the courses indicated until segregation of magma from residual crystals at M_1–M_4. The nature of the magma is determined by equilibria occurring at M_1–M_4 and not at F_1–F_4. Compare Figures 8-16, and 6-6. (From Contr. Mineral. and Petrol., **15**, 163, Fig. 12. Berlin-Heidelberg-New York: Springer, 1967, with permission.)

no solid solution; that is certainly not true for the systems considered here. Fractional crystallization in a system with olivines, pyroxenes, and plagioclase can produce liquid paths diverging considerably from the paths of equilibrium crystallization with the divergence increasing at lower temperatures; the liquids produced by fractionation in such a system will persist to temperatures which may be considerably lower than the equilibrium solidus temperature.

The liquid fractionation paths depicted in Figure 8-16 cannot be the reverse of the equilibrium melting paths implied in Figure 8-17 unless the minerals involved maintained constant composition during the processes. Green and Ringwood stated that the Fe/Mg ratios of olivines and pyroxenes in the successively studied basaltic liquids are similar to the Fe/Mg ratios of olivines and pyroxenes of mantle-derived ultramafic rocks, and this is the justification for treating the partial melting process as essentially the reverse of the fractionation process.

Figures 3-12 and 6-18 suggest that unusual conditions are required for the temperature to exceed the dry solidus of mantle peridotite. The most reasonable explanation involves the diapiric uprise of mantle material resulting from some gravitational instability as illustrated in Figure 8-17. This process has been depicted schematically in Figure 6-6, and the successive stages can be followed there as well as in Figure 8-17. The process that causes gravitational instability and initiates diapiric uprise is a matter for speculation. One possibility is that upward migration of juvenile water from the deep mantle could operate as shown in Figures 8-22 and 8-23.

The sequence shown in Figure 8-17 is:

1. Solid mantle material rises in diapiric form from a source S under adiabatic conditions.

2. Magma generation occurs when the temperature of the rising mass exceeds the solidus at a point F.

3. The partially melted mass continues to rise adiabatically with the degree of melting increasing as the temperature interval increases above the solidus probably with the liquid remaining in equilibrium with the residual unmelted crystals of the mush. The adiabat with liquid forming has a different slope.

4. Magma segregation occurs at some point M where the extent of partial melting is sufficient, 20 to 40%, and the liquid leaves the residual refractory crystals.

5. Magma rises upward as an independent magma body, cooling, crystallizing, and fractionating.

The critical factor is the composition of the liquid when magma segregation occurs, which is controlled by the depth and the temperature interval above the solidus. Figure 8-17 show the relative positions for the formation of liquids of various basaltic compositions taken from the fractionating scheme in Figure 8-16. The dashed lines above the solidus show the percentage of melting of the mantle.

The linear pattern of melting assumed by Green and Ringwood for dry pyrolite is an idealized situation. In systems where the percentage of melting has actually been measured, the pattern is step-like rather than linear as indicated in Figure 8-19b (Robertson and Wyllie, 1971). The dry melting curve here is my estimate based on the phase relationships and the results reported by Ito and Kennedy for a natural peridotite (Figure 8-14a). They reported that at 20 kb there was little increase in liquid content between 1320 and 1600°C. The curve is similar to that suggested by O'Hara in 1968. He discussed the partition of elements between crystals and liquid in five stages of fusion without giving temperatures; these stages are shown in Figure 8-19b.

Effect of Water

Figure 6-18 shows the general pattern of the melting interval introduced below the

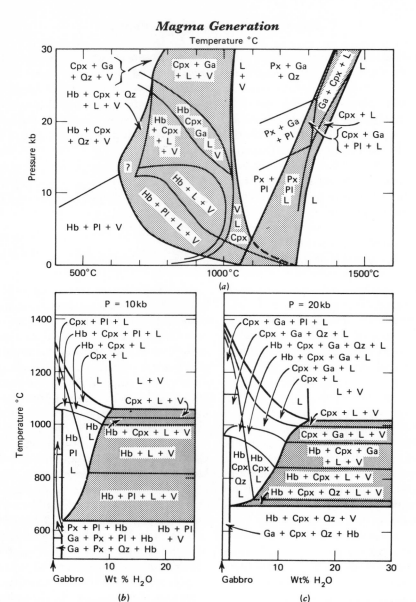

Figure 8-18. (*a*) Experimentally determined phase relationships in gabbro (Figure 8-14) and gabbro in the presence of excess water (unpublished data by Lambert and Wyllie; see Lambert and Wyllie, 1968, 1970, Hill and Boettcher, 1970). See Figure 6-12. The interval for the coexistence of amphibole (hb) and garnet (ga) is probably too narrow, because of the reluctance of garnet to nucleate; with seeded runs the garnet curve would probably be determined at lower temperatures. Zoisite, epidote, and kyanite or sillimanite are minerals whose presence was suspected but not proved at pressures greater than 10 kb. Abbreviations: ga — garnet, cpx — clinopyroxene, px — pyroxene, pl — plagioclase, hb — amphibole, qz — quartz, L—liquid, V—aqueous vapor phase. (*b*) and (*c*) Schematic isobaric *T-X* sections for 10 kb and 20 kb for gabbro-water, interpolated from the dry and water-excess results in Figure 8-18*a*. (Compare Figures 8-4*b* and 8-11). An important boundary for experimental determination is the vapor-absent liquidus below the field of L; changes in liquidus phases with pressure (Figure 8-14) and with water content may have a strong influence on fractionation products. A congruent reaction is depicted for amphibole in the vapor-absent region; an incongruent reaction is an alternative arrangement.

dry peridotite solidus by the addition of small amounts of water. Experimental results for peridotite-water are limited to reconnaissance experiments near the solidus, but we do have fairly complete data for the melting relationships of a gabbro which illustrate the problem involved. The melting intervals of the rock dry (Type I system) and in the presence of excess water (Type IV system) were compared in Figure 6-12. In Figure 8-18a the detailed phase relationships within the melting intervals are shown. This gabbro contains only 8.6% normative olivine and it becomes a quartz eclogite at high pressures.

Given the results shown in Figure 8-18a, isobaric sections through the water-deficient region can be constructed in the same way that Figure 8-11 was obtained from Figure 8-10 for the system granodiorite-water. The schematic results so obtained for gabbro-water at pressures of 10 and 20 kb are shown in Figures 8-18b and 8-18c.

There are significant differences between the phase relationships at 10 and 20 kb. Notice the distribution of quartz in Figure 8-18c for 20 kb; this ensures that the first liquids produced on partial melting in Types II, III, and IV systems are all silica-rich. For the Type II compositions at 10 kb (vapor-absent subsolidus; Figures 6-13 and 6-19b for an isopleth) the assemblage is transformed from hornblende eclogite to hornblende gabbro with increasing temperature, whereas at 20 kb the assemblage is a quartz-hornblende-pyroxenite. Note the melting pattern of Type II systems; liquid undersaturated with water is produced within the temperature interval where hornblende dissociates (*m'm* in Figure 8-4). In contrast, for the Type III system the first liquid is water-saturated however small the water content.

Information about peridotites and various basaltic compositions that is vital for comprehension of the petrogenesis of basaltic magmas and for evaluation of the effect of water on the physical properties of the mantle includes: (a) the percentage of liquid produced in the vapor-absent region as a function of pressure (depth), temperature, and water content and (b) the composition of the liquid as a function of the same variables. For a given water content the pattern of phase relationships can change from Type II (as in Figure 8-4b) to Type III (as in Figure 8-4a) where the hornblende breakdown curve passes below the solidus. The composition and water content of the liquids present near the mantle solidus are therefore markedly dependent on the stability of hydrous minerals such as amphiboles and phlogopite.

Amount of liquid. As a guide for experimentation D. H. Green has presented a comprehensive scheme for peridotite-water in the form of a petrogenetic grid which provides an internally consistent working model for mantle source composition, derivative liquids, peridotitic residues, and magmatic cumulates from the liquids at high and low pressures. This is based on reconnaissance experiments in water-deficient regions with peridotites and basaltic compositions. The model is illustrated in Figures 8-19a and 8-20.

Figure 8-19a shows the linear relationship assumed by Green and Ringwood for the percentage of melting above the solidus temperature of mantle peridotite. Compare Figure 8-17. The minerals coexisting with liquid for specific percentage intervals of melting are shown between Figures 8-19a and 8-19b. The other curve in Figure 8-19a is Green's estimate of the percentage of melting produced with 0.1% water at each temperature above the wet solidus for peridotite at a pressure of about 25 kb; the solidus is at 1080°C where amphibole breaks down (Figures 6-18b and 8-20). The position of this curve is based on the measured depression of the liquidus in the vapor-absent region by known amounts of water, but the data and method have not been published as yet. The corresponding contours for percentage of liquid developed as a function of pressure with 0.1% water present are plotted on Figure 8-20.

Green concluded that the near-liquidus role of orthopyroxene in water-bearing

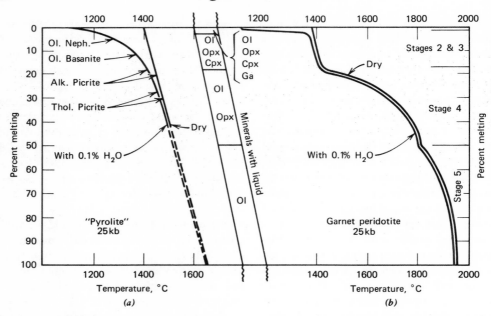

Figure 8-19. Fusion curves for peridotite dry, and in the presence of 0.1% water, showing the percentage liquid produced at successive temperatures above the solidus, and the minerals coexisting with the liquid in different parts of the melting interval (after Wyllie, 1971). (*a*) Dry curve from Green and Ringwood (1967), see Figure 8-17; curve with H_2O and estimated liquid compositions after Green (1970). (*b*) Dry curve estimated from partial experimental data by Ito and Kennedy (1967), see Figure 8-14*a*, with stages according to O'Hara (1968). The curve with H_2O contrasts with curve in *a*, and corresponds closely to Ringwood's (1969) estimate, except for the formation of a finite amount of liquid within a few degrees of the solidus in the *b* model, consequent upon dehydration of amphibole. (From Jour. Geophys. Res., **76**, 1328, 1971, with permission.)

nephelinitic magmas at high pressures may permit fractionation of magmas through olivine-rich basanites to olivine nephelinites at 20–25 kb and through picritic basanites and picritic nephelinites to olivine melilite nephelinites at about 27 kb. From these experimental studies he inferred that the highly alkaline, undersaturated magmas could form by partial melting of mantle peridotite in the presence of water with the general distribution as shown in Figure 8-19*a* for 25 kb and in Figure 8-20 for a range of pressures. The effect of 0.1% water on the magma generation scheme shown in Figure 8-17, as shown in Figure 8-20, is to lower the temperature of formation of the magma types shown in Figure 8-17 and to introduce the wide zones for the generation by fusion of the highly alkaline undersaturated liquids.

Green's curve for melting of peridotite with 0.1% H_2O indicates surprisingly high percentages of liquid at subsolidus temperatures. According to the curve in Figure 8-19*a* at the dry solidus temperature 0.1% of water is sufficient to produce about 17% melting and to dissolve all of the garnet and most of the clinopyroxene; Figure 8-20 indicates 25% of liquid at the dry solidus. Until this curve can be considered as securely based on experimental data I propose as an alternate pattern that shown in Figure 8-19*b*. A finite amount of H_2O-undersaturated liquid is produced within a few degrees of the solidus where hornblende breaks down (Type II composition, vapor-absent subsolidus in Figure 8-18*c*), and very little additional liquid is produced until the temperature approaches closely that of the dry solidus; at higher temperatures it follows the dry curve, rather than approaching it asymptotically as indicated by Green's curve between 30 and 40% melting. I have not attempted to estimate the actual percentages of melting because I do not have adequate data, and the percentages guessed are simply to illustrate a pattern which contrasts with that of Green. Distinction between these two patterns is significant because the amount of interstitial silicate liquid produced by traces of water below the dry melting temperature affects markedly the physical properties of the mantle, and the prospects that the liquid can escape from its host for independent uprise as a magma.

Composition of liquid. The composition of the liquid produced is in dispute. In 1955 A. Poldervaart suggested that magmas more silicic than basalt "may be formed in proportionately smaller amounts than basaltic magma by relatively low temperature partial melting of ultramafic material in the presence of high concentrations of water" (pp. 139–140). A series of experiments by I. Kushiro and associates in increasingly complex synthetic systems proves that the water-saturated liquid developed from Fo + En-bearing assemblages at high pressures is silica-saturated. Extrapolation of these results to the natural rocks leads Kushiro to conclude that equivalent liquids developed in the

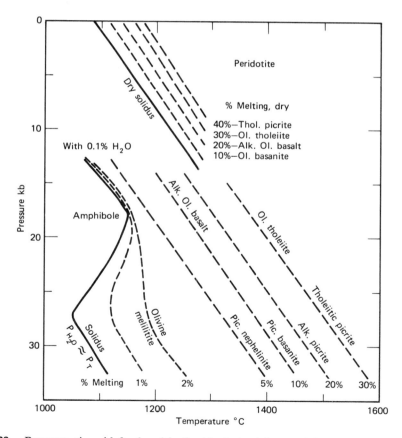

Figure 8-20. Petrogenetic grid for basaltic liquids derived by partial melting of mantle peridotite dry (Figure 8-17), and in the presence of 0.1% water (Figure 8-19*a*), according to Green (1971). (From Phil. Trans. Royal Soc. London, with author's permission.)

mantle are tholeiitic basalts or andesites; O'Hara made the same suggestion in 1965.

This interpretation is in direct conflict with the inferences of Green in Figures 8-19*a* and 8-20 that these liquids are nephelinitic. Green and Kushiro presented their respective interpretations in 1969. In his 1970 paper Green concluded that both sets of experimental data were probably correct and described preliminary experiments suggesting that at 22.5 kb the conclusions from Kushiro's synthetic systems cannot be extrapolated to a typical basalt. He doubted that a quartz-normative tholeiitic or andesitic liquid could coexist with olivine at water pressure greater than 10 kb.

Figure 8-21 shows that Kushiro's experiments have advanced a long way from the first runs with enstatite-water. The mineral components involved are Fo, En, Qz, Jd, Ne, An, and Ca-tschermaks molecule. The surface *VRBT* is the Fo-Opx liquidus boundary at 20 kb pressure under anhydrous conditions. The point *B* gives the composition

of the first liquid formed by partial fusion of an assemblage including Fo, Opx, Cpx, and an aluminous phase; it is probably in the Ne-normative region of the tetrahedron at 20 kb. The point *A* is the first liquid formed by partial fusion of Fo, Opx, Cpx, and Ga (a synthetic garnet peridotite) in the presence of water. This liquid is silica-saturated and Qz-normative at 20 kb and probably up to 25 kb. Kushiro informed me that he has since analyzed the liquid *A* using a microprobe confirming that its composition is andesitic. Above 30 kb, the first liquid *A* may be silica-undersaturated.

By analogy with our review of Figures 8-4 and 8-8 Figure 8-21 shows that however small the water content of the synthetic garnet peridotite, the first liquid formed is H_2O-saturated, with a composition at *A*. For water-deficient conditions the vapor dissolves in the liquid and with increasing temperature and decreasing P_{H_2O} the liquid composition follows a path close to *AB* but not coincident with it. The liquid composition thus moves

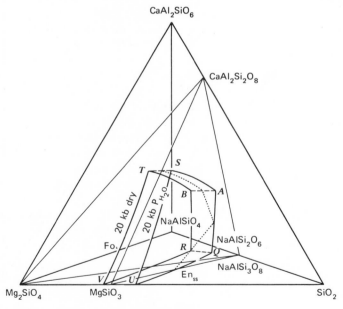

Figure 8-21. The positions of the forsterite-orthopyroxene liquidus boundaries in the system forsterite-nepheline-$CaAl_2SiO_6$-SiO_2-H_2O at 20 kb under water-saturated and dry conditions. The dotted line indicates a trace of the intersection of the plane $MgSiO_3$-$CaAl_2Si_2O_8$-$NaAlSi_3O_8$ with the volume *A-Q-U-S-T-V-R-B* (after Kushiro, 1970. Courtesy of Carnegie Institution).

further away from silica with progressive fusion, and when it crosses the plane En-An-Ab its composition becomes Ol-Hy-normative, still tholeiitic. If the liquid crosses the plane Fo-An-Ab it becomes Ne-normative. How much andesitic liquid can be generated and how far the liquid can move from A toward B depends on the water content.

This model assumes that no hydrous minerals are stable at the solidus: we are dealing with a Type III system (Figure 8-4a). If the pressure is appropriate for the stability of hornblende (or other hydrous mineral) then we have the situation in Figure 8-4b and the system may be either Type II or Type III depending upon the water content. If there is free vapor present as well as the hydrous mineral (Type III) the solidus temperature is lower than for the Type II assemblage with no vapor (Figure 8-4b). The apparent conflict of interpretation between Green and Kushiro may be resolved when all of the variables are considered. The pressure (depth) interval affects the composition of the water-saturated liquid and the stability of hydrous minerals; with hydrous minerals stable the water content defines the system as Type II or III. In different parts of a petrogenetic grid such as that proposed by Green (Figure 8-20) we have the following conditions, each of which may involve a liquid of different composition: (a) Type III system with no hydrous minerals, (b) Type III system with hydrous minerals, and (c) Type II system. The liquid produced in the Type II system is water-undersaturated and it may be significantly different from the water-saturated Type III liquid. The Type III liquid without hydrous minerals will have a composition different from a Type III liquid coexisting with amphibole. Kushiro's results at 20 kb suggest that Type III liquid could be andesitic and Type II liquid could be nephelinitic.

Initiation of diapiric uprise. Figure 8-17 shows a model for magma generation with diapiric uprise of mantle initiated by some unspecified gravitational instability. Figures 8-22 and 8-23 illustrate how juvenile water rising into dry peridotite from deep within the mantle (along the geotherm) would cause incipient melting in a layer ab just above the solidus for peridotite-water. This would lower its density and viscosity relative to the surrounding mantle, and it would tend to rise as shown in Figure 8-23b. It would rise adiabatically as in Figures 8-17 and 8-22 and when the layer ab reached the position cd it would be about 100°C hotter than the surrounding mantle. Continued uprise would carry it through the level near the dry solidus where significant melting occurs (Figure 8-19b), and in the layer at ef magma generation would occur in the normal way and the subsequent events could be described according to the dry model, Figure 8-17 for example.

Figures 6-18b, 6-19b, and 6-20 were used to explain the low-velocity zone in terms of incipient melting due to traces of water. If water rose upward from below the low-velocity zone, as in Figures 8-22 and 8-23a, the amount of interstitial liquid in the base of the zone would increase above the rising water. Increased liquid content would cause decreased density and viscosity, and diapiric uprise could thus be initiated. Magma generation caused by rising diapirs from the top of the low-velocity zone has often been suggested, but if the diapirs begin at deeper levels near the base of the zone, passing through material already partially melted, this could introduce considerable variety into the processes for distribution of trace elements and rare earth elements among the crystals and liquid of the rising diapir and the partially melted mantle in the low-velocity zone.

Basalt Petrogenesis

Current models for the generation and fractionation of basaltic magmas with an experimental basis include the following:

1. The compositions of primary basalts are controlled by the depth of partial fusion of mantle peridotite (Kushiro; Figure 8-1*d*).

2. Tholeiitic basalts are formed by olivine fractionation during uprise of a deep-seated parental picrite liquid with silica-poor alkalic liquids produced by deep-seated fractionation including eclogite fractionation (O'Hara, Yoder, Ito, and Kennedy; Figures 8-1*e* and 8-15).

3. The compositions of basalts are controlled by the degree of partial melting of dry mantle peridotite at the stage where the liquid is segregated, by the depth of segregation and the depth of subsequent fractionation (Green and Ringwood, Figures 8-16 and 8-17).

4. The presence of water in the mantle extends model (3) by introducing zones for small degrees of partial melting below the dry peridotite solidus (Green; Kay, Hubbard, and Gast, 1970; Figures 6-18*b* and 8-20). The low temperature hydrous liquids are silica-saturated according to Kushiro, silica-poor and alkalic according to Green.

These different interpretations emphasize the need for detailed experimentation under controlled conditions in water-deficient peridotite-basalt systems as illustrated in Figures 8-18*b* and 8-18*c*.

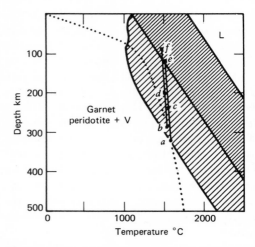

Figure 8-22. Peridotite-water isopleth from Figure 6-18*b*, and dotted oceanic geotherm from Figure 3-12. Shows the diapiric uprise of layer *ab* (Figure 8-23*a*) to successive positions *cd* (8-23*b*) and *ef* (8-23*c*), where basaltic magma is generated. Compare Figure 8-17. (From Wyllie, 1971, Jour. Geophys. Res., **76**, 1328, with permission.)

GENERATION OF BATHOLITHS AND ANDESITES

In 1950 Tilley drew attention to the problem of accounting for the voluminous andesites of the orogens. The plutonic rocks of batholiths approach chemical equivalence with andesites and their associated volcanic rocks and both associations are developed in the same tectonic environment. Although rocks in either association may have an origin independent of the other association, it makes sense to consider their petrogenesis jointly. It is generally held that at least some intrusions in batholiths represent andesitic magmas that failed to reach the surface for eruption.

The bitterness of the Granite Controversy subsided during the 1960's as many petrologists accepted an experimentally-based anatectic model.

We concluded that the normal product of anatexis in the crust is a mush composed of crystals and H_2O-undersaturated granite liquid. The temperatures required for the formation of liquid of intermediate composition, even under the most favorable conditions, are so high (Figures 8-10 and 8-11) that it seems improbable that andesite liquids could be generated in the crust. With this process excluded there remain the following possibilities for the origin of batholithic magmas and andesite magmas:

1. Crustal anatexis yielding a crystal mush of intermediate composition, composed of crystals with water-undersaturated granite liquid.

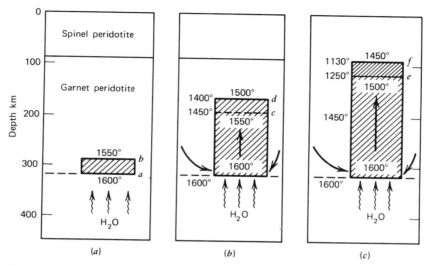

Figure 8-23. Schematic mantle sections showing the diapiric uprise of layer *ab* at the base of the low velocity zone into successive positions *cd* (*b*) and *ef* (*c*) on the adiabatic curves *ace* and *bdf* in Figure 8-22. Water migrating from the deep mantle increases liquid present in layer *ab*, causing diapiric uprise from the base of the low-velocity zone, and basaltic magma generation towards the top of the zone. (From Wyllie, 1971, Jour. Geophys. Res., **76**, 1328, with permission.)

2. Derivation from parent basaltic magma by fractional crystallization, assimilation, hybridism, or some combination of these.

3. Mantle anatexis producing primary andesitic magma.

4. Anatexis of crustal material under mantle conditions caused by downward transport of sediments and lithosphere in Benioff zones.

The first process provides a satisfactory explanation for many plutonic rocks of batholiths, but it cannot explain the eruptive andesites. This leads us to mantle conditions for the origin of andesites and possibly to mantle material which suggests in turn that mantle-derived material may be involved in the development of some batholithic intrusions.

The second process is the classic model, and the conditions under which basaltic magmas may yield andesitic derivatives have been discussed by E. F. Osborn. In 1959 he demonstrated that paths of crystallization for liquids in the system $MgO-FeO-Fe_2O_3-SiO_2$

varied according to the oxygen fugacity, and this has since been extended to more complex synthetic systems including plagioclase components. Applying the results to natural magmas leads to the conclusion that there are two reaction series as illustrated in Figure 4-5. In a closed system with oxygen fugacity remaining low the fractionation trend is towards liquid enrichment in Fe/Mg, whereas if the oxygen fugacity is maintained at a high level the residual liquids follow the calc-alkaline trend with enrichment in silica and alkali feldspar. In 1969 Osborn reviewed the problem in some detail and concluded that andesites were produced by fractional crystallization of olivine basalt magma in orogenic regions under conditions of high oxygen fugacity, produced by flow of water from the surrounding geosynclinal rocks into magma and migration away of hydrogen. According to H. P. Taylor in his 1968 comprehensive review of the oxygen isotope geochemistry of igneous rocks, the oxygen isotopes of andesites make it reasonable to assume that these were derived from a parent basaltic magma by some process of magmatic differentiation.

There appears to be little or no continental crust beneath some island arcs (Figures 5-1, 7-4b and 7-6b) which suggests that, in this environment, the andesites and associated lavas may be derived from the mantle. In 1959 when few petrologists paid much attention to the mantle J. T. Wilson argued that the andesites of island arcs were derived by partial melting of a small fraction of mantle peridotite between 70 and 700 km depth, and that they rose along fractures associated with deep earthquakes. Figures 8-19, 8-20, and 8-21 illustrate the current dispute about the composition of the liquid produced by partial fusion of peridotite in the presence of water; it may be Ne-normative or Qz-normative. Resolution of these different interpretations is important because of their implications with respect to the origin of andesites and the evolution of continents. We saw in Chapter 6 that the upper mantle is composed of peridotite and eclogite. T. H. Green and A. E. Ringwood presented experimental data in 1968 on a series of rocks with compositions between basalts and rhyolites, and they concluded that partial melting of eclogite or hydrous basalts at depth would yield andesitic magmas. They also reviewed hypotheses for the origin of the calc-alkaline suite. In a similar review O'Hara agreed with their conclusion noting that "Although the experimental evidence for this is based upon some exceptionally long linear extrapolations of liquidus temperatures (essentially a nonlinear function) through a minimum of fixed points the basic proposition that the partial melting product is silica-enriched cannot be in doubt." (1968, p. 101.) Thus primary andesite magmas can be generated in the upper mantle from eclogite and possibly from peridotite in the presence of water under some conditions. Origin from mantle material is favored by some trace element data and isotope data as reviewed by S. R. Taylor in 1969.

The fourth process has recently been hailed as the answer to all of the problems (Chapter 14). Partial fusion of sediments and rocks of the oceanic crust and upper mantle at high mantle pressures and varied temperatures occurs because of the downward movement of a slab of lithosphere at the Benioff zones. This produces magmas of basic, intermediate, and acid composition formed under dry or wet conditions, either within the down-going lithosphere or above it, for eruption at the surface or for intrusion into the cores of the young mountain chains developing adjacent to the oceanic trenches. There are plenty of variables to manipulate in order to obtain the solutions needed. What we lack are adequate constraints to facilitate selection among the solutions proposed. The study of trace elements and isotope evolution trends may eventually provide definitive evidence.

9. *Geosynclines and the Orogenic Cycle: Classical Views*

INTRODUCTION

For the most part this chapter is concerned with the classical treatments of the orogenic cycle, which include no satisfactory mechanism, published before plate tectonics introduced a new conceptual framework for orogenesis. Here we examine the geosynclinal concept as it was developed through a century by standard geological and stratigraphical methods of unravelling the history and paleogeography of a mountain chain. The recent models for mountain building are reviewed in Chapter 14.

In theories for the evolution of mountain chains, or the geotectonic cycle, the concept of geosyncline and orogenic cycle has been a dominant theme since 1859 when J. Hall realized that the Appalachian Mountains in New York State had been formed by elevation of a pile of sediments that were originally deposited in a subsiding trough. In 1873 J. D. Dana gave to such troughs the name geosynclinal which was later changed to geosyncline. This revolutionary concept of a great inversion of relief from depressed zone to folded mountain chain was followed by innumerable geological syntheses coordinating all of the subjects of geology and all of the processes outlined in Chapter 4.

The generalized sequence of the geosynclinal phase, the tectogenic phase, and the orogenic phase as traditionally presented in textbooks includes the following:

1. The accumulation of sediments in a subsiding trough, the geosyncline, along with marginal or submarine eruption of basic and ultrabasic lavas, including spilites and the ophiolite suite.

2. Folding, dislocation, and overthrusting of the rocks in the geosyncline.

3. Regional metamorphism and the emplacement of batholiths.

4. Uplift and the formation of marginal troughs with renewed sedimentation; widening of the geosynclinal zone, and repetition of (2) and (3).

5. Epeirogenic uplift with volcanic eruptions of basalts, andesites and rhyolites, and comagmatic plutonic intrusions.

6. Peneplanation.

The scheme has been modified repeatedly with some geologists extending the idea of a geosyncline to any subsiding region in which a significant thickness of sediments accumulates and others restricting the term only to troughs which give rise to strongly folded alpine-type mountain ranges. It has been generally assumed that orogenic activity implied the former presence of a geosyncline, and that high folded mountain ranges have formed only on the sites of geosynclines. With the revolution of the 1960's, revival of the theory of continental drift and development of the "new global tectonics" (Chapters 11 and 14), however, many geologists claim that the conventional cycle is obsolete and new models for mountain building involve movement and collision of crustal plates. The classical descriptive models are based on much geological data, and students should know this basis before being launched on a collision course.

CLASSIFICATION OF GEOSYNCLINES

Development of the concept of geosynclines is illustrated by various classification schemes. The ideas of Hall and Dana strongly influenced subsequent American thought which has differed in significant aspects from European views. The main differences between them arise because each group developed the concept with a different mountain range as the model. The standard example for American geologists is the Appalachian geosyncline and for European geologists it is the Alpine geosyncline as interpreted by Haug in 1900. Americans have often cited the coastal plain of the Gulf of Mexico as a standard example of a present day geosyncline, but Europeans have considered the Indonesian Archipelago as a standard example.

Both schemes are characterized by a great thickness of sediments, but whereas American geologists have contended that shallow-water sediments characterize subsiding basins European geologists recorded the occurrence of deep-water sediments and concluded that geosynclines were deep, elongated troughs. In the American view a progressively rising geanticline supplied sediment to the neighboring geosyncline in such a way that the rate of sedimentation just kept pace with the rate of subsidence. According to the European view this delicate balance was not maintained and the history and duration of a geosyncline depends on the relative rates of subsidence and sedimentation. Americans have considered geosynclines to occupy sites marginal to continents but Europeans have regarded them as forming either marginal to continents or between continental masses.

Tectonic Elements

The geosynclinal concept as originally developed in both America and Europe embodies a major paleogeographic distinction between one type of mountain chain, that characterized by strong folding, and the rest which can be defined as intracontinental or intracratonic. A geosyncline has a long history and it comprises several tectonic elements each distinguished by specific features and events in space and time. The classification of some of these individual elements as geosynclines has led to such broadening of the original concept that the term geosyncline has been applied to almost any tectonic element that involves subsidence and sedimentation. The main trends are illustrated in Table 9-1 which summarizes and compares various classifications.

Significant contributions were made by H. Stille who recognized cratons and orthogeosynclines as the two major crustal divisions. He subdivided cratons into hochkraton—stable continental crust, and tiefkraton—stable oceanic crust; and he distinguished between eugeosynclines and miogeosynclines. Eugeosynclines, found in the internal zones of the geosynclinal system (internides) most distant from the craton, are characterized by basic lavas and ophiolites; whereas miogeosynclines, found in the external zones (externides) of the system nearer the craton, are free or almost free of igneous activity. He also clarified the tectonic history of geosynclines. Epeirogenesis (undation) occurring over large areas for long periods of time produced geosynclines, geanticlines, and uplift of mountain chains with no

TABLE 9-1   Various Classifications of Geosynclines and Tectonic Elements
(Modified Version of Table 1 of Aubouin, 1965)

Stille 1935–40	Kay 1951	Krumbein and Sloss 1963; Badgley, 1965	Sinityzn and Peyve 1950	Aubouin 1965
Orthogeosynclines eugeosyncline miogeosyncline	*Orthogeosynclines* eugeosyncline miogeosyncline	Orthogeosyncline Miogeosynclinal transitional zone	*Primary geosyncline*	*Geosynclines* eu-furrows mio-furrows eu-ridges mio-ridges
				Back-deep Intra-deep
	Epieugeosyncline	Postorogenic basins	*Secondary geosynclines*	
Parageosynclines	*Intracratonal geosynclines* exogeosyncline zeugeosyncline autogeosyncline	*Intracratonic basins* marginal basin yoked basin interior basin	*Residual geosynclines*	Foredeep Intracratonic furrows Basins
	Taphrogeosynclines Paraliageosynclines	Rift valley Coastal geosyncline		Trenches
Hochkraton Tiefkraton	Craton	Craton stable shelf unstable shelf	Platform	

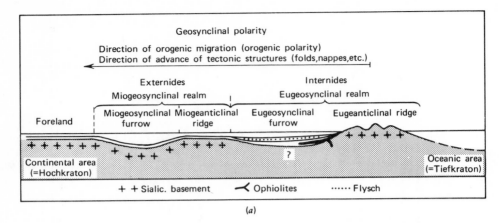

(a)

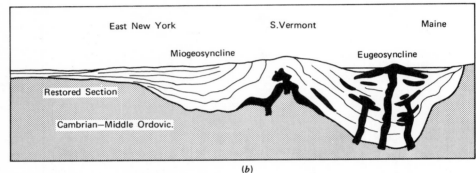

(b)

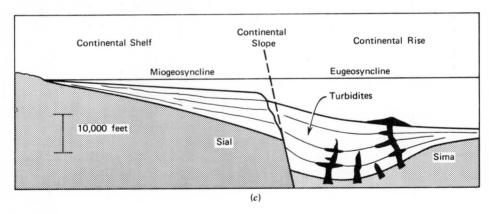

(c)

Figure 9-1. Reconstruction and interpretations of eugeosynclines and miogeosynclines. (*a*) Elementary couple (after Aubouin, 1965) showing conditions at the orogenic (terminal) stage of the geosynclinal period, when the emergent eugeanticlinal ridge was supplying detrital flysch material to the eugeosynclinal furrow (Figure 9-4). (Reproduced by permission, Elsevier Publishing Co.) (*b*) Restored sections of Cambrian and Ordovician in geosynclines from New York to Maine (after Kay, 1951, with permission of The Geol. Soc. Amer.). (*c*) Reconstruction of the section depicted in *b* at the end of Trenton time (after Dietz, 1963). This is analogous to the presumed existing situation off the eastern United States (Figure 9-6). (With permission from Journal of Geology, Copyright 1963 by the University of Chicago.)

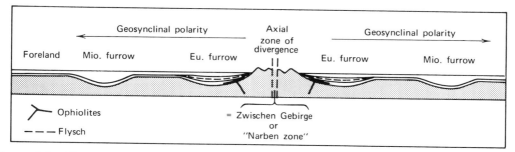

Figure 9-2. Characteristics of a divergent bicouple giving rise to a mountain chain of bilateral symmetry (after Aubouin, 1965). The two couples are depicted as they would appear during the orogenic (terminal) stage of the geosynclinal period (Figure 9-4). It is arbitrarily assumed for convenience that both couples have reached the same stage at the same time. (With permission of Elsevier Publishing Co.)

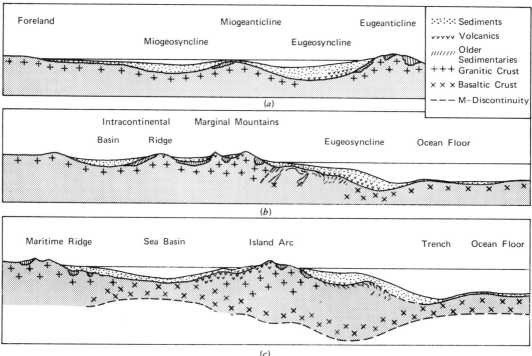

Figure 9-3. Tectonic frameworks (after Matsumoto, 1967). (*a*) Classical examples in Europe adapted from Aubouin (Figure 9-1*a*). (*b*) Based on Mesozoic examples in East Asia. (*c*) Based on Cenozoic examples on the west side of the Pacific. (From Tectonophysics, with permission.)

structural deformation. Orogenesis (undulation) was very limited in space and time producing intensely folded structures during brief episodes of deformation. In addition he prepared a detailed scheme for the correlation of igneous activity and metamorphism with the stage of evolution of the geosyncline.

The mountain chains rising from orthogeosynclines characterized by tight folds and nappes he termed alpinotype ranges, and other mountains arising from block faulting he termed germanotype ranges. Stille related germanotype mountains to a new class of geosyncline that was not subjected to orogenesis—parageosynclines. This was an extension and departure from the original concept of geosynclines, and he stated that parageosynclines were "second order" features and not "true" geosynclines.

Stille's classification was introduced into America by M. Kay, the first to work out the detailed development of systems consisting of a eugeosyncline and a miogeosyncline. His restored section of the Champlain and Magog belts from New York to New Hampshire is reproduced in Figure 9-1*b* from his 1951 memoir on North American Geosynclines. This memoir also contains Kay's subdivisions of Stille's parageosynclines into several groups classified on the basis of either their position in space relative to the craton or their distribution in time. Table 9-1 shows a comparison of these types with the names used for equivalent tectonic elements by European geologists. Kay's intracratonal geosynclines include exogeosynclines near the edge of the craton (fore-deep), autogeosynclines on the craton without associated highlands (classic basins), and zeugeosynclines which contain sediment from eroded highlands within the craton (introcratonic furrows). Kay also defined three late-cycle geosynclines formed in deformed eugeosynclines (back-deeps and intra-deeps), taphrogeosynclines or sediment-filled rift valleys, and paraliageosynclines along present continental margins (north coast of the Gulf of Mexico). Kay noted that the propriety of considering parageosynclines as geosynclines is debatable, and Table 9-1 shows that other American geologists Krumbein and Sloss and Badgeley prefer not to apply the term geosyncline to these tectonic elements.

According to the original concepts developed by Hall, Dana, and Haug the term geosyncline is restricted to a major paleogeographic feature. It is unfortunate, but not important, that the part of the Appalachian geosyncline originally studied by Hall has since proved to be a secondary feature. Aubouin classified as geosynclines only the major mobile belts, the orthogeosynclines and primary geosynclines of other authors. A geosyncline is a complex system of furrows and ridges.

Metamorphism and Igneous Activity

The development of metamorphic and igneous rocks in geosynclines was originally considered to be incidental and resulting simply from the downwarping of the crust and deep burial of sediments with consequent increase in pressure and temperature. It is now known that regional metamorphism and the associated granitic intrusions are bound up with the orogenic deformation of the geosynclinal sediments and not with the early downwarping of the geosyncline. In many examples the metamorphic zones cut across the tectonic units and are thus later than the major tectonic movements. The late-tectonic regional metamorphism is developed especially in the internal or eugeosynclinal belts and rarely in the external or miogeosynclinal belts. It was only in 1961 that A. Miyashiro drew attention to the fact that many geosynclinal zones, especially in the circum-Pacific region, contain paired metamorphic belts; one belt characterized by high pressure-low temperature metamorphism, often including glaucophane schists, and by abundant basic and ultrabasic rocks; and the other belt characterized by relatively low pressure-high temperature metamorphism with abundant syntectonic and late-tectonic and granitic intrusions. Feasible explanations for the metamorphic belts are provided by the plate tectonics model (Chapter 14).

It is obvious from the preceding discussions that the evolution of a geosyncline is marked at successive stages by specific types of igneous activity. The synthesis and classification proposed by Stille appear to be typical of the eugeosynclinal zones (internides). He recognized three main events which are correlated with Aubouin's scheme for geosynclinal evolution in Figure 9-4. The initial pretectonic magmatic episode produced basic submarine volcanic eruptions yielding the ophiolite suite and spilites; subsequent uplift and erosion may expose contemporaneous but deep-seated ultramafic rocks of the root-zones (Table 6-4). A synorogenic magmatic episode produces syntectonic "concordant" granitic intrusions associated with the regional metamorphism, and late-tectonic "discordant" granitic intrusions after the principal orogenic phase. The postmagmatic episode is subdivided into three parts: eruption of andesitic lavas, intrusions of granites and granodiorites, and finally eruption of basaltic lavas.

EVOLUTION OF GEOSYNCLINES

The evolution of a geosyncline and the formation of a mountain chain occupies a long period in geological history, and it is far more complex than indicated by the simple cycle outlined at the beginning of this chapter.

Geosynclinal Couples

Aubouin's book presenting a classical treatment of geosynclines was published in 1965. He reviewed the geology of the Mediterranean mountain chains of the Alpine cycle with specific reference to the Hellenides (including much of Albania, Yugoslavian Macedonia, and Greece) as a basis for comparison, and he concluded that they showed a characteristic pattern of organization and evolution. The fundamental unit is the eugeosynclinal-miogeosynclinal couple which is illustrated schematically in Figure 9-1*a*. According to Aubouin, the eugeosynclinal domain (internides) comprises a eugeosynclinal furrow and a eugeanticlinal ridge bordering the ocean. Between the internides and the continental craton (foreland) is the miogeosynclinal domain (externides) comprising a miogeosynclinal furrow and a miogeanticlinal ridge. The two couples are illustrated as if they had reached the same stage of development, but this is not always so: they can reach different stages at the same time. A similar bicouple is illustrated in Figure 9-1*b* by Kay's section of the Champlain and Magog belts in New York and New Hampshire.

These fundamental units may be paired, according to Aubouin, forming a divergent bicouple as illustrated schematically in Figure 9-2. This type of geosyncline produces a mountain chain with bilateral symmetry. The two eugeanticlinal ridges may have an intermediate hinterland between them, they may coalesce to form a single ridge, or they may be absent altogether so that the two eugeosynclinal furrows become one. Aubouin also illustrated a complex system of four eu-miogeosynclinal couples forming two divergent bicouples and one convergent bicouple; the axial zone of convergence corresponding to two miogeanticlinal ridges is situated between two divergent bicouples formed simply by repeating Figure 9-2.

Kay demonstrated that the geological development of North America was dominated by a series of eugeosynclinal-miogeosynclinal couples similar to those depicted in Figures 9-1*a* and 9-1*b*. He stated that no examples of these types of geosynclines are being formed today. R. S. Dietz, however, suggested that the Atlantic and Gulf Coast sedimentary prisms may be modern examples —an actualistic interpretation of the normal geosynclinal couple. Figure 9-1*c* is Dietz's suggested reconstruction of the same geosynclinal couple depicted by Kay in Figure 9-1*b*, drawn by analogy with the existing situation off the eastern United States. The eugeosyncline is considered to be the continental rise. This is a prism of sediments deposited on oceanic crust at the base of the continental slope largely by turbidity currents laying down turbidites (greywackes). The sediments lap onto the sialic continental slope. Plutonic and volcanic rocks invade the eugeosynclinal sediments. The continental shelf undergoes

TABLE 9-2 Sedimentary and Igneous Rocks Characteristic of Geosynclines
(Modified After Mitchell and Reading, 1969)

Atlantic Type		Andean Type		Island Arc Type		Japan Sea Type
Miogeosyncline	Eugeosyncline	Mountains	Trench	Islands	Trench	Margin of Restricted Basin
Continental Crust	Oceanic Crust	Continental Crust	Oceanic Crust	Intermediate Crust	Oceanic Crust	Intermediate, Modified Crust
Abundant A and B.	Common C. Rare D. Abundant E.	Rare A and B. Rare to abundant F. Abundant H. Common I.	Abundant C. Rare to common E. Common to rare G.	Locally abundant B. Rare C. Abundant F and G. Common I.	Abundant C. Common G.	Abundant A and E. Locally common B. C present if basin floor oceanic. Tuffs of F. Rare G.

Characteristic rock types:

A. Shallow marine and coastal plain clastic sediments
B. Carbonate sediments
C. Interbedded pelagic sediments, tholeiitic lavas, and ultrabasic rocks
D. Tholeiitic volcanic turbidites
E. Compositionally mature turbidites

F. Calc-alkaline volcanic rocks and minor intrusions
G. Calc-alkaline volcanic turbidites
H. Continent-derived coarse clastic sediments
I. Intermediate or acidic plutonic rocks

isostatic subsidence, and this permits a wedge of epicontinental deposits to build up on the marginal flexure of the continental shelf. These shallow-water sediments make up the miogeosyncline. Comparing Figure 9-1c with 9-1b, we see that there are three elements deleted from the classical couple to transform it into Dietz's actualistic couple. These are the outer half of the miogeosyncline, the island arc (eugeanticlinal ridge), and the tectonic borderland (miogeanticlinal ridge). Also a continental slope is added between the geosynclinal prisms. Dietz suggested that the onset of orogeny occurred when the oceanic sea floor became uncoupled from the continental crust and moved down beneath the continent impelled by sea-floor spreading (Chapters 12 and 14).

Two other types of tectonic profiles for geosynclinal systems have been constructed by T. Matsumoto following a review of the circum-Pacific orogenic system. Figure 9-3a is adapted from Aubouin's cross-section (Figure 9-1a). Figure 9-3b shows the reconstruction of a eugeosyncline at the continental margin of east Asia during the Mesozoic period, and Figure 9-3c is a profile showing sites of sediment accumulation in the region of island arcs and ocean trenches in the western Pacific during the Cenozoic period. Some of these such as the Marianas arc appear to be formed on the ocean floor far from a continent, whereas others, such as the Japanese arc, are superimposed on older, sialic basement structures. The rock assemblages characteristic of the geosynclinal environments illustrated in Figures 9-1 and 9-3 are listed in Table 9-2. Matsumoto concluded that the tectonic framework of the circum-Pacific mobile belt may have changed with time from the classical type of Figure 9-3a during the Paleozoic, through the Andean type in Figure 9-3b during the Mesozoic, to the island arc type in Figure 9-3c during the Cenozoic. Thus the complexity of the orogenic cycle becomes even more apparent when we add the dimension of time to the geosynclinal organization.

Geosynclinal Evolution According to J. Aubouin

Figure 9-4 summarizes Aubouin's scheme for the evolution of a eumiogeosynclinal bicouple corresponding to that of Figures 9-1a and 9-3a. This is based on his reconstruction of the geological history of the Hellenides, but he claims that it depicts a general pattern. J. Debelmas, M. Lemoine and M. Mattauer disputed this claim in 1967. They questioned the fitness of Aubouin's model to explain the Hellenides and concluded that it could only provide a very schematic picture of the tectonic style of Mediterranean Alpine ranges. They praised Aubouin's book as a source of valuable information but felt that the tectonic inventory was insufficient.

Figure 9-4a lists three major periods in the evolution of a geosyncline. The geosynclinal period (I), which is of much longer duration than the late-geosynclinal (II) and post-geosynclinal periods (III), is subdivided into the generative (A), the development (B), and the orogenic or terminal stages (C). The development stage is further subdivided into pre-flysch (i) and flysch periods (ii). Figure 9-4b shows the main events occurring at any time in each part of the geosyncline, and Figure 9-4c illustrates these events in six schematic geological sections corresponding to specific times in 9-4b. Note the time scale in Figures 9-4a and 9-4b.

Figure 9-4 illustrates geosynclinal polarity and in particular orogenic polarity. Different regions of the geosyncline experience similar events but at different times. The orogenic stage of the geosynclinal period (IC) begins first in the internal eugeanticlinal zone of the geosyncline and then migrates outward toward the miogeosynclinal zones. The time of transition from development stage (IB) to orogenic stage (IC) and from the geosynclinal period (I) to the late-geosynclinal period (II) thus depends on geographic position within the geosynclinal system.

For a period of about 120 m.y. after the initial subsidence the emission of ophiolites

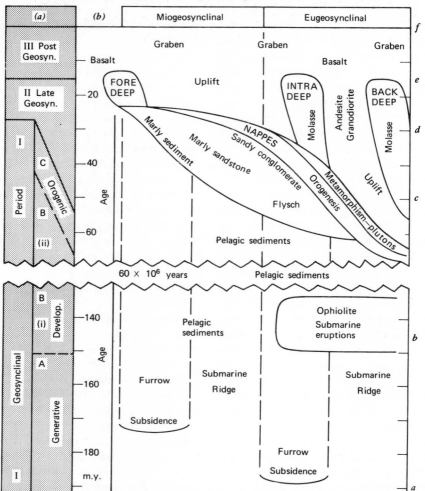

Figure 9-4. (a) Evolutionary pattern of a geosyncline, according to Aubouin (1965). Periods:— I. Geosynclinal, II. Late-geosynclinal, III. Post-geosynclinal. Stages: IA, Generative. IB, Development. IC, Orogenic (terminal). The development stage is divisible:—IB(i) pre-flysch period. IB(ii) flysch period. (b) Diagrammatic representation of geosynclinal polarity in a bi-couple (Figure 9-1a), according to Aubouin (1965). The orogenic stage (IC) begins at the eugeanticlinal ridge and migrates across the system from the interior to the exterior, preceded by the flysch deposits. This is illustrated in Figure c by generalized geological cross sections for specific times, a, b, to f. (c) Schematic geological cross sections showing the paleogeographical evolution and tecto-orogenic development of a geosynclinal couple as illustrated by the Alpine cycle, based on the Hellenides (after Aubouin, 1965). Sections a to e, geosynclinal and late-geosynclinal periods. Section f, post-geosynclinal period. (With permission of Elsevier Publishing Co.)

is the only activity in the geosyncline apart from the slow accumulation of deep-water, pelagic sediments both in the furrows and on the submarine ridges. Then the orogenic or terminal stage of the geosynclinal period (IC) begins with orogenesis and uplift of the

eugeanticlinal ridge. This is followed by rapid erosion, and terrigenous sediments are poured into the eugeosynclinal furrow from the rising highlands. These are poorly sorted sediments of greywacke type collectively termed flysch in the Alpine chains of the

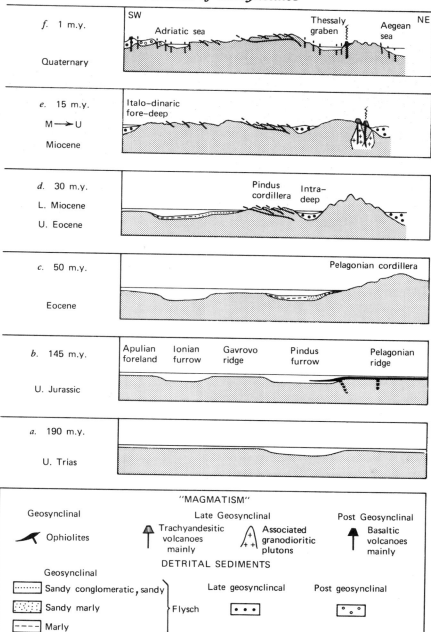

(c)

Mediterranean region. The deposition of greywackes marks the beginning of the flysch period of the development stage (IBii) which comes before the orogenic stage (IC); but Figure 9-4*b* indicates that in fact the flysch period does not begin until after the orogenic stage has commenced in the inner zones. With continued uplift of the eugeanticlinal ridge and the formation of a cordillera the flysch deposits pile up against the flank of the ridge and migrate across the furrow. As the flysch migrates, with the deposition of successively younger material from the interior to the exterior zones, the zone of orogenesis also migrates. Regional metamorphism and plutonism accompany and follow the orogenesis. Subsequent uplift is followed by erosion and the folded flysch itself becomes the source of new flysch that is deposited in the miogeosynclinal regions. Cross-section *d* shows that about 35 m.y. after the initial onset of orogenesis the sediments and ophiolites of the eugeosynclinal furrow have been thrust in great nappes over the miogeanticlinal ridge.

The late-geosynclinal period (II) follows immediately after the orogenic stage in any zone. Alongside the rising cordillera are formed narrow troughs, back-deeps, and intra-deeps (Table 9-1) which subside to receive unconformable detrital sediments derived from the young mountain ranges. These sediments are termed molasse in the Alpine chains of the Mediterranean region and clastic wedges elsewhere.

By about 50 m.y. after the beginning of orogenesis the complete width of the geosynclinal system is in the late-geosynclinal period of evolution (II) as shown by the cross-section *e* in Figure 9-4c. An additional trough, the foredeep, has developed between the geosynclinal system and the craton, receiving molasse deposits from the highlands rising from the geosyncline. This period is marked by the eruption of andesites in the tectonized internal zones accompanied by some basalts and rhyolites and by the intrusion of granodioritic batholiths probably associated with the volcanic activity.

The post-geosynclinal period (III) is characterized by vertical movements which often cut across the geosynclinal trends. They produce regional arching and tensional rift valleys or graben, often associated with basaltic volcanism. Simultaneously basaltic lavas are erupted on the foreland regions of the craton well beyond the geosynclinal belt.

Aubouin's book was published just before plate tectonics introduced many new concepts related to the significance and evolution of geosynclines. It would be an instructive exercise for students to attempt an interpretation of Figure 9-4 in terms of plate tectonics after reading Chapter 14. Can geosynclinal and orogenic polarity be interpreted in terms of plate collisions and subduction zones?

"Pacific" geosynclines of K. A. W. Crook

In 1969 K. A. W. Crook noted systematic differences between the geosynclines described by Aubouin which he called "Atlantic" geosynclines, and "Pacific" geosynclines which occur in the circum-Pacific region. According to Matsumoto (Figure 9-3) the tectonic framework of the circum-Pacific mobile belt may have changed with time, but Crook referred to something different. He used as examples the Paleozoic Tasman geosyncline of eastern Australia and the Mesozoic to Recent New Guinea geosyncline. He gave as characteristics of Pacific geosynclines:

1. They comprise a number of subparallel volcanic and nonvolcanic troughs and highs with the nonvolcanic elements near the craton.
2. The volcanic troughs contain volcanic-terrigenous flysch-like sediments; the volcanic troughs and highs may lack serpentinites; ophiolites have not been recognized.
3. Sediment in the nonvolcanic troughs is predominantly terrigenous flysch of sialic derivation.
4. Deformation structures are predominantly vertical and terminal tectonism proceeds outward from the craton.
5. Successive pairs of troughs do not display polarity.
6. They do not occupy the sites of older geosynclines and probably develop on a largely simatic floor.
7. They lie between sialic cratons and simatic ocean floors.

Crook concluded that although Atlantic and Pacific geosynclines are superficially similar there are differences between them which are significant and which appear to reflect their different geotectonic environments. Whereas Atlantic geosynclines may develop on the sites of older geosynclines, discordant to their trends, Pacific geosynclines appear to be newly formed on a largely simatic floor, marginal to sialic cratons. The Pacific geosynclines do not exhibit the sedimentary and igneous polarity characteristic of Aubouin's model (Figure 9-4c). The difference may be readily explicable in terms of plate tectonic theory if an Atlantic geosyncline is considered as two Pacific geosynclines which have collided. Figure 14-17 illustrates collision sequences.

EVOLUTION OF CONTINENTS

The hypothesis that continents have grown by a process of marginal accretion was introduced in Chapter 7. The essence of this hypothesis is contained in the concept of geosynclines and the orogenic cycle with the geosynclines developing marginally to continental cratons; the idea is inherent in the early writings of Hall and Dana. New volcanic material is added to the crust from the mantle, dispersed along with preexisting rocks by sedimentary processes, and the whole complex of rocks is then welded onto the continental plate by metamorphism and mountain building. There are many examples, however, where orogenic belts intersect each other rather than developing in successive, concentric arcs, and one school of geologists maintains that geosynclines form along zones of weakness within continental cratons. Another view is that the continents are not growing at all, but that continental masses are being destroyed by a process of oceanization; continental crust founders and is engulfed in basaltic magma (Chapter 15). In Chapter 14 we examine a fourth view that the oceans open and close with orogenesis and mountain building resulting from collisions of lithospheric plates. Although the size of a continental plate may be increased by marginal accretion resulting from collision at one stage, at another stage a continent may be separated into two plates by rifting and drifting.

One question relevant to all hypotheses of continental evolution is to what extent the continents are increasing in total volume by the addition of new material from the mantle. As noted in Chapter 7 the problem is complicated by the fact that during orogenesis a part of the existing crust is reworked and incorporated into the belts of younger material, and it is difficult to distinguish the regenerated older material from the younger material. We know that magmas from the mantle are added to the crust during the orogenic cycle (Figures 4-1 and 9-4), but we do not know how much of this material is derived from crustal rocks recycled through the mantle (Chapter 14). Strontium and lead isotope evolution trends should eventually solve these problems.

Detailed field mapping, structural studies, and petrological studies have much to contribute, and they provide the essential framework for the interpretation of isotope data. H. R. Wynne-Edwards in 1969 published Figure 9-5 which shows the geological evolution of an area comprising 13,000 square miles of the Grenville province of southwestern Quebec. This is based on systematic mapping and the recognition of structural and textural criteria for subdividing the plutonic and metamorphic rocks. Figure 9-5 merits careful study. The map legend at the right of the diagram representing the rocks as they are now is constructed from a series of steps, like building blocks, added to it at different times. Additions of material represented by the rise in steps up the "lithology" axis are shown for successive orogenies; the Kenoran about 2500 m.y. ago; the Hudsonian about 1750 m.y. ago; the Elsonian about 1400 m.y. ago; and the Grenville about 950 m.y. ago.

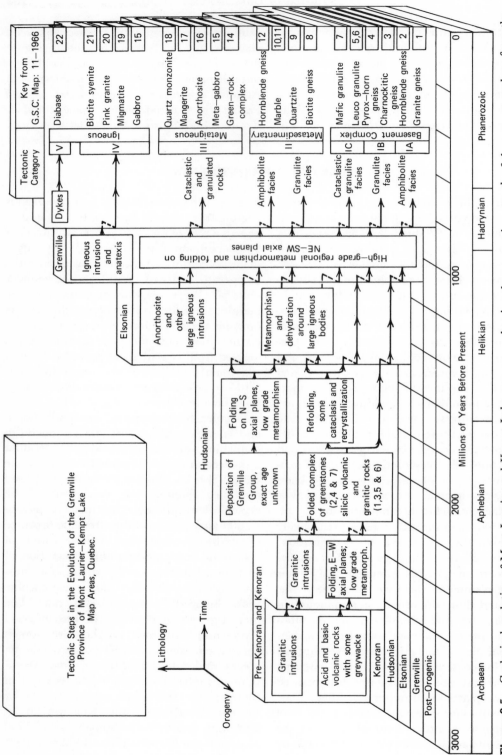

Figure 9-5. Geological evolution of Mont Laurier and Kempt Lake map areas showing the progressive growth of the present geology from the Archaean by a series of tectonic steps in which older rocks were reprocessed and new ones were added (after Wynne-Edwards, 1969, Figure 12, p. 80. With permission of the Geological Association of Canada.)

During each orogeny the older rocks were modified by folding, metamorphism, and sometimes melting, and generally updated by the redistribution of radiogenic products. The recycling process is represented by the steps along the "orogeny" axis. The arrows show the evolutionary paths followed by each block from the time it was added as new material. Rocks in the Grenville province thus show continued tectonic overprinting extending back 2500 m.y. to the Archean.

The Grenville tectonic cycle has been completed from initial deposition to erosion close to base level. The complete cycle includes about 1000 m.y. of sediment deposition and burial with active orogenesis and uplift occupying only 200 to 300 m.y. Most of the rocks of the Grenville province are much older than the Grenville orogeny and therefore they cannot be regarded as new contributions to the continental crust about 950 m.y. ago (Figures 7-13 and 7-14).

CONTEMPORARY GEOSYNCLINES

Criteria for the recognition of contemporary geosynclines include present physiography, whatever geological history can be deciphered from available evidence, and projections into the future. Geosynclines must be of appropriate dimensions and they should exhibit features such as those outlined in Figures 9-1 to 9-4. Mountain systems have arcuate and linear sections, lengths of thousands of kilometers, and widths of a few hundreds of kilometers. Deep troughs are characteristic of the generative and much of the development stage of the geosynclinal period, and thick piles of sediment or severe folding are characteristic of later periods. Our attention is therefore directed to the continental margins: the system of deep ocean trenches and the continental rises (Figure 2-2).

We know that sedimentary and volcanic deposits of geosynclinal dimensions are accumulating on and oceanward of modern continental margins, around island arcs, and in some small ocean basins (Figure 7-8). The characteristics of continental margins were outlined in Chapters 2 and 7. Figure 9-6 shows the distribution of three types according to A. H. Mitchell and H. G. Reading: Atlantic type margins lacking an ocean trench: Andean type margins with a mountain belt bordered by a submarine trench; and Island-arc type margins sepa-

rated by a small ocean basin from the island arc and associated ocean trench.

In their review of continental margins, geosynclines, and sea-floor spreading Mitchell and Reading distinguished five possible types of contemporary geosynclines which are related to the three types of continental margins. An Atlantic type geosyncline is shown in Figure 9-1c, an Andean in Figure 9-3b, and an Island-arc located on and around active island arcs as shown in Figure 9-3c. The small sediment-filled ocean basins on the concave side of many island arcs are the Japan Sea type of geosyncline (Figure 9-3c), and geosynclines of Mediterranean type occur in small oceans between continents. This subdivision of the sites of contemporary sediment accumulation is not a classification of geosynclines equivalent to those in Table 9-1. It is rather an actualistic interpretation of the environments where the classified geosynclines (Table 9-1) may be forming today. Figure 9-6 shows their global distribution. There is no reason to suppose that geosynclinal sequences which appear very similar in the geological record should all have formed in the same kind of environment. On the other hand detailed geological study of geosynclinal sequences should permit paleogeographic reconstruction of the former environment in terms of Atlantic, Andean, or Island-arc types, as previously mentioned

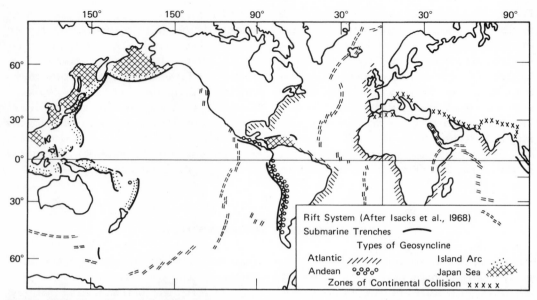

Figure 9-6. Positions of modern geosynclines in relation to world system of rifts and submarine trenches. See also Figures 2-2 and 2-3 (after Mitchell and Reading, 1969, from Journal of Geology with permission. Copyright 1969 by The University of Chicago).

for the circum-Pacific orogenic zone in connection with Figure 9-3. The associations and sequences of sediments and igneous rocks characteristic of each of these environments are summarized in Table 9-2.

Mountain building connected with these geosynclines may be one of three types; Andean, Island arc, or Himalayan. Mitchell and Reading related these to plate tectonic models. An Atlantic type geosyncline may change to a geosyncline of Andean or Island-arc type with the onset of orogeny. Andean and Island-arc type orogenies occur within and are associated with the development of their respective geosynclines. Himalayan type orogeny develops only if the continents drift (Chapter 14), and it then results from the collision of continents; it may affect any type of geosyncline because any one could be trapped between the moving continents.

Finally let us compare some aspects of the geology of a specific modern geosyncline with the evolutionary scheme of Aubouin shown in Figure 9-4. The impressive arcuate

sweep of the Sunda Islands and associated ocean trenches (Figure 9-7) has been linked with the concept of geosynclines since the beginning of this century. The islands are arranged in a double arc. The inner volcanic arc includes the large islands of Sumatra, Java, Bali, Lombok, Sumbawa, and Flores. This arc constitutes a cordillera with Mesozoic sediments partially covering an ancient crystalline basement. Unconformable late-Tertiary sediments also are present, and igneous activity has produced andesite lavas and granodiorite batholiths. The outer, nonvolcanic arc consists of smaller, scattered islands which are merely the points of emergence of a linear shelf. These include the Mentawi Islands, Timor, and Tanimbar Islands. Basement rocks are not exposed and the islands are composed of intensely folded sediments often of deep-water facies. In Timor an ophiolite nappe transported from the north occurs above the recently folded sediments. There is a fairly deep marine trough between these two arcs and the deep Indonesian trench lies outside the outer arc.

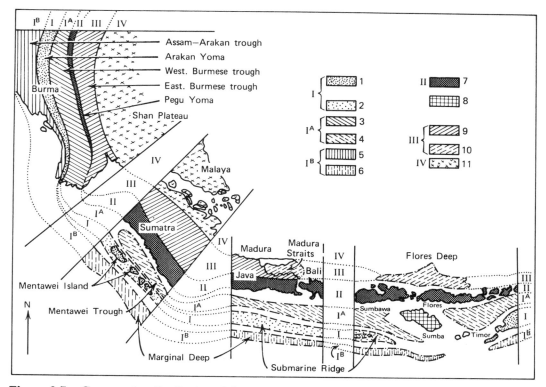

Figure 9-7. Comparative distribution of the structural zones of Indonesia and Burma, demonstrating the transition of structures of the Sunda Islands to the folded chains of Burma, which are quite clearly in the post-geosynclinal stage (Figure 9-4*a*) (after Aubouin, 1965, from Umbgrove, 1948). 1 = zone of strong Laramide folding; Miocene epochs of compression, weak in Burma, strong in other sectors; zone of strong negative anomalies of gravity. 2 = Submarine continuation of the same zone. 3 = Inner trough, accompanying zone I, filled up with Tertiary sediments in Burma sector. 4 = Submarine inner trough. 5 = Outer trough, accompanying zone I, filled up with Tertiary sediments in Burma sector. 6 = Submarine outer trough, marginal deep. 7 = geanticlines, Tertiary up to Recent volcanism; Miocene diastrophism moderate. 8 = Sumba Island; Upper Miocene folding; Tertiary volcanism; no recent volcanoes; interruption of zones I, I^A, I^B, and of zone of strong negative anomalies of gravity. 9 = Tertiary sedimentation troughs; moderate folding towards the end of the Pliocene. 10 = Submarine continuation of the same troughs. 11 = Regions above sea-level in Tertiary and Pleistocene times. (Reproduced by permission of Elsevier Publishing Co.)

The inner arc shows a positive isostatic gravity anomaly whereas the outer arc and the marine trenches are associated with an extreme negative anomaly.

A cross-section through this island arc shows similarities with the cross-sections *d* and *e* in Figure 9-4*c*. Aubouin stated that in passing from the inner arc to the outer trench, four features can be correlated directly with the schematic sections:

1. The inner arc with its late-Tertiary back-deep and andesitic and plutonic igneous activity corresponds to the cordillera of the eugeanticlinal ridge.

2. The marine trench between the island arcs corresponds to the intra-deep.

3. The outer arc represents another cordillera with material from the eugeosynclinal furrow being overthrust onto the miogeanticlinal ridge.

4. The Indonesian trench corresponds to the external miogeosynclinal furrow.

Aubouin stated that apart from the fact that there is no continental craton beyond the outer trench the present geology of the Sunda Islands corresponds very closely to the situation existing about 30 m.y. ago in Figure 9-4. The absence of continental craton adjacent to the miogeosynclinal furrow is an unusual arrangement according to other interpretations of geosynclinal couples (Figure 9-1). The inner volcanic arc and adjacent trench are in the late-geosynclinal period (II), the outer arc is in the orogenic stage of the geosynclinal period (IC), and the external trench appears still to be in the early flysch period of the development stage (IBii). The scale and curvature of the Mediterranean-Alpine system are very similar to those of the Sunda Islands.

The structural zones of the Sunda Islands can be extended northward into those of the folded mountain chains of Burma as illustrated in Figure 9-7. This suggests that these zones form part of a single geosyncline. According to Aubouin the alpine-type mountains of Burma are in the post-geosynclinal period (III) further advanced in evolution than the Sunda Islands. According to plate tectonic models the folded Alpine system resulted from collision producing a double system on the Asian continent. The Sunda Islands then represent only the northern side of this system with open ocean to the south and southwest.

This brief review of a contemporary geosyncline indicates that at any given point in time the stages of evolution of a geosyncline not only vary in cross-section but also along its length. Similar conclusions were reached by R. J. Roberts (1969) in connection with the eugeosynclinal zones of the western Cordillera of the United States during the Paleozoic and Mesozoic; there is evidence for repeated orogeny and epeirogeny in different parts of the system for more than 550 m.y. with considerable age variation for the activity both along and across the system. A geosyncline is thus a very complex assemblage of furrows and ridges subjected to similar events at different times in different places.

CAUSES OF SUBSIDENCE AND UPLIFT

The orogenic cycle incorporates most of the geological processes, and these are governed to a large extent by the occurrence and the rate of subsidence and uplift of segments of the Earth's crust. What is less certain is the cause of these major movements although it seems clear that they are second-order effects of major, global processes. The magnitude of the effects, in terms of depth or pressure variations and changes in temperature distribution, can be estimated from study of metamorphic rocks in conjunction with laboratory experiments on metamorphic reactions at known pressures and temperatures.

Subsidence of Geosynclines

Figure 9-8 summarizes in diagrammatic form some of the mechanisms that have been proposed to account for the formation of a geosynclinal trough. It has often been claimed that the development of a geosyncline is a self-generating process as shown in Figure 9-8a. The weight of accumulated sediments is presumed to load the crust sufficiently to produce continued downsinking. Removal of magma from the depths and its eruption at the surface could also contribute to subsidence (Figure 9-8b). The operation of subcrustal currents producing downbuckling of the trough and the formation of a tectogene as depicted in Figure 9-8c is the process that has been invoked most frequently during the last 40 or 50 years. Figure 9-8d and Chapter 14 show a layer of the crust or lithosphere being carried down

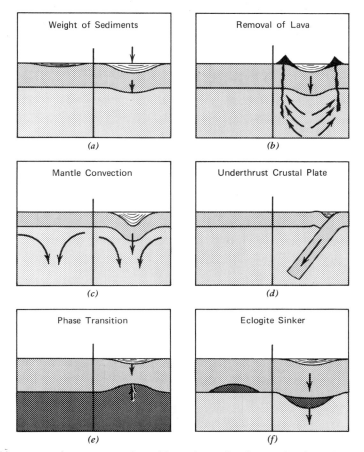

Figure 9-8. Diagrammatic representation of hypotheses for the mechanism of producing subsidence for a geosynclinal trough.

into the mantle as a result of sea-floor spreading (Chapter 12) producing a trough at the surface. According to the phase transition hypothesis cooling at depth causes the conversion of gabbro into eclogite (Chapter 5) with contraction of the deep rocks and the resultant subsidence at the surface. This is illustrated in Figures 9-8e and 12-25 and discussed in more detail in Chapter 10. Figure 9-8f shows a crustal section of gabbro overlying peridotite with a layer of eclogite forming from the gabbro. The denser eclogite would tend to sink dragging down a section of the crust and forming a depression at the surface.

According to Figures 9-8c and d, the vertical movements associated with mountain building are a consequence of horizontal movements occurring beneath the crust, whereas according to Figures 9-8e and f the vertical movements develop independently of lateral movements and the controlling forces are related to changes in heat distribution and local mass readjustments in response to gravity. These represent two viewpoints that are strenuously maintained by different groups of geologists and geophysicists. The intense folding in orogenic belts is developed either as a direct result of compression arising from lateral movement of crustal slabs or as an indirect result of vertical movements with consequent instability and gravity tectonics; compression is then a secondary effect of the vertical movements.

Metamorphic Rocks as Guides to Geosynclinal Conditions

Figure 4-1 shows that subsidence and burial of sediments in the orogenic cycle causes metamorphism, and Figure 4-4 shows the general range of depths (pressures) and temperatures for metamorphic rocks. Figure 9-9 is a more detailed version of the crustal part of Figure 4-4 showing the distribution of the main metamorphic facies in terms of depth and temperature.

The relative positions of the metamorphic facies have been known ever since P. Eskola introduced the concept in 1915. N. L. Bowen described a petrogenetic grid in 1940 suggesting that laboratory measurements at high pressures and temperatures would permit calibration of the metamorphic reactions that form the boundaries of metamorphic facies. Apparatus became available for such experiments in 1949 (Chapter 8) and in 20 years sufficient accurate experimental data has been gathered to provide diagrams such as Figure 9-9. According to F. J. Turner, who presented an experimental appraisal of critical metamorphic reactions in his 1968 book, Bowen's abstract concept has now become an instrument increasingly capable of effective use in calibrating temperature and pressure gradients in metamorphic terranes.

Figure 9-9 shows Turner's preferred estimates of the positions of selected metamorphic facies in terms of depth and temperature based on the measured positions of mineralogical reactions inferred to have occurred in the formation of various metamorphic rocks. Figures 4-4 and 7-11 show estimated continental geotherms for comparison with Figure 9-9.

If sediments are buried in a geosyncline they will be subjected to increases in pressure and temperature as they follow a geotherm. They become metamorphosed according to the position of the geotherm and the changes in temperature distribution caused by the subsidence of cool sedimentary rocks or by the uprise of magma and they subsequently become uplifted and exposed by erosion. The sequence of metamorphosed rocks in the roots of each mountain chain is unique,

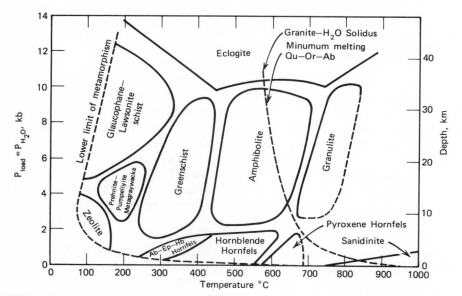

Figure 9-9. Tentative scheme of metamorphic facies in relation to total pressure (assumed equal to P_{H_2O}) and temperature; all boundaries are gradational (after Turner, 1968). See Figure 4-4. (From "Metamorphic Petrology". Copyright McGraw-Hill Book Company, 1968. Used with permission.)

and the mineralogy of the rocks contains the imprint of the range of pressures and temperatures occurring in the sequence during its metamorphism. From study of the mineralogy the positions of the rocks on Figure 9-9 can be located. In some terranes the rock sequence is from greenschist, through amphibolite, to granulite facies, and in others blueschists (glaucophane schists) and eclogites are formed.

The development of granulite facies rocks requires temperatures considerably higher than those indicated by the continental shield geotherm (Figure 3-12) suggesting that regional metamorphism is accompanied by higher than normal heat flow from the mantle. On the other hand the formation of blueschists requires very high pressures or depths of burial with very low temperatures. Any scheme of global tectonics must provide an explanation for the extraordinarily wide range of temperatures to which metamorphic rocks have been subjected at equivalent depths in different geosynclines.

Theories of Global Tectonics

There have been five main global theories: (a) the Earth is rigid and contracting, (b) the Earth is rigid and expanding, (c) the Earth is mobile with the continents drifting over the mantle, (d) the Earth is mobile with the ocean floors spreading apart probably because of convection in the mantle, and (e) the Earth is mobile with rigid plates moving over the asthenosphere. Theories (c), (d) and (e) are not independent of each other (Chapter 14).

In the nineteenth century the contraction theory was accepted as the cause of folding, thrusting, and mountain building. In its classical form the contraction theory holds that the Earth is contracting because it is cooling, and the outer relatively cool and rigid zone is therefore compressed. Lord Kelvin's celebrated model of an Earth cooling from a molten state had to be modified with the discovery of radioactivity and the study of the distribution of radioactive elements within the Earth, but the contraction theory persists and most textbooks have been written with this concept as their foundation. Applications to global and regional tectonics were reviewed in the 1959 book by J. A. Jacobs, R. D. Russell, and J. T. Wilson. Evidence of worldwide rifts in the ocean basins indicative of tension has made this theory less attractive.

One of the pioneers of the expansion hypothesis, O. C. Hilgenberg, suggested a crustal expansion rate of 0.17 cm/year up to the early Mesozoic, followed by a faster rate

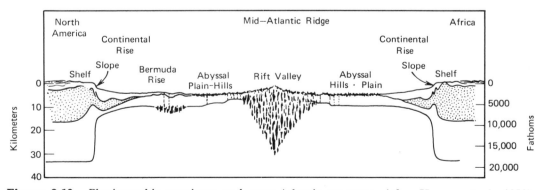

Figure 9-10. Physiographic provinces and trans-Atlantic structure (after Heezen et al; 1959). Based on scattered seismic-refraction measurements in the North Atlantic which have been projected along province boundaries. The topographic profile was pieced together from continuously recorded echo-sounding profiles from New York to Spanish Sahara. (With permission of the Geological Society of America.)

of 1.1 cm/year. Decrease in G, the gravitational constant, would produce expansion. L. Egyed published papers advocating a uniform rate of radius increase of about 0.05 cm/year based on an inferred phase transformation in the Earth's core, and estimates that the areas of the crust covered by shallow seas have decreased with time. The theory was revived in the 1950's by several sets of independent interpretations some of which are illustrated in Figure 12-10. Carey's ideas are outlined in Chapter 11. The worldwide rifts mentioned above were interpreted by B. C. Heezen as rents in the crust resulting from expansion (Figure 9-10). Reviews of the expansion theory are given by Irving (1964, p. 288–292) and Holmes

(1965, p. 965–975). Available evidence does not support this process as a cause of global tectonics.

Continental drift is reviewed in Chapter 11, sea-floor spreading in Chapter 12, and plate tectonics in Chapter 14. The occurrence of mantle convection would not preclude either contraction or expansion of the Earth.

The relative significance of vertical and horizontal movements as driving forces in tectonics is still in dispute (Figure 9-8), but advocates of horizontal movement of lithospheric plates have certainly dominated the literature since 1967. Vertical movements are reviewed in Chapter 10 and horizontal movements in Chapters 11-14.

10. Tectonic Significance of Phase Transitions

INTRODUCTION

There is no doubt that parts of the outer shell of the Earth have undergone repeated subsidence and uplift, and this is the reason for the operation of most of the geological processes outlined in Chapter 4. Without subsidence there would be no great piles of sedimentary rocks, and without uplift there would be no mountains or exposed metamorphic rocks. The vertical movements may involve large blocks of the crust or elongated belts of the crust. There is some question as to whether these movements are a secondary effect caused by mantle convection and lateral movement of the lithosphere or whether they are caused directly by volume changes at depth within the crust and mantle. Volume changes may be caused by localized heating or cooling, by phase changes such as melting and magma crystallization, or by solid-solid transitions such as those involved in the transformation of gabbro to eclogite.

In the preceding chapter we noted that one possible cause for subsidence in geosynclines was conversion of gabbro to eclogite (Figure 9-8). In Chapter 5 we noted that in tectonically active regions the geophysical evidence concerning the nature of the crust-mantle boundary was consistent with the existence of material involved in this phase transition (Figures 5-4 and 5-18). Furthermore in Chapter 6 we concluded that phase transitions occur at several depths within the mantle and specifically within

the upper 100 km of the mantle. Migration of a phase transition zone at depth must have some effect on the elevation of the surface. Thermal or pressure perturbations at depth may initiate changes of phase, and these changes, involving latent heat, must in turn exert considerable influence on the temperature distribution within the earth.

Successive papers have reviewed various aspects of the problem, and we will examine some of these in historical sequence because this provides a convenient development from simple to increasingly complex models. In all models reviewed the problem was simplified by treating the phase transition as univariant rather than divariant. It appears from Figure 5-17 that the gabbro-eclogite transition may be considered to approximate a pair of univariant reactions. The effects of expansion and contraction at depth resulting from changes in temperature distribution were described by H. H. Hess for the peridotite-serpentinite transition and by J. F. Lovering for the gabbro-eclogite transition. Lovering discussed conversion of gabbro to eclogite or of eclogite to gabbro with resultant contraction or expansion respectively. Contraction at depth would produce subsidence at the surface, and expansion at depth would produce uplift at the surface. G. C. Kennedy considered in addition the effects of isostasy, and he pointed out that crustal thickening caused by downward migration of the Moho would be followed by isostatic uplift, whereas crustal thinning would be followed by isostatic subsidence.

The isostatic effect is illustrated schematically in Figure 10-1. Figure 10-1a is a standard representation of blocks of wood (B) floating in isostatic equilibrium in a fluid (E). The surface level of the fluid is maintained constant in all figures by some method not illustrated. The lower boundaries between the blocks B and the fluid E represent the Moho. We will change conditions for the middle block with the other two remaining fixed for reference. We can represent downward migration of the Moho by placing an additional block of wood beneath the central block as shown in Figure 10-1b. The thickened, light crust is floated up to higher

levels as shown in Figure 10-1c in order to restore isostatic equilibrium. Similarly in Figure 10-1d migration of the Moho upward can be represented by cutting off a piece of the middle block and removing it. The thinner, light crust therefore loses buoyancy and sinks for restoration of isostatic equilibrium as illustrated in Figure 10-1e.

Kennedy also considered the effects of loading or unloading on the surface with consequent pressure perturbations at depth. This led him to the problem of the effects of sediment deposition, which produces first a pressure perturbation followed by a temperature perturbation because of the thermal blanket effect of a column of sediments. It is this problem which has occupied the attention of subsequent workers.

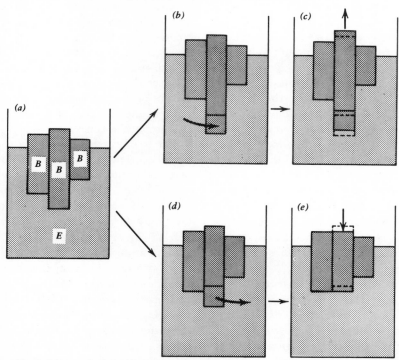

Figure 10-1. Illustration of isostatic adjustment resulting from motion of a phase transition boundary. Wooden blocks, B, floating in fluid, E, represent crustal basalt in isostatic equilibrium in mantle eclogite. The level of the fluid is maintained constant in all vessels by a device not illustrated. (a) Original condition. (b) Additional block of B is inserted below the center block; this displaces fluid, and is equivalent to conversion of eclogite to basalt and thickening of the crust. (c) The center block floats up until it reaches a new equilibrium position. (d) A portion of the centre block of (a) is cut off and removed; this is replaced by fluid, and is equivalent to conversion of basalt to eclogite and thinning of the crust. (e) The center block sinks down until it reaches a new equilibrium position.

EFFECT OF THERMAL PERTURBATIONS

Peridotite-Serpentinite Model of H. H. Hess

Hess drew attention to the tectonic implications of serpentinization of mantle peridotite in 1955. This is a chemical reaction rather than an isochemical phase transition, and it requires the addition of water and possibly the migration of other constituents in solution. The precise chemistry of the serpentinization process remains a matter of debate. Hess assumed that serpentinization of peridotite was accompanied by expansion but this too remains debatable. It was shown in Figure 5-7 that with the present temperature distribution in the Earth the upper mantle in many tectonic environments lies within the stability field of serpentine, so that if water were available the mantle peridotite would become serpentinite. The possible consequences of this are illustrated in Figure 10-2.

Figure 10-2*a* shows the position of the reaction curve for the dehydration of serpentine, which is taken as the dehydration boundary for serpentinite and the initial

position of the geotherm beneath the oceanic section shown in Figure 10-2*b*. The point of intersection, *A*, provides the level in the mantle of Figure 10-2*b* above which the peridotite lies within the stability field of serpentinite. If water and other volatiles slowly leak from the deep mantle along a favorable zone then peridotite would be converted to serpentinite above the level *A* as shown in Figure 10-2*c*. The volume increase so produced would cause the crust to rise with the amount of uplift increasing as the serpentinization worked gradually upward. If the top of the serpentinization zone reached the base of the crust the additional water would then be ejected through the crust with no further tectonic effect. If warming occurred at depth the geotherm in Figure 10-2*a* would move to higher temperatures and the intersection point *A* would migrate toward *B*. This would cause dehydration of serpentinite at the base of the zone as illustrated in Figure 10-2*d*, and the water would migrate upward and escape through the crust. The decrease in volume so produced would cause subsidence at the surface. Figure 10-2*e* shows

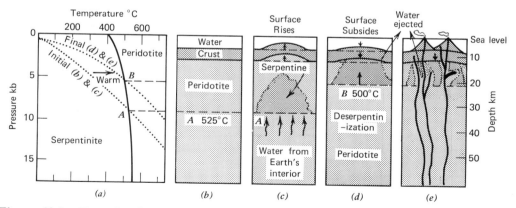

Figure 10-2. Tectonic effect of serpentinization of a peridotite mantle. (*a*) Peridotite-serpentinite reaction boundary based on Figure 5-7*a*. Initial geotherm intersects boundary at *A* for the sections (*b*) and (*c*), and warming at depth shifts the intersection point to *B* for (*d*) and (*e*). (*b*) Initial section, showing mantle peridotite at temperature below 525°C that is in the stability field of serpentinite, if water should become available. (*c*) Effect of migration of water upward across boundary *A*. (*d*) Effect of warming at depth, and deserpentinization. (*e*) Effect of volcanic activity in producing localized dehydration. (With permission of The Geological Society of America. (*b*) to (*e*) after Hess, 1955.)

that local volcanic activity could also cause deserpentinization with resultant subsidence at the surface.

Hess discussed applications of this kind of process to oceanic features with specific reference to the mid-Atlantic ridge and the guyots of the Pacific Ocean. The amount of serpentinite formed, and therefore the amount of expansion produced, is dependent on the amount of water available. Hess' later development of the idea of serpentinization of peridotite beneath the midoceanic ridges and its lateral migration forming the crust of the ocean basins is reviewed in connection with sea-floor spreading in Chapter 12. Hess also suggested that the elevation of the Colorado Plateau probably results from an expansion reaction, and that possibly serpentinization of peridotite below the Moho might account for this epeirogenic movement.

Gabbro-Eclogite Model of G. C. Kennedy

Figure 10-3 illustrates Kennedy's explanation for the uplift of plateaus. The initial geotherm intersects the basalt-eclogite phase transition boundary at point A, which produces the crustal section in Figure 10-3b.

Warming at depth changes the position of the geotherm with the point of intersection moving down to B and producing the crustal section in Figure 10-3c. Migration of the Moho from A to B causes conversion of eclogite to basalt with a volume expansion of at least 10%. This expansion causes uplift at the surface. The thickened crust is then uplifted further by isostatic adjustment as illustrated in Figures 10-3d, 10-1b, and c. This causes the Moho to be moved upward as well which means that the point of intersection B in Figure 10-3a also has to change. This means in turn that the temperature distribution at depth must change again. Now we begin to see that the problem involves many factors. One most important factor neglected so far is the latent heat of the gabbro-eclogite phase transition. Heat is absorbed when eclogite is converted to gabbro, so that as soon as migration begins from A toward B the phase change tends to oppose the regional warming trend and to maintain the Moho in its initial position.

The reverse process of cooling at depth would cause the Moho to migrate upward with conversion of basalt to eclogite and resultant contraction and subsidence at the

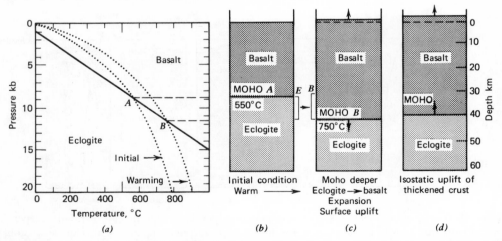

Figure 10-3. G. C. Kennedy's model illustrating qualitatively the uplift of plateaus. (*a*) The initial geotherm intersects the basalt-eclogite phase transition boundary at A. With warming at depth, the geotherm moves and the point of intersection migrates to B. (*b*) Crustal model corresponding to initial condition in (*a*). (*c*) Crustal model corresponding to later condition in (*a*). (*d*) Later isostatic uplift of the thickened crust.

surface. The effect of isostasy would then produce heat, opposing the cooling effect and opposing migration of the Moho. Obviously the rate of heat transfer to and from the phase boundary is an important parameter in this problem.

EFFECT OF PRESSURE PERTURBATIONS

The effect of a change in pressure at depth can be illustrated to a first approximation using Kennedy's example of persistent mountain ranges. Figure 10-4 illustrates a geotherm intersecting the phase transition boundary at point A and producing the crustal section in Figure 10-4b. This section includes a mountain range with average height above sea level of 4 km. Kennedy pointed out that as the mountains are eroded the pressure at the Moho beneath the mountains decreases causing downward migration of the Moho with deepening of the light roots that float the mountains upward again. This is illustrated schematically in Figure 10-4. Suppose that the mountain range could be removed by erosion instantaneously without producing any change in the position of the geotherm. This would provide a pressure decrement at depth of about 1 kb. The depth scale in Figure 10-4 is fixed by the geometry of the Earth, and the pressure scale therefore moves downward after erosion. At each depth the pressure becomes about 1 kb less than it was before erosion. The phase transition is a function of pressure rather than depth, and the position of the phase transition after erosion has moved deeper by about 1 kb as shown in Figure 10-4a. The point of intersection corresponding to the Moho therefore moves to point B giving the crustal section in Figure 10-4c. This requires conversion of eclogite to gabbro, expansion, and uplift. Again we have neglected the opposing effect of the heat absorbed by the phase transition which would tend to transfer the geotherm to lower temperatures and thus move the intersection point B to lower pressures and lower temperatures along the

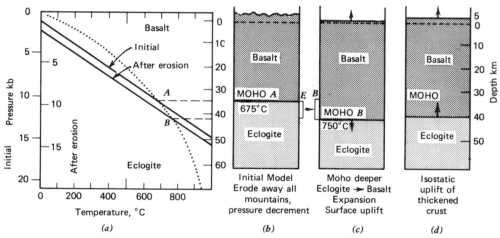

Figure 10-4. G. C. Kennedy's model illustrating qualitatively the rejuvenation of mountains and the permanence of continents. (a) and (b) Initial condition, with the Moho intersection at A. The mountains are completely eroded away, with no temperature change at depth, causing the phase transition boundary to move downward in (a) to a new intersection point B. The Moho thus moves to new position in (c). (c) Crustal section after erosion, showing uplift caused by conversion of eclogite to basalt. (d) Later isostatic uplift of thickened crust.

new phase transition boundary. The additional uplift resulting from isostatic adjustment is illustrated in Figure 10-4*d*. Thus the effect of erosion is to rejuvenate the mountains.

The deposition of sediments in a subsiding basin provides pressure increments at depth, with results opposite to those illustrated in Figure 10-4 as shown in Figures 10-5*a*, *b*, and *c*. The problem was outlined in a general way by Kennedy. The first effect of pressure loading by sediments would be the opposite of that of unloading by erosion which is illustrated in Figure 10-4. Initially the pressure at the Moho would increase as sediments displaced water causing upward migration of the Moho, conversion of gabbro to eclogite, and therefore continued subsidence at the surface. The column of sediments behaves as a thermal blanket, however, and the problem involves both pressure and temperature perturbations.

SEDIMENT DEPOSITION: SUBSIDENCE FOLLOWED BY UPLIFT

It is generally considered that the deposition of large thicknesses of sediments such as those occurring in geosynclines cannot be explained by the classical theory of isostasy; but the sequence of subsidence, sediment deposition, and uplift becomes explicable if there is a phase transition involved in the region of the crust-mantle boundary. The problem of determining the changes in depth of a phase transition boundary in response to both temperature and pressure perturbations is difficult because it involves a nonlinear condition at the phase boundary. Solutions to the problem have been attempted by MacDonald and Ness (1960), Wetherill (1961), van de Lindt (1967), and Joyner (1967). O'Connell and Wasserburg (1967) used analytic expressions applying to the initial part and the final part of the motion of a phase boundary.

Dynamics of Motion of a Phase Boundary after R. J. O'Connell and G. J. Wasserburg

The detailed contribution of O'Connell and Wasserburg in 1967 provides insight into the characteristics of the problem of sediment loading at the surface with a phase transition boundary at depth. They presented approximate analytic solutions with both impulsive and continuous loading. The effects of isostasy, sedimentation, and erosion were considered as modifiers of the essential process of the response of the phase transition to changes in pressure. We should be aware of their conclusions before proceeding with review of the various models developed by others between 1960 and 1967.

They covered first the steady state behavior and then the dynamic problem. They obtained numerical results for various models introducing the effects of thermal blanketing, the time at which this would cause reversal of the motion of the phase boundary, and the effects of isostasy. They concluded that the important parameters affecting motion of the phase boundary were the latent heat of the phase transition, the difference in slope between the geotherm and the phase transition curve, and the effect of isostasy. Of minor importance are convective heat transport and the distribution of heat sources. In the sedimentation model it is the rate of removal of heat that governs the rate of movement of the phase boundary. The redistribution of the latent heat liberated may produce significant changes in temperature distribution within the earth, extending to considerable depths and therefore depending upon deep-seated thermal conditions (Figure 10-5). The long-term motion of the phase boundary depends primarily upon the overall geometry of the model and the boundary condition at depth.

The Models of G. J. F. MacDonald and N. F. Ness and G. W. Wetherill

There are two ways to examine the motion of a phase boundary resulting from sedimentation. One is to consider steady-state or equilibrium configurations, and the more difficult approach is to determine the motion of the phase boundary as a function of the sedimentation rate and the thermal properties of the material involved in the phase transition. MacDonald and Ness used both approaches (1960), and Wetherill (1961) determined the effect of several variables by using steady-state calculations. His results differed from those of MacDonald and Ness for reasons discussed in his paper.

We can use Wetherill's model in Figure 10-5 to illustrate the problem before considering the models in detail. The first effect of sediments accumulating in a trough would be to increase the pressure at depth with very little change in temperature. The pressure scale and the phase transition curve would then move upward relative to the fixed depth scale as shown in Figure 10-5a. Figure 10-5c shows a hypothetical stage where original 1 km water in the trough has been completely replaced by 1 km of sediments. This would produce a pressure increment of about 0.14 kb and, neglecting isostasy and heat effects, the phase boundary would therefore migrate upward to the point of intersection B. This would cause conversion of gabbro to eclogite, contraction, and resultant subsidence of the trough. Thus the short-time effect of rapid sedimentation, as in a geosyncline, would be one of sinking. The sediments accumulating in the trough, however, are of low thermal conductivity and

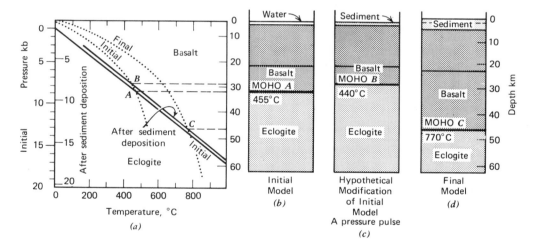

Figure 10-5. G. W. Wetherill's steady state crustal sections showing the effect of deposition of 6 km of sediments in a trough originally 1 km deep. (*a*) Initial condition: assumed phase transition boundary is intersected by the geotherm at *A*, corresponding to crustal section in (*b*). After deposition of 5 km sediment the final equilibrium situation provides intersection *C* with the phase transition boundary (migrated upward after sediment deposition) and crustal section (*d*). (*b*) Initial crustal section. (*c*) Hypothetical modification of initial model: 1 km of water in (*b*) is replaced by 1 km sediment with no change in temperature at depth. Phase transition boundary in (*a*) moves upward (contrast Fig. 10-4) producing new intersection point *B*. This would cause the thinned crust to sink in order to restore isostatic equilibrium; compare Fig. 10-1 (*d*) and (*e*). (*d*) Final equilibrium situation after deposition of 5 km of sediments. The effect of the sediment as a thermal blanket causes heating at depth, with the equilibrium geotherm intersecting the equilibrium phase transition boundary (only approximate in (*a*)) at point *C*, with crustal thickening.

they are richer in radioactive materials than the surrounding rocks. The new sediments act as a thermal blanket with their own heat source, and given sufficient time the temperature at depth will slowly rise causing the phase discontinuity to reverse its motion and migrate downward as illustrated in Figures 10-3 and 10-5. This causes conversion of eclogite to gabbro, expansion, and uplift. The crust thickens and the column of sediments in the geosyncline rises to form a mountain range as shown in Figure 10-5d, the final equilibrium state after deposition of 5 km of sediment.

Figures 10-5b and d show the equilibrium states calculated by Wetherill before and after deposition of 5 km of sediments. Figure 10-5a shows the initial position of an assumed phase boundary and the calculated geotherm for the initial crustal model adopted by Wetherill. This is a trough of water 1 km deep above a two-layer crust with total thickness 30 km; the depth to the Moho, the phase change discontinuity, is therefore 31 km (Figure 10-5b). This model is in

isostatic equilibrium. After deposition of 5 km of sediments with properties shown in Figure 10-5d, Wetherill calculated that for isostatic and thermal equilibrium the Moho was depressed to a depth of 46.7 km below the surface, and the surface of the sediments was elevated 1.7 km above sea level. The final temperature distribution was as shown in Figure 10-5a. For convenience to prevent overcrowding of the diagram the final position of the phase transition curve in Figure 10-5a is made to coincide with that drawn for the hypothetical intermediate stage.

MacDonald and Ness (1960) studied the motion of the phase change boundary with respect to time for several crustal models. The results of their time-dependent study of "Model C" are shown in Figure 10-6b by the dashed lines. The position and slope of the phase transition curve in Model C are very similar to those used by Wetherill (Figure 10-5a). Sedimentation begins initially in a basin 3 km deep, with the phase boundary at a depth of 30 km. The initial effect of

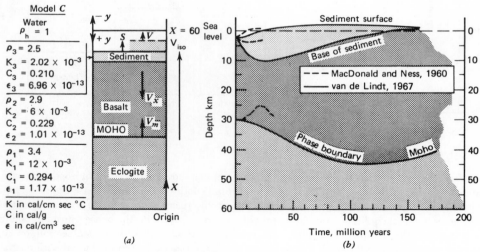

Figure 10-6. Time-dependent solutions for sediment deposition in a trough. (a) Crustal section with physical properties of Model C of MacDonald and Ness. The dynamic model is from van de Lindt, and the meaning of the velocity symbols and distance symbols is explained in the text. (b) The motions of the sediment surface, base of sediments, and phase transition boundary at depth calculated for Model C in (a). The calculations of MacDonald and Ness neglected isostasy whereas van de Lindt included effects of isostatic adjustment. (From Jour. Geophys. Res., **72**, 1289, 1967, with permission.)

sedimentation causes the phase boundary to rise, but the trough becomes filled with sediments in about 8×10^6 years. After 21×10^6 years the phase discontinuity has migrated downward again, and the sediments have become uplifted to a height of 240 m above sea level. MacDonald and Ness drew attention to the three principal time constants involved in the problem: the rate of sedimentation, the rate for obtaining thermal equilibrium, and the rate of establishment of isostatic equilibrium. Rates of sedimentation can be estimated from geological observations. They assumed that the time constant for attainment of isostatic equilibrium is small compared with the time required to attain thermal equilibrium, and they estimated that the time scale for approaching thermal equilibrium would be of the order of 100×10^6 years. The time interval of 21×10^6 years plotted in Figure 10-6b is thus not long enough for equilibrium conditions to be attained. Because of the mathematical difficulties, MacDonald and Ness neglected in their calculations the effects of isostasy and the heat carried by mass transport as the layers moved up or down. Wetherill concluded that both of these factors are likely to be important, and that in order to make further progress in solving the problem it was necessary to obtain time-dependent solutions with these effects considered; it was six years before such solutions were published (Figures 10-6 and 10-7).

Time-Dependent Solution of W. J. van de Lindt

W. J. van de Lindt (1967, 1968) used the crustal model illustrated in Figure 10-6a to calculate the movement of the Moho and the base and surface of the sediments as a function of time. Figure 10-6a provides a fairly simple picture of the various factors, velocities, and relative velocities involved in the problem, so we will examine his treatment of this model in some detail. The numerical results shown in Figure 10-6b are for Model C of MacDonald

and Ness. The origin of the coordinate system is chosen below the Moho and the system is at rest compared to the upper mantle; however, it is moving with respect to a coordinate system fixed in the center of the Earth, because of the isostatic adjustment taking place as a result of the surface loading by sediments. The assumption that this adjustment is instantaneous will not cause large errors as the time necessary to reach isostatic equilibrium is about 10^4 years, whereas the time constants involved in heat conduction are of the order of 40×10^6 years.

The depth of the surface of the Earth beneath the water (constant sea level is assumed throughout) is y, and if the surface rises above the water y is negative. The rate of sedimentation or erosion is taken as proportional to y. The sedimentation speed is s; that is, a layer of thickness s is added to the surface per unit time. The speed with which the surface of the sediment approaches the surface of the water is v. The speed of isostatic adjustment is v_{iso}, and at the depth of isostatic equilibrium in the mantle the pressure will remain unchanged although sedimentation occurs, giving

$$dP/dt = \rho_s s - \rho_h v + \rho_1 v_{iso} = 0 \quad (10\text{-}1)$$

$\rho_s s$ is the weight of the sediment added, $\rho_h v$ is the weight of displaced water, and $\rho_1 v_{iso}$ is the amount of mantle material added to the vertical column because of the isostatic adjustment. If the upward velocity of the Moho with respect to the mantle material is v_m, and v_x is the downward velocity of the material immediately above the Moho, then

$$v_m \rho_1 = v_m \rho_2 + v_x \rho_2 \quad (10\text{-}2)$$

and hence v_x is determined as a function of v_m.

The speed v of the Earth's surface is given by:

$$v = v_{iso} + s - v_x$$

$$= v_{iso} + s - v_m \frac{\rho_1 - \rho_2}{\rho_2} \quad (10\text{-}3)$$

From Equations 10-1 and 10-3 v and v_{iso} can be expressed in terms of s and v_m, which gives

$$v = \frac{(\rho_1 - \rho_s)}{(\rho_1 - \rho_h)} s - \frac{\rho_1(\rho_1 - \rho_2)}{\rho_2(\rho_1 - \rho_h)} v_m$$

$$\text{for} \quad y \geqslant 0 \quad (10\text{-}4)$$

If the Earth's surface is above water there is no longer any displacement of water and the equation becomes

$$v = \frac{(\rho_1 - \rho_s)}{\rho_1} s - \frac{(\rho_1 - \rho_2)}{\rho_2} v_m$$

$$\text{for} \quad y < 0 \quad (10\text{-}5)$$

The velocity of the Moho is a function of rates of heat transfer as indicated in a general way in our discussions of Figures 10-3, 10-4, and 10-5. This problem was tackled by van de Lindt as follows. He assumed that the density, ρ, specific heat, c, heat conductivity, k, and radioactive heat source strength, ϵ, in each layer is constant, although they may vary from layer to layer as shown in Figure 10-6a. In each layer, i, the heat conduction equation is valid

$$\frac{\partial T}{\partial t} = \frac{ki}{\rho_i c_i} \frac{\partial^2 T}{\partial x^2} + \frac{\epsilon_i}{\rho_i c_i} - v_i \frac{\partial T}{\partial x} \quad (10\text{-}6)$$

where T is the temperature and t the time. The first term on the right hand side is the diffusion term, the second is the radioactive heat source term, and the third represents the heat transport due to mass transport. The velocity v_i is zero for the material below the Moho, and it will have the same value for all layers above the Moho. At all interfaces between layers with different physical properties, except the Moho, the interface conditions are

$$T_i = T_{i+1} \quad (10\text{-}7)$$

and

$$k_i \left(\frac{\partial T}{\partial x}\right)_i = k_{i+1} \left(\frac{\partial T}{\partial x}\right)_{i+1} \quad (10\text{-}8)$$

which describes the continuity of heat flux across the boundary. It is assumed that the heat flux from the interior of the Earth is known and constant in time at the origin giving

$$-k \left(\frac{\partial T}{\partial x}\right)_{x=0} = Q \quad (10\text{-}9)$$

At the Moho the boundary conditions are

$$T_1 = T_2$$

and

$$k_1 \left(\frac{\partial T}{\partial x_1}\right) - k_2 \left(\frac{\partial T}{\partial x_2}\right) = \rho_1 r v_m \quad (10\text{-}10)$$

where r is the latent heat of the phase transition. This assumes slow rates of sedimentation and only a small temperature rise at the Moho, otherwise r would have to be replaced by $(r - c\Delta T)$.

When slow sedimentation starts the reaction curve for the phase transition moves from A as illustrated in Figure 10-5a, and the Moho attempts to follow to B while the latent heat released moves the geotherm to higher temperatures tending to maintain the Moho at the deeper level. Diffusion begins to take place, and the Moho eventually moves to its new position of equilibrium. The rate of temperature increase is

$$\left(\frac{dT}{dt}\right)_{\text{Moho}} = g \left(\frac{dP}{dt}\right)_{\text{Moho}} \quad (10\text{-}11)$$

where g is the slope of the reaction curve, (dT/dP). The total rate of change of pressure at the Moho is given by

$$\left(\frac{dP}{dt}\right)_{\text{Moho}} = \rho_s s - \rho_h v - \rho_1 v_m \quad (10\text{-}12)$$

and substitution of (10-4) into (10-12) and the result into (10-11) gives the rate of temperature rise at the Moho as

$$\left(\frac{dT}{dt}\right)_{\text{Moho}} = g \left[\rho_s - \rho_h \frac{\rho_1 - \rho_s}{\rho_1 - \rho_h}\right] s$$

$$- g \left[\rho_1 - \rho_h \frac{\rho_1(\rho_1 - \rho_2)}{\rho_2(\rho_1 - \rho_h)}\right] v_m$$

$$\text{for} \quad y \geqslant 0 \quad (10\text{-}13)$$

If y is negative there is no longer a displacement of water and Equation 10-13 becomes

$$\left(\frac{dT}{dt}\right)_{\text{Moho}} = g(\rho_s s - \rho_1 v_m) \quad \text{for} \quad y < 0 \tag{10-14}$$

These are the equations necessary for solution of the problem, and van de Lindt solved it on a digital computer by numerical means. His results for Model C are shown in Figure 10-6 and contrasted with the very different results obtained by MacDonald and Ness for the same model. The principal cause of the difference is the larger amount of sediment deposited, and this was caused in turn by taking into account the isostatic adjustment. The results of van de Lindt are closer to the steady-state results of Wetherill which were also obtained by isostatic considerations (Figure 10-5).

The original trough with a depth of 2.7 km becomes filled to sea level in about 1.8×10^7 years after which the land rises above sea level. The maximum thickness of .the sediment is about 10 km. Notice that the initial movement of the Moho in response to the rapid sedimentation is not upward as predicted in the simple explanations (Figure 10-5c) and in the results of MacDonald and Ness (Figure 10-6). Thinning of the crust for about 20×10^6 years, however, does indicate upward movement at the phase boundary relative to the layers above. Similar calculations for other models by Joyner (1967) do indicate initial upward movement of the Moho. The downward movement of the Moho which accompanies thickening of the crust produces upheaval of the sediments, and the maximum depth to the Moho in Figure 10-6 is 45 km (compare Figure 10-5d). The maximum elevation of the sediments is about 3 km. After about 120×10^6 years the Moho starts to rise again aiding the effects of erosion in lowering the elevated sediments and introducing the prospect of oscillatory motions of the Earth's surface about sea level. Using a modified computer program van de Lindt extended the time scale to

400×10^6 years, with the results shown in Figure 10-7a. A second interval of sediment deposition occurred after about 210×10^6 years with elevation of the sediment surface above sea level occurring again about 30×10^6 years later. This was followed by more than 150×10^6 years with little change; the sediment surface remained about 1.5 km above sea level during this erosion cycle.

The effects of varying the values for the parameters listed in Figure 10-6 were calculated and illustrated by van de Lindt, and he concluded that the values of density and thermal conductivity of the sediments are important whereas the influence of the slope of the phase transition curve and the latent heat turn out to be small.

O'Connell (1968) offered a critique of van de Lindt's analysis, and van de Lindt (1968) corrected some errors in his original paper. He stated that none of his computations were affected by the errors. O'Connell also noted that in contrast to the conclusion of van de Lindt he and Wasserburg had shown that the difference in slope between the phase transition curve, the temperature distribution in the Earth, and the latent heat of the phase change are the parameters that primarily determine the initial dynamic response of the model. He suggested that van de Lindt did not detect the importance of these factors because of the specific models investigated. Van de Lindt replied that for his model and within the range of parameters investigated the velocity of the Moho is relatively insensitive to these factors, but he did not claim that this was true for all models. MacDonald and Ness had previously concluded that the amplitude of movements of a phase boundary depended most critically upon the relative slopes of the phase transition reaction curve and the geotherm.

Time-Dependent Solution of W. B. Joyner

All of van de Lindt's models had initial basin depths of 2.7 km. Joyner (1967)

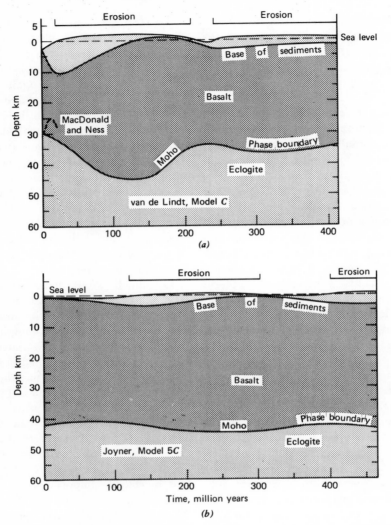

Figure 10-7. Time-dependent solutions for sediment deposition in a trough, illustrating cycles of sedimentation and erosion. Compare Fig. 10-6. (*a*) Solution of van de Lindt in Fig. 10-6(*b*) extended for twice the time. (*b*) Solution of Joyner for deposition in shallow-water. (From Jour. Geophys. Res., **72**, 4977, 1967, with permission. See also Figure 10-6.)

published the results of finite-difference calculations for the same problem using similar basic assumptions but different techniques of implementation and obtained results that he considered to be in general agreement with those of van de Lindt if allowance is made for the differences in parameters. Joyner's models had initial basin depths of 0.5 or 1.5 km and the sedimentation rate was

assumed to be constant in most models.

Joyner examined specifically the problem of thick accumulations of sediments deposited in shallow waters of epicontinental seas. Basins with initial depths of 2.5 to 3 km approach the depth characteristics of ocean basins, and they are incompatible with initial crustal thicknesses of 30 km. Such models are useful, however, in establishing the validity of the phase change hypothesis to explain thick sedimentary deposits. Joyner selected a phase transition line based on the diagram of Yoder and Tilley (Figure 5-14a), which has a more gentle slope, (dP/dT), than those used in the models discussed in this section (Figures 10-4 and 10-5). He obtained results for other transition curves indicating that the behavior of the models was quite insensitive to the slope of the transition, which he explained because the effect of changing the slope was compensated by changing ΔS in order to satisfy the Clapeyron equation for the transition. His results indicated that the thickness of sediments deposited depends strongly on the initial water depth and is rather insensitive to variations in deposition rate and the other parameters.

The cyclic results obtained by Joyner for his model 5c are compared in Figure 10-7b with those of van de Lindt. In this model Joyner used a reaction curve with a slope of 66°C/kb which is similar to the curves used in Figures 10-4 and 10-5. The initial crustal thickness is 41 km and the basin, initially 500 meters deep, fills with 3 km of sediments in 118×10^6 years. Contrast this with the deep water model of van de Lindt in Figure 10-7a, where 10 km of sediments accumulate much more rapidly in a basin initially 2.7 km deep. In Joyner's model the crust thins slightly during sediment deposition, and this is followed by slight thickening as the phase change migrates downward after about 100×10^6 years. A period of erosion lasts for 185×10^6 years, during which all of the sediments are eroded away, and then a second period of deposition occurs and the basin receives 3 km of sediments again. The second deposition period lasts for about 100×10^6 years and is followed in turn by a second period of erosion. During the second deposition period the water depth never exceeds 170 meters while 3 km of sediments are deposited. Joyner examined a number of different models, and in all of them deposition was followed by uplift with uplift cycles lasting as long as 200 to 300 million years.

Oscillatory Movements

The results reviewed above confirm that if there is a phase transition in the depth range 30 to 60 km, with characteristics approximating that of the gabbro-eclogite phase transition, then the process of sedimentation will cause migration of the phase transition boundary or zone, with continued subsidence occurring at the surface followed later by uplift of the sediments. Oscillatory motions can occur in this way. Any phase transition at depth will presumably have similar capabilities, the response at the surface varying according to the nature and depth of the phase transition.

Figures 10-5, 10-6, and 10-7 illustrate some quantitative results for various models. In their detailed analysis O'Connell and Wasserburg derived analytic approximations to the actual solution of the problem which permits specific geophysical models to be considered without the need of obtaining numerical solutions for each one. They concluded that for a phase change at a depth of about 40 km the initial upward motion of the phase boundary would be reversed after about 20 million years. It is this time interval which limits the thickness of sediments that can accumulate in a basin and not the initial depth of the water. Therefore when a basin is not filled within 20 million years the sediment thickness depends almost exclusively on the sedimentation rate. For several examples they calculated the maximum thicknesses of sediments deposited in basins with initial depths of 3 km or 5 km and the maximum final elevations. Calculated sediment thicknesses ranged from 8.6 km to 21 km, and the maximum final elevations ranged from 1.0 km to 5.7 km assuming no erosion.

The differences among these various numerical results and the others reported in the papers reviewed are neither unexpected nor discouraging. MacDonald and Ness pointed out in 1960 that a judicious choice of thermal constants could lead to any desired amplitude for the movements. Rather it is encouraging that the various calculations can yield amplitudes and time scales that are in general agreement with geological observations and deductions. Whatever scheme of global tectonics one prefers the effect of phase transitions at depth on near-surface geologic processes must be taken into consideration.

GRAVITY AND PHASE TRANSITIONS: CENTRIFUGE MODELS OF H. RAMBERG

If the material within the Earth undergoes a phase transformation it may become more or less dense than the material around it. As we saw in the preceding section the contraction or expansion at depth will change the surface elevation. In addition superimposed and modifying this effect there may be vertical movement of large bodies of material relative to the surrounding rocks. Under the influence of gravity the material changed through phase transformation will tend either to sink if it has become more dense or to rise if it has become less dense, provided that the mechanical properties of the rock layers will permit the movement. The significance of such processes is illustrated by two examples. If gabbro at the base of the crust (density 3.0) is transformed into eclogite (density 3.5) it is considerably more dense than the rocks around it and the mantle peridotite (density 3.3) below it. Gravity tends to make it sink (Figures 9-6f and 12-25a). Partial melting of mantle peridotite producing a mush of crystals with interstitial basaltic magma is accompanied by a density decrease. Gravity then drives the crystal mush upward if it can overcome the mechanical strength of the overlying rocks. In Chapter 6 we interpreted the existence of the low-velocity zone in the upper mantle as being due to the presence of interstitial liquid; the low-velocity zone is also of lower density than the mantle above it. The low-density layer is then a potential source of material rising buoyantly under the influence of gravity, as proposed in several models of magma genesis (Figure 6-6, Chapter 8).

In a series of papers since 1963 H. Ramberg has considered the role of gravity in the movement of material within the crust and upper mantle and the tectonic effects produced by such movements. This work, both theoretical and experimental, was brought together in a 1967 monograph. The analytical part of the volume treats rocks as continua in a mechanical sense and uses the theories of fluid dynamics and strength of materials. Unfortunately most rocks are not simple in their mechanical behavior, and the complex geometrical structure of realistic tectonic systems usually defies a rigorous theoretical analysis. Ramberg therefore investigated the effect of gravity on various geometrical structures by constructing dynamic scale models. The Earth's gravity field was imitated at the appropriate scale by the centrifugal force in a large-capacity centrifuge. This permitted the construction of models using more viscous materials than those used in previous model studies.

In previous small-scale model studies conducted in the normal gravity field it was necessary to use mechanically weak and soft materials in order to achieve collapse and flow. These studies have provided valuable information, but they can never correctly imitate gravity-generated features in large structures. In centrifuged models the centrifugal force per unit mass can be made several thousand times stronger than the gravitational force per unit mass, and scale can be maintained when using materials several thousand times stronger and correspondingly more viscous than the materials used in noncentrifuged models of the same size. Materials used in the latter type of experiments include asphalt, heavy oil, soft wax, and wet clay. Materials used by Ramberg include wax, modelling clay, and various putties. The scarcity of material with suitable properties remains a problem if consistent model ratios are to be maintained. Any pair of materials gives a defined model ratio of density, but their strength and viscosity ratios would usually be out of proportion for the defined model ratio. The more complex the structure modelled and the greater the number of rheological properties significant for the tectonics involved the less likely is it that the scale model will be reasonably consistent. Nevertheless the approach is a considerable advance over other model studies, and the structural patterns produced bear striking similarities to some crustal structures. This suggests strongly that gravity is an effective force in deep-seated tectonics with many resultant effects in the superstructure.

The procedure is to prepare a model using layers of different materials with the bottom of the supporting caps in the centrifuge model shaped parallel to the equipotential surface during rotation; the centrifugal force is then directed perpendicular to the layers. Because of the rigidity of the materials used the structures adopted because of deformation during rotation are preserved when rotation ceases, and at the end of a run the model is removed and sliced perpendicular to the original layers for inspection.

Dome Models

It is generally agreed that gravity has played a dominant role in the emplacement of salt domes, batholiths, and mantled gneiss domes. The low-velocity, low-density layer in the upper mantle may be a source for similar gravity-generated dome-like bodies. Figure 10-8a shows a diagrammatic cross-section through an initial model of silicone putty with density 1.12 g/cm^3 between layers of painter's putty with density 1.85 g/cm^3. There is a small bulge in the buoyant layer which initiated the doming. Figure 10-8b shows a cross-section through the model after a run of four minutes at 700 g in the centrifuge. This shows the typical geometric shape of domal structures in

general with a trunk region (T) rising from the root region and spreading out over the more dense material in the form of a hat (H). The arrows show the marginal sink of the more dense layer. Various stages in the development of such a structure have been examined by centrifuging for a shorter period. At a later stage in the evolution the hat portion may become detached from the trunk forming a surface layer or a sill below a higher level strong layer.

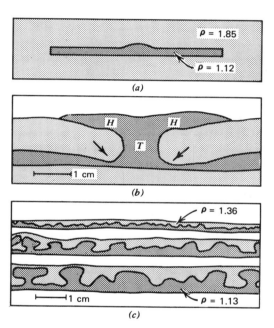

Figure 10-8. Results of centrifuged models illustrating the role of gravity in the formation of tectonic structures (after Ramberg, 1967). (a) Model 026 prior to run in centrifuge. Buoyant layer of silicone with initiating bulge, embedded in painter's putty. (b) Model 026 after run of 4 minutes at 700 g in the centrifuge, showing dome formation. (c) Profiles through three models with inverted density stratification run in centrifuge: for 2.5 min at 800–2000 g, for 1.5 min at 2000 g, and for 5 min at 800 g–2000 g. The dark grey bottom layers have a density of 1.13 g/cm^3, the light top layer a density of 1.36 g/cm^3. Both materials are silicone (viscosity of the order of 10^5–10^6 poises). (With permission of Academic Press.)

According to theory the diameter of domes and the spacing between them should be proportional to the thickness of source layers in models. In the three examples shown in Figure 10-8c the thickness of the domes and the spacing between them is roughly proportional to the thickness of the layers. This was found to be true for many of the models run.

The models illustrated in Figure 10-8 are built of materials representing rocks. It is impossible to include material representing magma in these models because the scaled viscosities turn out to be impossibly low. Even water is much too viscous to represent the silicate liquids. Water at 20°C has a viscosity making it a model material suitable for representing obsidian at 800°C; obsidian of course is a viscous glass and not a magma. Ramberg illustrated sections of layered models which started with small chambers at the bottom filled with a solution of $KMnO_4$. In most models, the solution passed right through the overburden quite rapidly and was extruded at the surface leaving a brown stain along its path of ascent. Passage of the solution strongly disturbed the layering at several levels. The buoyant rise of a solution, with properties representing an approach towards a fluid magma, produces geometric patterns quite unlike the domes developed in models with less contrasted viscosity of buoyant layer and overburden. The geometry of plutonic intrusions is thus clearly related to the physical properties of the intruding material compared with those of the country rock.

Models of Subsiding Sheets

The uprise of a buoyant mass of rock or magma must be balanced by the subsidence of more dense material. Figure 10-9 shows the structures developed when dense horizontal layers sink through less dense material. Figure 10-9a shows a central double layer of putty, $P + D$, suspended in a layer of lighter material, S. This rests on a denser layer and the whole model is covered by a thin over-

burden of modelling clay, M, and silicone putty alternating with layers of modelling clay, L. Figure 10-9b shows a sketch of the cross-section through the model after centrifuging. As the more dense double layer $P + D$ began to sink through the lighter layer, S, it assumed first an anticlinal shape and then the complementary uprise of the lighter material produced a central dome structure which burst through the sinking layers. These were then separated into the two portions shown. The general arrange-

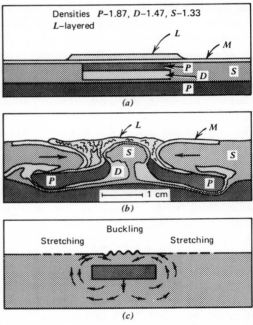

Figure 10-9. Results of centrifuged scale models illustrating the role of gravity in the formation of tectonic structures (after Ramberg, 1967). (a) Sections through a model of unstable density stratification prior to run in centrifuge. Model composed of painter's putty, $\rho = 1.87$ g/cm³; silicone, $\rho = 1.33$ g/cm³; dark silicone, $\rho = 1.47$ g/cm³; silicone putty with sheets of modelling clay; silicone putty, $\rho = 1.35$ g/cm³; modelling clay. (b) Section through model in a after run in centrifuge. (c) Flow lines around a heavy body sinking through a viscous mass. Deformation in competent crust indicated. (Figures 9-8f and 12-25a.) (With permission of Academic Press.)

ment of flow lines around a subsiding body sinking through a viscous mass is depicted in Figure 10-9c. The effect of similar flow lines is evident in the structure shown in Figure 10-9b. Layer D has flowed out sideways from beneath the heavier layer P and then upward and inward over the top of the layer despite the fact that its flow has been through the less dense material S in defiance of gravity. Sinking of the heavy layer P has produced very marked effects on the surface layers, M and L. The layer M was first dragged downward as P subsided, and then when two portions of P were forced apart as the buoyant dome S burst through it the sequence S-M was overturned. The layered overburden, L, was dragged down by the sinking slab, and

severely compressed and folded by the inward flow of layers S above the subsiding layers. These surface strata although much too light to sink by themselves have been pulled down to considerable depth in narrow wedges immediately overlying the heavy layer P. Figure 10-9c shows that the deformation in a competent crust overlying a sinking body would involve compression and buckling immediately above the sinking layer with stretching of the layer on either side.

The geological applications of Figures 10-8, 10-9, and the many other models illustrated in the book were reviewed by Ramberg. We shall have cause to remember these figures when we consider the effects of horizontal movements in Chapter 14.

PHASE TRANSITIONS AND MANTLE CONVECTION: ANALYSIS BY J. VERHOOGEN

We have examined the effect of temperature and pressure perturbations on the migration of a phase boundary at depth, and the effect of gravity in causing vertical movements of bodies with density different from that of their surroundings. Now we must consider what happens if a moving body reaches a level where a phase boundary or phase transition zone exists. The problem has been examined in the context of whether phase transitions in the upper mantle will enhance or hinder mantle convection and opinions are divided. In 1965 J. Verhoogen concluded that phase transitions in the upper mantle probably form no serious obstacle to mantle convection. The examples which follow are based on Verhoogen's treatment.

Figures 10-10a, 10-11a, and 10-12a show a schematic phase diagram for a univariant transition phase 1 $\rightleftharpoons$ phase 2; the heavy dashed lines represent the geotherm for the depth interval in the upper mantle that is represented. The diagrams to the right of the phase diagrams represent sections through the same depth interval of the mantle. In the following discussion "phase 1" and "phase 2"

refer to the complete mantle mineral assemblages containing these two phases respectively. The temperature of the mantle at any depth is given by the geotherm, and temperatures are written alongside the mantle sections. The point of intersection of the geotherm with the univariant curve in the phase diagram gives the depth within the mantle column of the boundary between the more dense phase 1 and the overlying phase 2. We are going to consider the movement of bodies of material through this stable, phase-stratified mantle.

The moving bodies are represented by the circles in the mantle sections, and the mineral assemblages present at specific depths are indicated by open circles for phase 2, closed circles for phase 1, and part-filled circles for a body consisting of both phases 1 and 2. The changes in temperature as a function of position at depth of the moving bodies are plotted on the phase diagram as heavy lines with arrows showing the upward or downward direction of movement. At any depth the temperature of a moving body can be read from the phase diagram; these temperatures are written alongside the circles in the mantle sections. For each phase diagram numbers have been assigned to the temperatures specified in the sequence of increasing temperature, and the appropriate points on the phase diagram have been designated by these numbers. This shows directly in the mantle sections whether the moving bodies at any depth are hotter or cooler than the mantle surrounding them. The symbol with the higher subscript is the higher temperature; T_4 is warmer than T_2.

Consider first a small mass of phase 1 at point 5 in Figure 10-10a. This is represented by the closed circle at temperature T_5 just at the phase boundary in the mantle section of Figure 10-10b. Let this mass be displaced upward from point 5 to a level with lower pressure as shown by the rising column of the mantle section in Figure 10-10b. This moves it into mantle composed of phase 2 and therefore phase 1 in the body begins conversion into phase 2. This transition (from more dense to less dense phase) is endothermic, and assuming that the reaction runs adiabat-

ically the temperature of the moving body decreases. As long as phase 1 is undergoing transition to phase 2 the body is constrained to follow the path of the phase boundary. The simultaneous decrease in pressure and temperature of the body consequent upon its upward displacement is therefore given in Figure 10-10a by the path 5-2. If the transition is completed at point 2 then the rising body consists of phase 2 at temperature T_2 within mantle of the same phase at a higher temperature T_4. If displacement were continued upward as indicated in Figure 10-10b, then the path followed would be an adiabat such as 2-1 in Figure 10-10a. At point 1 the body at temperature T_1 is cooler and therefore it remains more dense than the surrounding mantle of the same phase at higher temperature T_3. The system is thus stable with respect to upward movement of phase 1 starting from point 5. The density difference between the rising body and the surrounding mantle tends to reverse any upward motion.

Consider now a small mass of phase 2 at point 5 in Figure 10-10a, represented by the

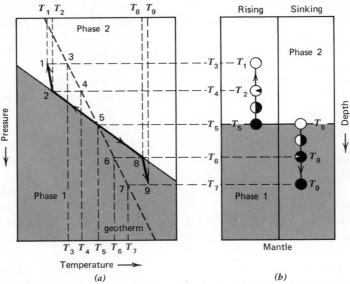

Figure 10-10. Mantle convection and phase transitions; an illustration of treatment by Verhoogen (1965). Shows the depth-temperature paths followed by material moving from position 5. (a) Transition boundary between phase 1 (more dense) and phase 2 (less dense) in mantle material, and a geotherm. (b) Mantle sections along the geotherm in a. Temperature variations and phase transitions of material moving up or down from the position 5 (Figure a) on the transition surface in the mantle.

open circle at temperature T_5 just at the phase boundary in the mantle section of Figure 10-10b. Let this mass be displaced downward from point 5 as shown in the sinking column of the mantle section, and the body then follows the path 5-8-9 in Figure 10-10a, with phase 2 changing to phase 1 as shown in Figure 10-10b. At all points between 5 and 8 the density of the sinking body is less than that of the surrounding mantle and its temperature is higher. The tendency therefore is for reversal of the downward movement.

Therefore the mantle system with a univariant phase transition occurring at the level shown in Figure 10-10b is completely stable with respect to vertical movements beginning at the level of the phase transition boundary.

Figure 10-11 illustrates the situation when a small mass of phase 1 is displaced upward from point 8 on the geotherm at temperature

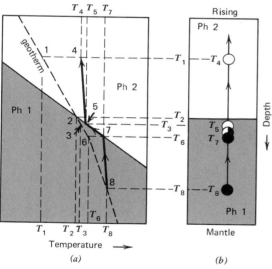

Figure 10-11. Mantle convection and phase transitions; an illustration of treatment by Verhoogen (1965). (*a*) Transition boundary and geotherm as in Figure 10-10a. Shows the temperature-depth path followed by material moving upwards from position 8. (*b*) Temperature variations and phase transitions of material moving up through mantle from position 8 in Figure *a*.

T_8 well below the level of the phase transition boundary in the mantle column. The temperature of the rising body will fall as shown by the adiabatic line 8-7 until it reaches the transition boundary in the phase diagram (Figure 10-11a) at some distance below the boundary in the mantle (Figure 10-11b). At point 7 the rising body at temperature T_7 will begin to transform from phase 1 to phase 2, surrounded by cooler mantle at temperature T_6 which remains as phase 1. The density of the rising mass thus decreases even further relative to the surrounding mantle and upward movement is enhanced. During the endothermic reaction the path of the rising body is constrained to follow the line 7-5 as shown in Figure 10-11a until all of phase 1 has been converted to phase 2. The body then consists of phase 2 at temperature T_5 surrounded by more dense mantle of phase 1 at a lower temperature T_3; the body therefore continues to rise following an adiabatic line 5-4 with density remaining less than that of the surrounding, cooler mantle.

Provided that the phase transformation is completed before the mass reaches point 2 in Figure 10-11a then the density of the rising body remains always less than that of the mantle and upward motion can continue right through the phase transition level. If the phase transformation were not completed by the time the body reached point 2, then its path would continue along the phase boundary to higher levels beyond 2 producing a situation where the body consisting of phase 1 and phase 2 would be surrounded by mantle of phase 2 only, at a higher temperature. The body would then be of greater density than the mantle and the upward motion would be opposed.

Verhoogen estimated the depth at which upward movement would have to start (point 8 in Figure 10-11a) in order that the transformation be completed by point 2. The results are strongly dependent upon the thermodynamic properties of the phase transition, and upon the relative gradients of the geotherm and the adiabatic curve. One set of assumptions gave a vertical distance between points 8 and 2 of about 400 km, and another gave a value of about 2000 km. The inhibition of upward movement is small if ΔS of the reaction is small and if the univariant reaction curve is very steep. Similar considerations are applicable to the downward movement of a body starting at point 1 in Figure 10-12a and following initially the adiabatic path 1-2.

Most phase transitions in the mantle are likely to be divariant or multivariant rather than univariant (Chapter 6), and Verhoogen also analyzed the conditions for movement of material through a divariant transition interval. He showed that for vertical movement of material through such a zone within the mantle the adiabatic gradient need be only slightly greater than the gradient of the initial geotherm.

Verhoogen concluded that although phase transitions act as filters tending to stop small perturbations, they permit large, deep-seated vertical movements to pass through. The analysis assumes that reaction rates are such that the moving bodies are able to transform into the equilibrium phase assemblage stable for the local pressure-temperature conditions. According to Verhoogen transformation rates under upper mantle conditions are not likely to hinder convection.

METASTABLE PHASE TRANSITIONS CAUSE EARTHQUAKES ACCORDING TO J. G. DENNIS AND C. T. WALKER

In their first paper on metastable phase transitions in 1965 Dennis and Walker expressed dissatisfaction with the elastic rebound theory of earthquake source mechanism, and proposed instead that energy was stored chemically and released abruptly in a spontaneous phase transition from metastable material. Their scheme is illustrated in Figure 10-12 using the same approach adopted for Figures 10-10 and 10-11.

Consider the downward movement of a body of mantle material from position 1 in Figure 10-12a. This will follow the adiabatic path 1-2 with temperature becoming progressively less than that of the surrounding mantle so that its increase in density will encourage continued downward movement. According to Verhoogen's evaluation of this problem, when the body reaches point 2 its transformation to phase 1 begins and its path then moves along the line 2-6 (compare path 8-7-5 for the rising body in Figure 10-11a). Dennis and Walker suggested that because of the high activation energy for solid state reactions involving breakdown of the lattice the probability of transition at or near equilibrium, at temperature T_2, was

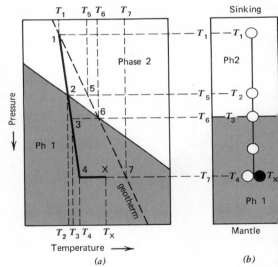

Figure 10-12. Metastable phase transitions in the mantle; an illustration of the scheme proposed by Dennis and Walker (1965) and Walker and Dennis (1966). (a) Transition boundary and geotherm as in Figures 10-10a and 10-11a. Shows path of material moving downwards from position 1, and passing metastably through the phase transition boundary. (b) Mantle section showing temperature variations and phase transitions of material moving down from 1 in Figure a.

infinitesimal and that the body would continue down the adiabatic curve 2-3-4 as metastable phase 2 surrounded by mantle in the form of phase 1 at a higher temperature. They stated that with increasing temperature (as along 2-3-4) the probability of nucleation of phase 1 would increase; at point 4 this occurs and they suggested that the heat liberated during the exothermic reaction would accumulate, accelerating the rate of nucleation and reaction to such an extent that an implosive transition would occur. This is represented in Figure 10-12 by the body of phase 2 at temperature T_4 at point 4 changing at this depth to phase 1 at some higher temperature T_x.

In the preceding section we referred to Verhoogen's consideration of transformation rates. He concluded that the transitions in the mantle would probably run at rates comparable with the velocities of vertical movements. If this is true then the path 1-2 could not continue far in the direction 2-3 in Figure 10-12a without nucleation of phase 1 beginning; how far along the path 2-3-4 the metastable material would have to travel before it had the capability of a spontaneous implosion is not known. Walker and Dennis recognized this problem in a second paper in 1966 after they learned that the olivine-spinel transition in Fe_2SiO_4 composition was completely and readily reversible in the laboratory at temperatures greater than 700°C. Therefore they suggested that downward migration in the upper mantle must begin at levels where the temperature was considerably less than 700°C, and that spontaneous transformation must occur when the temperature of the sinking body reached a critical value in the region of 700°C; this limit corresponds to T_4 in Figure 10-12.

In order to present a limiting scheme that they considered workable they depicted downward migration of mantle from beneath the continental crust at a temperature of about 500°C (T_1) reaching a depth of about 300 km at about 600°C corresponding to point 2 where the normal temperature T_5 was about 1350°C. The phase transition in the mantle (point 6) was passed at a depth of almost 400 km; and the maximum depth of spontaneous transformation of the metastable material was 700 km at a temperature of 700°C (T_4), where the temperature of the surrounding mantle was about 1650°C (T_7). The difference in temperature of about 950°C between sinking body and surrounding mantle at a depth of about 700 km seems rather extreme (difference between T_4 and T_7), and maintaining mantle phase 2 in a metastable condition through a depth of about 400 km (between points 2 and 4) in the temperature range 500–700°C (T_2 to T_4) also seems rather unlikely. For less extreme examples their proposal that a moving body of metastable phase 2 could undergo spontaneous transformation somewhere along the path 2-3-4 above and below the location of the phase transition in the mantle at 6 remains a possibility. Whether such a spontaneous transformation would amount to an implosion remains somewhat more speculative.

There is another problem, and that is the lower density of the moving body than the surrounding mantle along the path 3-4. Here we have metastable phase 2 surrounded by more dense phase 1, and this will oppose continued downward migration. Therefore we cannot expect the limit of point 4 to extend much below point 3. The 300 km depicted as a maximum in their 1966 model is surely excessive. It seems that this proposed mechanism can only be effective if there is some force driving the less dense, metastable material into the more dense mantle. Figure 14-8 illustrates possible mechanisms for sinking lithosphere slabs. It is left as an exercise for students to consider whether or not these mechanisms and the temperature distribution in Figure 14-13 are consistent with the proposal of Walker and Dennis.

11. *Continental Drift: Debate of the Century*

INTRODUCTION

The development of the geological sciences has been shaped by a series of great controversies. These include the Neptunists versus the Plutonists during the period 1775–1825, Catastrophism versus Uniformitarianism between about 1775 and 1835, and the vigorous debate about continental drift during the present century—Mobilists or Drifters versus Stabilists or Fixists.

The idea of a mobile Earth, with the crust floating on a molten interior, was familiar to geologists by 1900, but the concept of continental drift did not gather a large following until 1915 when Alfred Wegener published the first edition of his book *The Origin of Continents and Oceans*. Lateral migration of the continents provided an alternative mechanism to that of contraction for compressing the rocks of mountain ranges.

The evidence presented by Wegener and his followers failed to convince many scientists of the validity of continental drift, and a heated controversy continued between 1910 and 1960. Then the concept of sea-floor spreading was proposed by H. H. Hess with tectonic implications propounded by R. S. Dietz. This was supported by a variety of new evidence from paleomagnetism, geochronology, and marine geology and geophysics. The concept received tremendous impetus when papers published in 1966 and 1967 indicated that measurements of three different features of the Earth all change in the same ratios: these are the ages of magnetic polarity reversals in lava sequences, the depths of reversals of remanent magnetization in deep-sea cores, and the widths of the linear magnetic anomalies parallel to midoceanic ridges. The relationships between a time scale and distances are translated into a velocity of crustal spreading away from the midoceanic ridges. Converts flocked to the Mobilist camp, but some steadfast Fixists such as Harold Jeffreys and the Russian school led by V. V. Beloussov maintained that the theory has no foundation.

The theories of continental drift and sea-floor spreading were combined in schemes of plate tectonics where thin slabs of rigid lithosphere (including continent and ocean crust) moved over the less viscous asthenosphere in the upper mantle. By 1970 with the results of the Joides deep-sea drilling project presented as additional support for sea-floor spreading Earth scientists were coping with a ruling theory. It becomes almost futile to argue against a ruling theory, especially one which appears to solve so many problems. But other problems do remain and in 1970 A. A. Meyerhoff published the first in a series of papers arguing against the drift theory.

The successive discoveries and steps leading to the formulation and modification of the theory of plate tectonics contain many lessons for students; in the next four chapters we shall examine the developments in some detail.

THE DEBATE UNTIL 1950

Continental drift is an old idea formulated initially to explain the striking parallelism of the Atlantic coasts. The geometric fit of the continental margins, however, was not acceptable by itself as critical evidence that the Atlantic Ocean was once closed. Indeed the close fit has been considered as evidence either favorable or unfavorable to continental drift depending on the viewpoint of the scientist. Those against continental drift maintained that if a catastrophic event powerful enough to split a continental mass had occurred, then the newly generated continental margins would be stretched and severely deformed and they could not possibly fit so well.

Alfred Wegener and others compiled an impressive list of evidence supporting continental drift, based on aspects of geography, geodesy, geophysics, geology, paleontology, and paleoclimatology. Counterarguments were erected opposing each line of evidence, and it seemed that the theory could not be proved or disproved. The theory suffered from a lack of quantitative data, and the type of evidence put forward was perhaps psychologically unacceptable to many geologists.

Opponents of drift following the lead of eminent geophysicists such as Harold Jeffreys argued against the theory partly on the grounds that there was no satisfactory explanation as to why the continents had drifted and, furthermore, that the known physical properties of the Earth were such that the proposed lateral migration of the continents was impossible. Proponents of drift, on the other hand, argued that geological facts should not be ignored simply because there was no explanation available for them. Then other experts disputed the geological

"facts." The arguments continued in this indeterminate vein, like some medieval philosophical controversy, until a stalemate was reached in the 1940's. Just about everything that could be argued for and against continental drift had been written not once but many times and the debate faded for lack of additional evidence.

The Theory of Continental Drift

All theories of continental drift require that before the Mesozoic the continents were grouped together into a single block or into two blocks. Wegener proposed the name *Pangaea* for the single continent which he believed existed in the Late Carboniferous as illustrated in Figure 11-1a. The shaded areas on the continents represent shallow seas. According to Wegener *Pangaea* began to split up during the Jurassic period, with the southern continents moving either westward or toward the equator, or both. South America and Africa began to drift apart during the Cretaceous a little more than 70 m.y. ago. Figure 11-1b shows the distribution of the continents during the Eocene about 45 m.y. ago. Opening of the North Atlantic he thought was mainly accomplished during the Pleistocene, and Figure 11-1c shows that Greenland and Norway only started to part company about 1.5 m.y. ago. The Indian peninsula, which was originally long (Figure 11-1a), drifted northeastward and was compressed into fold mountains against the Asian continent; and similarly the European Alps and the Atlas Mountains of North Africa were caught between Africa and Europe, and can be interpreted as an extension of the Himalayan chain. Wegener suggested that as the front margins of the

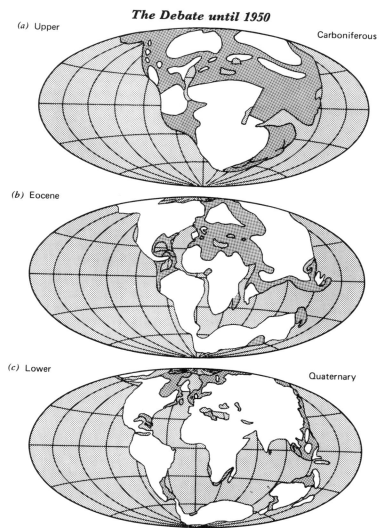

(a) Upper   Carboniferous

(b) Eocene

(c) Lower   Quaternary

Figure 11-1. Stages of continental drift according to Wegener (1915; reprinted version, 1966). Compare Dietz and Holden (1970). (From "The Origin of Continents and Oceans" by A. Wegener, Dover Publications, Inc., New York. Reprinted with permission of the publisher.)

moving continents met the resistance of the ocean floor, they became compressed and folded into mountain ranges. In this way he explained the western cordillera of the Americas and the mountain ranges of New Zealand and New Guinea. His explanation for the tapering ends of Greenland and South America and for the island arcs including the Antilles and those on the east of Asia is that these represent stragglers left behind in the wake of the moving islands. Detached fragments formed the long strings of islands.

The leading proponent of two primordial continents was A. L. du Toit who maintained that from the middle of the Paleozoic era to the beginning of the Tertiary the continent of Laurasia in the northern hemisphere was separated from the continent of Gondwana-land in the southern hemisphere by Tethys, a huge, geosynclinal ocean from which the Alps and Himalayan mountain chains eventually emerged. This grouping of the continents is consistent with the major structural features of the Earth as they exist at present.

Figure 2-3 shows the distribution of Tertiary mountain ranges in the form of two great rings; one surrounding the northern continents of North America and Eurasia (Laurasia) and the other surrounding the southern land masses of South America, Africa, Arabia, India, Australia, and Antarctica (Gondwanaland). The major structural features of the Earth thus consist of:

1. The Laurasian group of continents with the intervening basins of the North Atlantic and Arctic Oceans.

2. The Gondwanaland group of continents with the intervening basins of the South Atlantic and Indian Oceans.

3. The Pacific Ocean basin.

Evidence Cited for Continental Drift

The geological evidence which was debated for half a century has been reviewed many times before and we need consider only the outline of the controversy. There are four main lines of evidence that the continents were once joined together: (a) paleoclimatology, (b) paleontology, (c) the geometrical fit of the continents, and (d) the matching of stratigraphy and truncated structures across oceans.

The anomalous paleolatitudes deduced from paleoclimatology in the 1920's provided strong evidence for continental drift. Climatic zones in the late Paleozoic have been determined from the distribution of glacial deposits, desert sand and deduced wind directions, salt deposits, coal deposits, ancient coral reefs, and other fossil plants and animals indicative of climate. The paleoclimatic zones so deduced suggest either that the continents have moved relative to their present latitudes, or that there have been major climatic fluctuations with time at specific latitudes with stationary continents. (A. A. Meyerhoff reached a different conclusion in 1970; see Chapter 15). The concept of polar wandering was introduced in the nineteenth century to account for variations in climatic and biological sequences in different parts of the southern continents. The distribution of Permocarboniferous tillites in Australia, India, South Africa, and South America has been cited as evidence that these continents were in contact, or at least in close proximity, during this period.

Fossil plants and animals believed to be incapable of crossing deep water are distributed in limited geographic regions on both sides of the Atlantic and on continents separated by other oceans. Their distributions suggest that there were former land connections between Europe and North America, and between South America, Antarctica, Australia, India, and Africa. The land connections apparently required in earlier times appear to be nonexistent after the Jurassic. There were vigorous paleontological discussions about the reality and location of hypothetical land-bridges, which were supposed to provide temporary land connections across oceans from one continent to another. The theory of continental drift makes the land bridges unnecessary.

Attempting to fit together the continents marginal to the Atlantic Ocean has been a popular pastime since the last century. Figure 11-1a shows Wegener's reconstruction for the carboniferous period before the onset of drifting. Although most scientists are impressed by the approximate fit of the Atlantic coastlines others remain sceptical.

If the continents were once joined then the structures, ages, and petrology of the rocks must match across the joins. There are remarkable similarities between the stratigraphic and lithological columns for Brazil and southwest Africa from the Silurian through Cretaceous periods and several structural links between groups of rocks in Africa and South America have been proposed. There have been many attempts to reconstruct a single predrift geosyncline from the Caledonian belts of Norway, East Greenland, Scotland, Ireland, and the Appalachian chain. During the great debate the reality of the claimed structural fits across the

Atlantic Ocean was disputed by many geologists.

Mechanisms Postulated for Drifting the Continents

A major problem facing proponents of continental drift was to find a mechanism capable of producing the breakup and drifting apart of the continents. None of the proposed mechanisms was considered convincing by opponents of the theory, and I suspect that those who proposed them were not fully convinced either.

In the nineteenth century it was suggested that the Atlantic Ocean opened up early in the history of the Earth as a direct result of the loss of the moon from the site of the Pacific Ocean. The moon was also involved in the hypothesis of F. B. Taylor, who proposed in 1910 that the moon was captured by the Earth only in Cretaceous times and was at first much nearer to the Earth than it is today. This produced tidal forces which increased the rate of the Earth's rotation, and also caused the continents (Laurasia in the north and Gondwanaland in the south) to slide away from the polar regions towards the equator. The great Tertiary mountain ranges, and mountainous loops and arcs, were pictured as being raised up in front of the sliding continental masses. Ocean basins were produced from the stretched and broken regions left behind or between the sliding continents. A major weakness of the theory is that if mountain building is correlated with capture of the moon it gives no explanation for the orogenic cycles prior to the Tertiary. Also tidal forces strong enough to move the continents on such a scale and to cause the uplift of great mountain ranges, would surely have slowed the Earth's rotation almost to a standstill within a very short time.

Wegener described the drift of the continents as *Polflucht*, the "Flight from the Poles," and he suggested that this resulted from the gravitational attraction between the continents and the Earth's equatorial bulge. He also postulated a general drift to the west reasoning that the differential attractions of the sun and moon would cause the continents to lag behind the rotation of the Earth as a whole. Both of these forces are quite inadequate to overcome the frictional resistance to motion of the continents although Wegener maintained that even such small forces, if maintained for millions of years, would eventually displace the continents.

In his book, *Our Wandering Continents*, dedicated to Wegener in 1937, Du Toit proposed a model whereby a continent slides over the mantle by the action of gravity. Marginal loading of a continental block by a geosynclinal pile causes the continent to become slightly tilted and gravity causes it to slide toward the ocean. The central part of the continent is thus subject to tension which causes rifting, permitting the uprise of magma which contributes to further splitting and arching of the rifted region, which in turn makes the continent slide further outward. This original hypothesis lacked quantitative evidence, and the idea received little attention. Recent interpretations of sea-floor spreading, however, include mechanisms driven by gravity (Figure 14-8c).

In the nineteenth century we already find the essence of the ideas of polar wandering, mantle convection, and sea-floor spreading which have become topics of major concern to geologists and geophysicists since 1950. The study of paleoclimates and paleontological sequences early led to the conclusion that the geographic poles had shifted relative to the positions of the continents, but attempts to trace the locus of polar displacements through geological time always led to contradictions. The concept of polar wandering was generally distrusted, and it has been difficult to place any precise meaning on the term. It is now usually used to convey the idea that while the geographic poles remain fixed relative to the rotating Earth, an outer shell of the Earth becomes decoupled from the mantle and shifts as a whole relative to the poles. Others have understood by polar wandering a turning over of the entire body of the globe relative to its axis of rotation. Another interpretation, coming with the advent of paleomagnetic studies, implied that the magnetic poles wandered relative to the axis of rotation or geographic poles. Polar wandering can occur without continental drift, which implies movement of one continent relative to another, but continental drift cannot occur without polar wandering. If continental drift occurs then different continents will have different paths of polar wandering.

In 1881 Osmond Fisher published a book, *Physics of the Earth's Crust*, which postulated the existence of convection currents in the Earth's fluid interior, with uprise beneath the oceans causing expansion at the surface by the addition of volcanic rocks in the median position and descent of the currents beneath the continents. As the oceans expanded the continents contracted to form fold mountains.

Fisher's ideas were regarded as wild speculations and therefore ignored. The idea of convection in the mantle had been suggested before and it became popular later, but the Earth's interior is now known to be solid.

The first to suggest that continental drift might be explained in terms of convection within a solid mantle appears to have been Arthur Holmes. The model that he described in 1928 (published in 1931) is remarkably close to the concept of sea-floor spreading developed in 1960 by Hess and Dietz; although it could more aptly be termed sea-floor stretching. Figure 11-2 shows Holmes' diagrams for the convective-current mechanism using gravitational energy and thermal energy within the Earth to engineer continental drift and the development of new ocean basins. In Figure 11-2a, we have a thick basaltic crust with an overlying sialic layer forming the continents. A current ascending at *A* spreads out laterally beneath

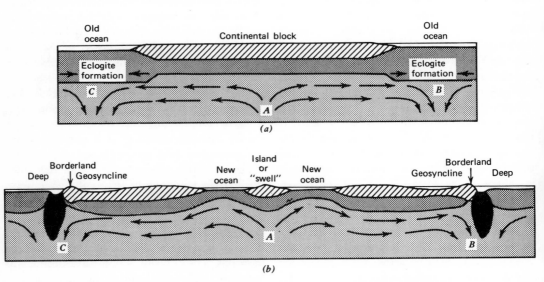

Figure 11-2. Convection current mechanism for causing continental drift (after Holmes, 1931). (*a*) Convection currents in the mantle create a region of tension in the continent, and convert basalt to eclogite where the currents descend. (*b*) The continent is split apart, and the continental masses are carried by the mantle convection currents, leaving a stretched and broken region between them; this becomes new ocean, and the size of the ocean grows with time as basaltic magmas are erupted above the ascending limb. (Compare with model for sea-floor spreading in Figures 12-22, 12-23, and 13-9.) Eclogite formed at *B* and *C* sinks (Figures 9-8(*f*), 10-9(*c*), and 12-25(*a*)), thus making room for the advancing continents.

the crust and extends the continent producing a stretched region or a disruptive basin between two main blocks. These are carried apart on the backs of the currents as depicted in Figure 11-2*b*. The migration is accomplished by removal of the obstruction of the old oceanic crust at *B* and *C* where the currents turn downward. Cooling at *B* and *C* causes conversion of basalt to eclogite. The heavier eclogite sinks thus making room for the advancing continents. The foundered masses of eclogite at *B* and *C* join in the main convective circulation melting at depth to form basaltic magma and rising again in ascending currents as at *A*. The basaltic magma heals gaps in the disrupted continent and contributes to the new ocean floor.

Holmes stated that the formation of a new ocean floor would involve the discharge of a great amount of excess heat. Figure 11-2*b* shows a fragment of sial left behind within the new ocean as an example of how Iceland may have formed, but this Holmes considered as only a local effect; normally, the newly exposed ocean floor consisted of basaltic material only.

Whereas Wegener and others had considered the continents as active elements drifting over and through the substratum, Holmes introduced the concept of passive continents being carried as if on a conveyor belt. This concept, and his use of eclogite as a sinker (Figure 9-8*f*), have become essential parts of recent theories.

DEVELOPMENTS SINCE 1950

The debate about continental drift stagnated in the 1940's, but the controversy was revived in the 1950's by the work of P. M. S. Blackett and S. K. Runcorn on paleomagnetism. Their results indicated that the former positions of the magnetic poles had changed relative to the continents, and continental drift is a process that could explain these changes. This new evidence led many geophysicists to consider the theory of continental drift seriously, while many geologists remained unimpressed by, and suspicious of, this new approach. We shall review the evidence in Chapter 12.

During the same period exploration of the ocean floor by marine geologists and geophysicists such as B. C. Heezen and M. Ewing showed that the midoceanic ridges were more nearly continuous than previously suspected, and that the ocean basins contained far less sediment than previously assumed. The mid-Atlantic ridge was found to be remarkably parallel to the continental borders of the Atlantic Ocean. These observations led geologists to conclude that the ocean basins were relatively young, and that some kind of upwelling occurred beneath the rifted oceanic

ridges. The formulation of these ideas into Hess's concept of sea-floor spreading is reviewed in Chapter 12.

During the 1960's former lines of evidence for continental drift were made more acceptable to scientists trained in physics and chemistry by the use of computers and isotopes (a) to match geometrically the borders of continents on opposite sides of oceans, and (b) to match structural provinces in terms of the isotopic ages of rocks. Examples of these approaches are reviewed below.

Other developments during the 1960's are reviewed in detail in Chapters 13 and 14.

These include:

1. The crucial interpretation of linear magnetic anomalies in terms of dated polarity reversals of the Earth's magnetic field and spreading from the oceanic ridges.

2. The interpretation of seismic data along ridges, island arcs, and transform faults.

3. The interpretation of sediment cores recovered from the deep ocean floor during the Joides program.

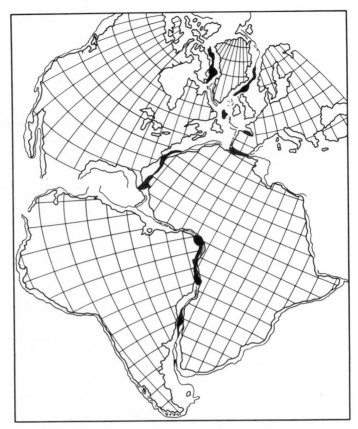

Figure 11-3. Computer fit of the continents around the Atlantic Ocean (after Bullard et al., 1965, with permission). Black areas show overlap of continental shelves. Drake et al. (1968) proposed an alternative fit based on a change in the pattern of magnetic anomalies which occurs on both sides of the Atlantic Ocean (north but not south). They fitted together the boundaries between the marginal "quiet magnetic zones" and the normal, higher amplitude anomaly pattern. Vogt et al. (1970) reviewed this and other hypotheses for the quiet magnetic zones. (Royal Society, London.)

Geometrical Fit of the Continents

Continental fits such as proposed by Wegener (Figure 11-1) are not accepted by all scientists. This is clearly shown by a remark of H. Jeffreys in 1964 according to E. C. Bullard: "I simply deny there is an agreement." In order to place the geometrical fitting of continental masses on a more objective basis Bullard and his associates therefore examined the fit of the continents around the Atlantic Ocean by numerical methods, and they found remarkably good

fits at the 500 fathom (about 915 m) contour which lies on the steep part of the continental slope. A computer was used first to fit South America to Africa with the result shown in Figure 11-3. Overlaps and gaps are indicated. The largest misfit, of 270 km, is at the Niger delta which is a recent addition to the continental edge (since the Tertiary). A similar computer procedure brought together from around the North Atlantic the continental masses of North America, Greenland, and Europe. The fit, shown in Figure 11-3, has omitted Iceland (composed of Tertiary and

Recent igneous rocks). The root-mean-square errors for these fits are 30 to 90 km. The fit of the southern block to the northern block is much poorer, and distortion of the continents to avoid overlaps requires rotation of Spain to close up the Bay of Biscay as shown in Figure 11-3. The reconstruction shows large gaps in the Caribbean and the Mediterranean, and most of Mexico and Central America has been omitted. Adjustments will have to be made to include the Paleozoic rocks known in Oaxaca, Mexico. The root-mean-square misfit is about 130 km. Note the relative rotations of the continental masses required in order to make the fit; this is shown by the lines of latitude and longitude in Figure 11-3.

There have been many estimated fits of Australia to Antarctica, mostly of the sketch-map variety because the problem of fitting the continents around the Indian Ocean is more difficult than that for the Atlantic Ocean. Using a corollary of the technique adopted for construction of Figure 11-3, W. P. Sproll and R. S. Dietz determined a good computerized fit for Australia against Antarctica in 1969, with a root-mean-square misfit of only 51.9 km at the 1000 fm (about 1,830 m) contour on the continental shelf. The fit is geologically permissible. A year later they published a result for the juxtaposition of Africa and Antarctica, seeking the Gondwana supercontinent. Also in 1970 A. G. Smith and A. Hallam published a reconstruction of the complete Gondwanaland, by bringing South America, Africa, Arabia, Australia, Antarctica, India, Madagascar, and New Zealand together by computer fit at the 500 fathom isobath, with the constraint that they investigated only configurations that were geologically reasonable. Their solution for the Africa-to-Antarctica fit differs in some respects from that of Dietz and Sproll, because the continent edges were fitted at different depths. The reconstruction is remarkably close to that published by Du Toit in 1937 when far less data were available to guide the matching of geological structures.

Du Toit's reconstruction is used in Figures 7-15 and 11-5.

In 1970, R. S. Dietz and J. C. Holden used the new geometrical and geological fits to repeat Wegener's reconstruction of *Pangaea* (Figure 11-1) with cartographic precision. They presented maps illustrating the breakup and dispersion of the continents during the past 180 m.y. Absolute geographic coordinates were assigned for the continents as well as for the active ocean rift zones and the oceanic trenches as they migrated to their present positions. They also extrapolated present day plate movements to predict the appearance of the world 50 million years from now.

Matching Age Provinces on Continental Reconstructions

Computer fits of continental margins such as that shown in Figure 11-3 give a less subjective framework than estimated fits for the comparison of structural trends from one continent to another, and recent work involving age measurements is beginning to provide more precise data. A good example is the matching of geological age provinces in West Africa and Northern Brazil, reported by P. M. Hurley and others in 1967 and illustrated in Figure 11-4.

West Africa is divided into two major age provinces, with K-Ar and Rb-Sr age determinations generally in the range 2000 m.y. in Ghana, the Ivory Coast, and regions to the west; and in the range 550 m.y. in the eastern part of Dahomey, Nigeria, and regions to the east. The sharp boundary between these provinces, shown by the dashed line in Figure 11-4, appears to head southwestward from a point near Accra. If Africa and South America had been together according to the fit in Figure 11-3 at the time the boundary was formed, the boundary would have entered Brazil just east of São Luis. Age analyses were therefore made on specimens collected near São Luis to see if this boundary did extend into South America. Both K-Ar and

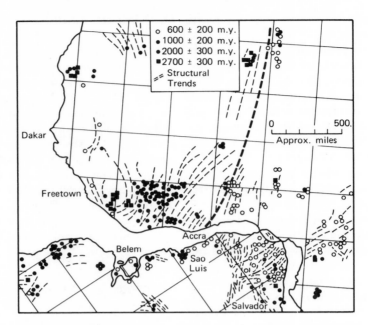

Figure 11-4. West Africa and South America shown fitted together according to the reconstruction of Figure 11-3. In West Africa the 2000-million-year Eburnean age province (solid circles) adjoins the 550-million-year Pan-African age province (open circles); the boundary between them is shown by the heavy dashed line. If Africa and South America were once joined together, this line would have entered Brazil near São Luis. The age measurements for Brazil appear to show the same age provinces as those in West Africa, with the boundary at the predicted location. There may be a similar correlation between West Africa and the east coast of Brazil north of Salvador (after Hurley et al., 1967, Science, 157, 495–500. Copyright 1967 by the American Association for the Advancement of Science. Reproduced by permission).

whole-rock Rb-Sr age measurements were made on the same samples where possible in order to obtain added information on the history of the basement rocks. The results show that the same age boundary does appear at almost exactly the predicted location. For specimens near the boundary, the K-Ar ages are in the range 410 to 640 m.y., and the whole-rock Rb-Sr ages are still in the 2000 m.y. range. Slightly further to the east the whole-rock age also has dropped to 665 m.y. The measurements plotted suggest the possibility of another trans-Atlantic age-boundary correlation further south towards Salvador where a 500 m.y. province gives way to base-ment exposures of 2000 m.y. age overlain or intruded by rocks of various ages down to 500 m.y. Geological studies have subse-quently confirmed the existence of a Pre-cambrian tectonic province in this part of Brazil, perpendicular to the coast, which is repeated in Gabon, Africa, in the position expected from the geometric fit of Figure 11-3 (Allard and Hurst, 1969).

Similar links have been noted between Europe and North America. The Lower Paleozoic carbonatites and alkalic rocks asso-ciated with the St. Lawrence graben system in Canada have been correlated with the similar rocks in Greenland and Scandinavia. These rocks appear to belong to a single alkalic rock province, defined by a rift system extending at least from central Canada to eastern Sweden on the predrift reconstruc-tion of Figure 11-3. Age determinations indicate that this rift system was active throughout its length about 565 m.y. ago (Doig, 1969).

Age determinations for massif-type anorthosites range from 1100 to 1700 m.y., but they cluster about 1300 ± 200 m.y. The global distribution of the anorthosites takes on a special significance when plotted on predrift reconstructions of the continents such as Figure 11-3 and Du Toit's reconstruction of Gondwanaland (Figure 11-5). The anorthosites then appear to lie within broad belts, at least one connecting America and Europe in Laurasia, and another traversing the continents forming Gondwanaland. N. Herz (1969) thought that the distribution and reasonably consistent ages of the anorthosites suggested that they were the product of a unique cataclysmic event or a thermal event that was normal only at an early time in Earth history. Of interest to us at present, however, is the fact that these anorthosite belts become apparent only if we assume that the continents have drifted apart since the anorthosite emplacement.

Recent compilations of the ages of basement rocks on the continents lends support to the former existence of two ancient continental cratons. Figure 11-5 shows again the predrift reconstruction of the continents, and the shaded areas are the continental blocks having apparent ages greater than 1700 m.y. The ancient cratons are not scattered uniformly throughout the predrift continental crust, and they occupy areas that can be enclosed within two rather smooth ovoid boundaries. One area is centrally located within the supercontinent of Laurasia, and the other similarly within Gondwanaland. These areas are chopped up by younger transcurrent, geologically active belts, and encircled almost entirely by belts of younger continent (Figure 7-15). This grouping of the ancient continental cratons makes it difficult to conceive of a series of drift motions, with splitting of the continents followed by gathering together again, earlier than the break-up that apparently occurred during the last 200 m.y. This evidence thus favors

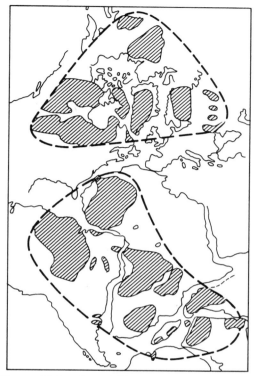

Figure 11-5. A predrift reconstruction in which the continental blocks having apparent ages >1700 million years (hatched areas) appear to be in a coherent grouping within two restricted regions. These blocks are transected and circumscribed by belts of younger rocks. Compare with Figure 11-4. It seems unlikely that during the time between 1700 and 200 million years ago the continents were scattered and drifting, only to be reassembled with this degree of ordering at 200 million years ago. Instead, there appear to have been nonmoving ancient nuclei and continental accretion up to the time of the great drift. (After Hurley and Rand, 1969.) Compare Figure 7-15. See Dietz and Sproll (1970), Smith and Hallam (1970), Schopf (1970), Dietz and Holden (1970). Copyright 1969 by the American Association for the Advancement of Science, Reproduced by permission.

the idea that continental drifting is a unique event which occurred only recently in geological time.

Twenty years detailed study of global tectonics convinced S. W. Carey that the continents could not be fitted together as claimed by Wegener on a Paleozoic Earth (Figure 11-1) if it had the same radius as the Earth today. In a symposium with contributions published in 1958 he showed how this could be done if the Permian Earth had a radius about 0.75 times the present radius, and a surface area about one half its present area. This is equivalent to an average expansion rate of about 0.5 cm/year. Carey adopted a model with an expansion rate increasing with time, however, as shown in Figure 12-11 in order to account for the large dispersions of the continents after their inferred break-up during the Mesozoic Era.

In 1969 R. Meservey also suggested that continental drift might require a large expansion of the Earth's interior. If the continents are moved simultaneously backward in time from their present positions to the relative positions shown in Figure 11-5, the perimeter of the Pacific Ocean must increase in order for it to pass over the Earth's circumference, and then decrease as the continents are reassembled on one side of the Earth. The size of the Pacific Ocean would increase from its present 35% of the Earth's surface to more than 50%. Meservey concluded that the evidence was strong that the combined length of the linkages around the Pacific perimeter decreases as we go back in time at least to 80 million years ago. If this conclusion is valid then there is no topologically possible transformation of the continents on an Earth of the present size that can explain continental drift. The apparent paradox could be resolved if the Earth had expanded during the last 150 million years.

REVOLUTION IN THE EARTH SCIENCES

As the evidence documenting the movement of lithospheric plates accumulated in the literature during 1966, 1967, and 1968 review articles began to appear in which proponents of the Mobilist view point announced a revolution in Earth sciences. J. Tuzo Wilson considers the revolution to be similar to, and as significant as, that which changed the approach to chemistry about 1800, that which occurred in biology about a century ago with the introduction of Darwin's theory of evolution, and that which occurred in physics when classical views were replaced by modern. The revolution, embracing the essential ideas of continental drift that had been debated for half a century, shows a promise of advancing the study of Earth sciences from the stage of data-gathering into the stage of formulation of a precise, comprehensive theory of global geology and use of the theory to make predictions. The new global tectonics challenges all of the past tectonic theories based on fixist or stabilist concepts. In particular the geosynclinal theories reviewed in Chapter 9 require re-evaluation within the new conceptual framework.

The December 1968 issue of *Geotimes* printed (a) an exposition of the revolution by Wilson, (b) a letter from V. V. Beloussov maintaining that the theory of continental drift and the concept of sea-floor spreading should serve only as working hypotheses along with others such as oceanization of the continental crust, and (c) a reply from Wilson restating his contention that the revolution will unite branches of the Earth sciences, formerly fragmented, into a new unified science of the dynamic Earth. I recommend this as required, critical reading for all students.

Whether or not textbooks should be rewritten and education reorganized in accordance with the developing ideas of global tectonics, as Wilson suggests, there is no doubt that the conventional geological subjects have benefited from critical reexamination in the context of global geology. This approach has certainly given geologists a different frame of reference for interpretation of their data, and all indications are that during the 1970's most of the classical and rather descriptive approaches to geology will be tested and modified in terms of the movements and interreactions of continents and oceanic plates. Wilson maintains that because the Earth is a single system the new theory of global tectonics should be learned first, and the traditional subjects can then be studied and more easily understood with the principles of Earth behavior providing the basis.

In the next three chapters we examine in detail the evidence which heralded this revolution.

12. *Paleomagnetism, Polar Wandering, and Spreading Sea Floors*

INTRODUCTION

Interest in continental drift was revived during the 1950's by paleomagnetic studies. Interpretations of fossil magnetism in rocks suggested that through geologic time the position of the Earth's magnetic axis had migrated relative to its rotational axis. The results of paleomagnetism and of marine geology and geophysics set the scene for Hess's formulation of the concept of sea-floor spreading. The subsequent discovery that the polarity of the magnetic field had reversed periodically provided a means for dating the linear magnetic anomalies that figured so prominently in the development of the concept (Chapter 13). The hypothesis of mantle convection as the driving force for horizontal motions at the Earth's surface was revived in various forms.

THE 1950'S: PALEOMAGNETISM AND POLAR WANDERING

All rocks exhibit magnetic properties one of which is a fossil magnetism or natural remanent magnetism (NRM), which was acquired during the formation of the rock and possibly modified afterward. The direction of magnetization was in response to the prevailing magnetic field, and NRM therefore provides the possibility of determining the direction of ancient geomagnetic fields and, in principle, the intensity of the field. The fossil magnetism is due to only a small proportion of minerals in rocks—the accessory iron oxides and sulfides. The intensity of magnetization of the rocks is low for this reason, and because magnetization occurred in a weak magnetic field. The magnetization of rocks has been studied for many years and by 1930 several important points had been clarified. It had been established that the Earth's magnetic field had not changed greatly in the recent geological past. The great majority of observations have been published since 1950, and it was not until then that the potential application of paleomagnetism to paleogeographic problems was put into practice. An interesting account of the historical development of this approach was given in 1967 in a book *Debate About the Earth* by H. Takeuchi, S. Uyeda, and H. Kanamori. E. Irving in 1964 published a book, *Paleomagnetism*, which summarized and reviewed all available paleomagnetic data.

Remanent Magnetism in Rocks

Remanent magnetism, or remanence, is that magnetization remaining in a substance in zero applied field. Thermoremanent magnetization (TRM) is the remanence acquired upon cooling through a certain temperature interval in the presence of a magnetic field. A mineral is ferromagnetic only at temperatures below its Curie point usually about 500°C. The magnetic minerals crystallize from lavas or magmas at temperatures greater than the Curie point, but they do not become magnetized until the temperature falls below 500°C. It has been shown experimentally that igneous rocks acquire a strong thermal remanent magnetism as they pass through the temperature interval 500 to 450°C. The magnetic minerals mainly responsible are within the system FeO-Fe$_2$O$_3$-TiO$_2$. TRM is many times stronger than that which could be induced in a rock at room temperature. It is not only strong but stable. Therefore igneous rocks provide excellent fossil magnets remaining unaffected by most of the later disturbances that may occur.

A small proportion of the detrital fragments in sediments are magnetic, and these have a tendency to align themselves in the direction of the prevailing geomagnetic field during deposition. The rock so formed then acquires detrital remanent magnetization, sometimes called depositional remanent magnetization (DRM). Many factors may disturb this alignment when the particle hits bottom or during compaction and cementation. It is therefore convenient to recognize depositional DRM acquired due to particle alignment during sedimentation, and post-depositional DRM acquired by particle rotation after deposition but before consolidation. An inclination error arises in many sediments; the declination of the magnetized sediment agrees with the magnetic field, but the inclination is often less than the field inclination. The difference may reach 25°. A bedding error is introduced if the surface of deposition is tilted. DRM is about a hundred times weaker than TRM and less stable.

Many rocks contain magnetic minerals which were formed at low temperatures by chemical processes or during metamorphism at temperatures below the Curie point. The growth of magnetic minerals produces chemical remanent magnetization (CRM) in

the rocks, and the direction of magnetization then corresponds to the prevailing field at the time of alteration and not at the time of formation of the rock.

The intensity of primary magnetization may decay with time, which is called viscous demagnetization or viscous decay. Also a new magnetization may be acquired at temperatures below the Curie point over long time spans, and this is called viscous remanent magnetization (VRM). Viscous decay and VRM in a prevailing field different from that of the original NRM will tend to obscure the primary magnetization.

If a rock is anisotropic the TRM may be deflected away from the prevailing field towards a direction of "easy" magnetization. Stresses arising from cooling or tectonic causes may also affect the remanence. Sometimes noticeable components of secondary or temporary magnetization are added to a sample between collection and measurement in the laboratory.

Paleomagnetic Measurements

Paleomagnetic surveys are made in rock units of known geological age such as a set of sedimentary beds, lava flows, or intrusive rocks which might span a time interval of 10^3 to 10^6 years. The meaning and accuracy of paleomagnetic interpretations is completely dependent on a knowledge of the geology of the samples. An adequate sampling scheme is essential for satisfactory results. Samples oriented with respect to the geological structure are collected so that their attitude when formed is known. The direction and intensity of NRM is then measured. As discussed in the preceding section NRM is usually complex; the primary or original magnetization may have been modified by secondary magnetizations. These components are resolved by using various techniques of which demagnetization in stages is the most important. The unstable components of NRM are removed by alternating magnetic fields (magnetic cleaning), by heat (thermal cleaning), by chemical treatment, or by combinations of these techniques. The more detailed the tests conducted, the more certain are the paleomagnetic results. All paleomagnetic studies must be evaluated in terms of the stability and reliability of the reported remanent magnetization.

The basic assumptions in paleomagnetic studies are that the Earth's magnetic field has always approximated a dipolar field (Chapter 3) and that the mean direction of permanent magnetization of the rocks at the place of observation represents the mean direction of the ancient geomagnetic field during the time of formation of the rocks. There is good evidence that the first assumption is valid at least as far back as mid-Tertiary and furthermore that the magnetic poles remained close to the present geographic poles in this interval. The second assumption is not necessarily valid because the geomagnetic pole position measured is that corresponding to the time of magnetization, which may be later than the time of formation or the physical age of the rock. Consideration of the fossil magnetism with respect to some geological feature of the rock unit, such as a fold, often permits a conclusion about the acquisition of NRM relative to the times of formation of the rock unit and of the geological feature of the unit.

The final value obtained for each sample is a geographic direction with an angle of dip approximating to that of the ambient geomagnetic field at its time of magnetization. This is plotted on an equal-area or stereographic projection as a point, and points are plotted for every sample measured in the rock unit. Figure 12-1 shows examples of NRM directions measured in four separate rock units. Figure 12-1a shows a close grouping of single determinations from 13 samples of recent sediments. They cluster around direction F of the present dipole field of the Earth. Twenty-one samples from the Miocene sediments of the Arikee Formation in South Dakota show a high dispersion of directions in Figure 12-1b. Figure 12-1c is an

example of a smeared distribution in Triassic sediments of Sidmouth, England, with the directions strung out approximately along the great circle passing through the direction F of the present axial dipole field of the Earth at Sidmouth and the poles M_S^- and M_S^+ of the axis of stable magnetizations determined for Triassic sediments elsewhere in England. Close groupings of directions are often obtained from observations from the same lava flow as shown in Figure 12-1d.

In a paleomagnetic study a question always remains about the different components contributing toward the NRM and their timing relative to the age of the rock. If the measured directions are closely grouped and divergent from the present field as in Figure 12-1d, this indicates the absence of viscous components imposed recently and of CRM from recent weathering. Consistency tests are of greatest significance if applied to results for rock types of different origins and composition. Figure 12-2 shows good agreement between results obtained for three groups of Upper Triassic rocks of New Jersey: red sediments, dolerite intrusions, and basalt lava flows. The mean directions do not differ significantly among the groups. This indicates the absence of inclination error in the sediments and implies that the effect of cooling stresses in the igneous bodies was negligible. The mean direction is considered to be reliable.

The directions recorded for samples in a rock unit represent former magnetic fields as they existed through an appreciable time interval, 10^3 to 10^6 years. In view of the secular variations known to occur in the Earth's magnetic field (Chapter 3) it is not surprising that the directions never agree exactly. Average values have to be calculated from sets of observations such as those shown in Figures 12-1 and 12-2. Statistical tests for the accuracy of calculated mean directions are described in Irving's book. It is necessary to provide some measure of the dispersion of observations about the mean direction and to estimate the accuracy with which the

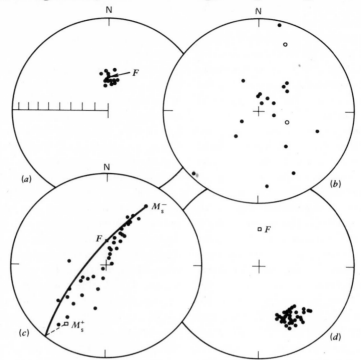

Figure 12-1. Examples of observations (all NRM directions) from a single locality. Directions with positive (negative) inclination are plotted as dots (circles) on a polar equal area net. (*a*) Payette Formation, Glenns Ferry, Idaho, United States. (*b*) Arikee Formation, South Dakota, United States—the scatter in directions is very great ($k = 3$). (*c*) Triassic marls from Sidmouth, England—the great circle through the poles M_s^-, M_s^+ of the axis of stable magnetization and the direction F of the geocentric axial dipole field is marked. (*d*) Cenozoic lava flow, Australia (after Irving, 1964, Figure 4.1, with permission of John Wiley and Sons).

mean is defined. The 95% circle of confidence is used to compare two mean directions. If the circles do not intersect the two mean directions may be judged as differing significantly (Permian and Triassic circles in Figure 12-6*b*). When the circles overlap a further test is required. The mean paleomagnetic pole is defined relative to the present geographic pole for a specific sampling unit covering an interval of time and not for the entire Earth at an instant of time. Thus it is not the ancient equivalent of the present geomagnetic pole. Estimates of the positions of mean paleomagnetic poles are the basic information needed for study of the variations in the Earth's field and for the comparison of results from different rock units on

different continents. No significant shift of paleomagnetic poles relative to the present geographic poles has been detected since mid-Tertiary times, but evidence indicates that they could have shifted considerably earlier. Paleoclimatic studies compared with paleomagnetic results, however, seem to indicate that the ancient magnetic poles have remained approximately coincident with ancient geographic poles as might be anticipated from ideas about the origin of the Earth's magnetic field (Chapter 3).

Interpretation of Paleomagnetic Pole Positions

Mean paleomagnetic directions are ex-

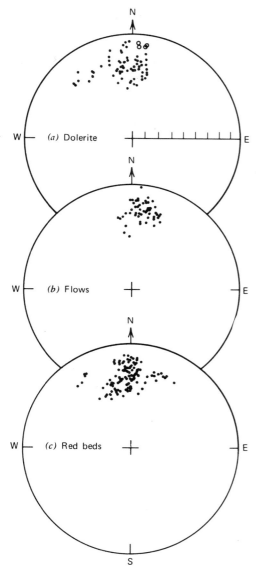

Figure 12-2. Paleomagnetic directions in the Newark Series of New Jersey. The directions in the red sediments are those of NRM. The NRM of the igneous rocks contains viscous components directed along the present field. These have been minimized by magnetic cleaning in 150 oe (peak) alternating magnetic field and the directions after this treatment are given here. Equal-area projection (after Irving, 1964, Figure 5.1, with permission of John Wiley and Sons).

pressed in polar coordinates in terms of the angle of declination from the present meridian

and the angle of dip from the horizontal (compare Figure 3-13*b* for declination and dip of the present magnetic field). The paleomagnetic pole of a rock unit from a given land mass tells us the direction of the land relative to the paleomagnetic pole and the paleolatitude of the rock unit relative to this pole (Figure 3-13*a*). The land mass could be anywhere on a small circle centered on the paleomagnetic pole.

It was in the 1950's that paleomagnetic studies were extended back in time beyond the Tertiary, and results obtained from Triassic red beds in England by P. M. S. Blackett's group were the first to reopen the question of the possibility of continental drift. Clegg, Almond, and Stubbs (1954) found that the geomagnetic field direction in the Triassic diverged about 30° from the present north geographic pole, and it had a dip of about 30° in contrast to the 65° dip at the present time. This proved that the paleomagnetic pole and England had moved relative to each other, but provided no information about the relative movements of England, the paleomagnetic pole, or both relative to the present geographic coordinate system. They assumed that the geomagnetic poles had remained approximately coincident with the present geographic poles, and explained the change in angle of declination as a rotation of England through about 30° since the Triassic. The change in dip was interpreted in terms of a change in latitude, England having migrated northward since the Triassic (note the dips at various latitudes in Figure 3-13*a*). This was startling new evidence for continental drift.

Additional data were gathered for different land masses during different geological periods, and the method of presentation preferred by Blackett's group is shown in Figure 12-3. This shows for various geological periods the ancient latitudes and orientations of Europe, North America, India, and Australia with respect to the paleomagnetic pole, which is assumed to have remained coincident with the present north geographic

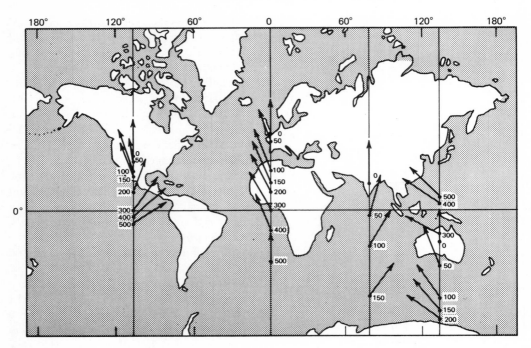

Figure 12-3. Paleolatitudes and orientations of North America, Europe, India, and Australia with respect to the magnetic pole at various geological periods. The numbers indicate the ages of points in millions of years. The paleomagnetic pole is assumed to have remained coincident with the present geographic pole (after Blackett et al., 1960, as redrawn by Takeuchi et al., 1967, with permission of the authors. Reproduced by permission of Freeman, Cooper & Co.).

pole. Results for each continent are given on a line of longitude passing through a town located approximately at the center of the continent, and the ancient latitude and orientation of each continent are given in terms of the reference town. No ancient longitudes are given because they are indeterminable, although this does not mean that there has been no longitudinal movement. Figure 12-3 contains all of the information that can be provided by the measurement of paleomagnetic poles.

S. K. Runcorn and his colleagues examined the data with the initial assumption that the land masses remained fixed in respect to the present geographical coordinates. For England and other European countries, they located the average positions of the paleomagnetic poles for each geological period since the Precambrian and plotted them on a

world map. The older the period the farther removed was the paleomagnetic pole from its present position, as shown in Figure 12-4a. Figure 12-4a can be considered as a diagram presenting in graph form the relative positions of Europe and its paleomagnetic pole at different periods, with no implications for the absolute movements of poles or continents. This is a path of apparent polar wandering, which may reflect polar wandering, continental drift, or some combination of both. The data from a single region are insufficient to illustrate what their relative contributions may have been. The figure shows in a different way the same kind of data illustrated for Europe in Figure 12-3. The study of paleoclimatology, however, had earlier produced a locus of polar wandering for the geographic north pole, which is reasonably similar to the locus of the paleomagnetic poles shown in

Figure 12-4. This suggests that the geographic pole and the paleomagnetic pole had moved together relative to the present geographic coordinates. In 1954 Runcorn and his colleagues interpreted the paleomagnetic data for Europe in terms of true polar wandering without invoking appreciable continental drift.

The locus of apparent polar wandering based on North American samples is shown in Figure 12-4b. It is quite similar to that obtained for Europe but not identical. The paths shown in Figure 12-4 are from Runcorn's 1962 book, and they are based on considerably more specimens than the first paths for Europe and North America which were published in 1954 and 1956 respectively. Yet according to Runcorn they are not different in broad outline from the early paths. The two paths are compared in Figure 12-5. Their approximate parallelism suggests that true polar wandering has occurred. Runcorn and his colleagues concluded that there is a consistent distance of about 30° of longitude between the two curves from Precambrian to Triassic. Therefore a polar wandering hypothesis alone is inadequate, because if only the poles had moved each continent would have an identical locus of polar wandering. These paleomagnetic data are satisfied, however, if continental drift according to Wegener's scheme is added (Figure 11-1). If North America were moved 30° of longitude to the east the two curves would become effectively coincident from the Precambrian to the Triassic and the Atlantic Ocean would be closed. If the continents had moved apart in the Triassic the coincidence of the polar paths from the Triassic to the present is also explained.

Extension of paleomagnetic research to other continents revealed that each continent had a different locus of apparent polar wandering. Whereas it is fairly straightforward to bring together the paths shown in Figure 12-5 by closing the Atlantic Ocean, it becomes much more difficult to obtain a solution for the relationship between polar wandering and continental drift when so many curves are involved. Not only continental drift but continental rotation must be invoked (Figure 12-3). Thus the collection of additional paleomagnetic data raised additional problems of interpretation.

Alternate interpretations have been proposed. Not everyone was convinced about the statistical validity of the distinction between the two paths for Europe and North America (Figure 12-5), and it was suggested by F. H. Hibberd in 1962 that a single path of polar wandering would satisfy most of the data from all continents with all discrepancies between this and previously published loci resulting from secondary magnetization (see also Figure 12-6). Figure 12-1 shows that secondary magnetization can produce rather drastic effects. The assumption of a dipolar field prior to the Tertiary has been challenged. Some consider it more reasonable to assume the existence of a multipolar field than multiple loci of polar wandering.

Despite these reservations the paleomagnetic results by the end of the 1950's were forcing geologists and geophysicists into a reconsideration of continental drift. This led to a symposium organized for the Royal Society of London in 1964 by P. M. S. Blackett, E. Bullard, and S. K. Runcorn which was published as a special volume in 1965. In his introduction Blackett noted that it was advances in two virtually new subjects, the study of magnetism of rocks and the study of the floors of the oceans, that were helping to overcome the former widespread objections to the concept of continental drift. In the final discussion, after 320 pages of text, M. G. Rutten stated that remembering the violently opposed continental drift discussions during the 1920's, the

"papers of this symposium have shown that, apart from paleomagnetic data, nothing much has been changed . . . It still depends on which part of the geological data one finds most strongly heuristic, if one is a 'drifter' or a 'fixist' . . . The only way to remain fixist

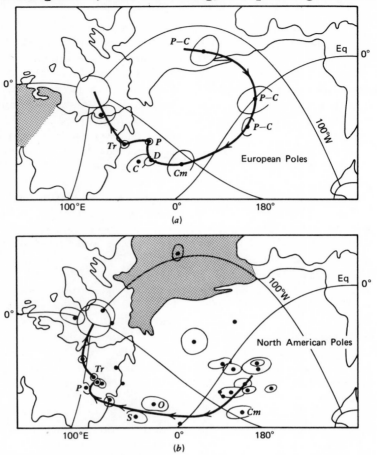

Figure 12-4. Polar wandering curves based on measurements on rocks in (*a*) Europe, and (*b*) North America. Key: *P-C*, Precambrian; *Cm*, Cambrian; *O*, Ordovician; *S*, Silurian; *D*, Devonian; *C*, Carboniferous; *P*, Permian; *Tr*, Triassic (based on Figure 19 of Runcorn, 1962, by permission of Academic Press, Inc.).

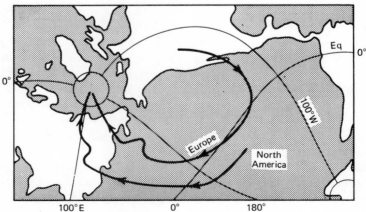

Figure 12-5. Comparison of polar wandering curves for Europe and North America, redrawn from Figure 12-4 (after Runcorn, 1962, by permission of Academic Press, Inc.).

now, is to disbelieve paleomagnetism, a position which becomes more and more awkward as its methods tend to become better substantiated. In future we shall have to base all of our geological theoremata on the data supplied by paleomagnetism."

PALEOMAGNETISM SINCE 1960: MAGNETIC REVERSALS

Polar Wandering and Continental Drift

Figure 12-5 summarizes the paleomagnetic results of the 1950's which suggested that true polar wandering and continental drift had both occurred. Improvement of techniques and the accumulation of more data permitted additional tests for consistency (Figure 12-2), with revision and reevaluation of the significance of the apparent polar wandering curves. Several compilations and reviews were published. The review by A. Cox and R. R. Doell in 1960 represents the state of the art at the end of an exciting decade. Irving's book of 1964 includes more results and shows how rapidly data accumulated.

The accumulation of data did not solve all problems and lead to general acceptance of the thesis portrayed in Figure 12-5. This is demonstrated in striking fashion by three reviews published almost simultaneously. In a symposium volume edited by Runcorn in 1967 J. Hospers and K. M. Creer reviewed essentially the same data for North America and Europe, and Creer extended his synthesis worldwide to include all paleomagnetic data available. Hospers concluded that the data suggest true polar wandering at least since the Cambrian and cast serious doubt on the reality of continental drift. Creer prepared apparent polar wandering curves for seven continents, superimposed the Upper Paleozoic portions for five of them, and concluded that the circumstantial evidence for drift now appears to be overwhelming. In 1968 I. A. Rezanov examined the data in Irving's book, together with all available data for the U.S.S.R., and concluded that paleomagnetic data are still so unreliable that they cannot be used as evidence either for or

against the relative drift of continents. We cannot spare the pages to attempt resolution of such conflicting interpretations; all we can do is to expand slightly this outline of their conclusions.

Irving reexamined the apparent polar wandering curves of Figure 12-5, using all pole estimates available by late 1963 that fulfilled his minimum reliability criteria. Instead of the previous simple pattern, Figure 12-6a shows what Irving referred to as rather curious changes in the relative longitudes for the pole paths of North America and Europe-northern Asia. The North American path begins to the west of the Eurasian path in the Cambrian. They cross around Silurian times and again between Triassic and Cretaceous times. Irving concluded that the differences between the Lower Paleozoic curves may or may not be fortuitous, and that many more results are needed before this feature can be adequately discussed. He emphasized that it would be many decades before an adequate coverage would be available. On the other hand there is little doubt, on the basis of the 1963 evidence, that the longitude difference in the Permian poles is significant and incidentally the most marked in the diagram; 95% circles of confidence for several of the mean poles are shown in Figure 12-6b.

Figure 12-6a shows in addition to Irving's plotted poles and paths six poles determined by Hospers using paleomagnetic triangulation. The procedure used is to take widely separated sampling sites from the same continent using rocks of the same age. The paleomagnetic pole for the continent at this time lies at the point of intersection of the paleomeridians drawn through each site. If everything is as it should be the pole positions

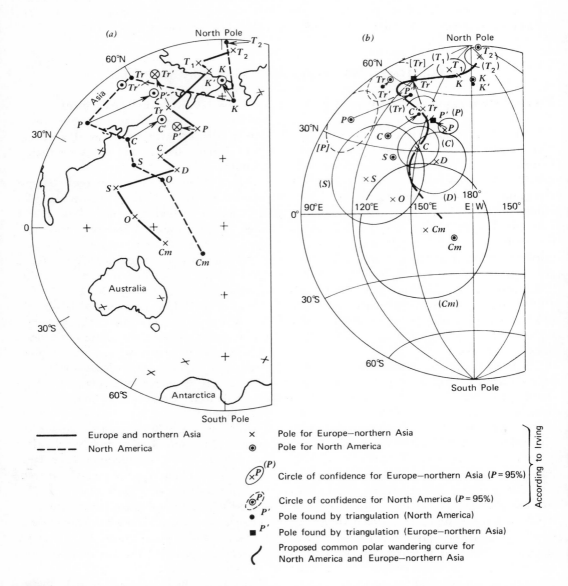

Figure 12-6. Polar wandering curves after Hospers (1967). (*a*) The Cambrian and younger pole positions for Europe and North America are shown, according to data evaluated by Irving in 1964. Separate paths for each continent are shown, and these paths intersect each other twice. Contrast Figure 12-5. The poles that Hospers located by triangulation are distinguished from the others by symbols with a prime (e.g. *Tr'*). The difference between Irving's poles and the triangulated poles is shown by arrows. Key: T_2, Upper Tertiary; T_1, Lower Tertiary; *K*, Cretaceous; *J*, Jurassic; *Tr*, Triassic; *P*, Permian; *C*, Carboniferous; *D*, Devonian; *S*, Silurian; *O*, Ordovician; *Cm*, Cambrian. (*b*) Hospers' preferred single polar wandering curve common to Europe-northern Asia and North America, based on all previous data and the new triangulated data. (Reproduced by permission of John Wiley and Sons.)

on the meridians calculated in the usual way from the inclination (dip) should coincide with the intersection point determined by triangulation. In order to test the reliability of Irving's poles Hospers compared them with the six triangulation poles. For North America he found that results from Carboniferous and Permian samples (sediments) show a statistically significant difference, whereas Triassic and Cretaceous data (including igneous samples) do not. For the Eurasian poles he found no statistical difference in Permian samples but a significant difference for the Triassic results. Hospers reviewed the many sources of systematic error in paleomagnetic inclinations measured in sediments, and concluded that these errors are likely to decrease the inclination and therefore provide a calculated pole position too far from the sampling site (Figure 3-13a). He therefore considers the paleomagnetic intersections to be more reliable. Figure 12-6a shows that the new pole positions found by triangulation bring the separate polar wandering paths closer. Although Irving had concluded that the longitude difference between Permian poles is significant, Hospers's largest revision closes the gap between the two Permian poles.

Hospers presented a single polar wandering curve for North America and Europe-northern Asia using his six preferred triangulation points and the other poles from Irving. Figure 12-6b shows the curve passing among the pole positions transferred from Figure 12-6a, together with some of Irving's 95% circles of confidence. The curve is nowhere outside the circles of confidence for the poles used and is within a few degrees of the triangulated poles. Hospers concluded that on this common polar wandering curve pole positions for the Upper Tertiary, Cretaceous, and Triassic times may be fixed with considerable confidence and for Permian and Carboniferous times with less accuracy. The Silurian and Cambrian poles are uncertain, and any comparison of great circle distances between Eurasian and North American

poles is meaningless for Cambrian and Silurian times. The rate of polar wandering indicated is about 90° in 600 m.y. or an average of about 1 or 2 cm per year. Creer felt that Hospers's conclusions were not justified. He stated that Hospers's preference for triangulated poles over Irving's calculated mean poles only makes sense if it can be shown that the inclinations of the remanent magnetism of the rock formations studied are likely to be systematically low, and this is not so.

Creer's first step in the synthesis of worldwide paleomagnetic data was to construct apparent polar wandering paths for the seven continents: South America, Africa, Australia, Europe-Russia, North America, India, and Australia. These were based for the most part on Irving's data compilation with inclusion of all other reliable data. For many of the poles data were sparse. No Cambrian poles and only one Ordovician pole was used. The principal paleomagnetic argument supporting the drift hypothesis is the divergence of apparent polar wandering curves from the geographic poles when drawn on the present globe, and Creer found marked divergence of the curve for Europe from those for Australia and India.

More data are available for Europe-Russia than for any other land mass, and one might assume therefore that the pole path was reasonably established. Nevertheless Creer modified the curve for Eurasia quite significantly by introducing a remagnetization hypothesis to explain some peculiarities about NRM of Devonian red-beds. He presented an up-to-date version of the original polar wandering curve for Eurasia (Figure 12-4a), which he based on essentially the same data as the curve in Figure 12-6a. This is shown in Figure 12-7a for the south poles with standard error circles; the curve is not reconcilable with curves for South America and Africa, and Creer resolved this problem by means of evidence that two significantly different paleomagnetic poles could be derived from studies of Devonian

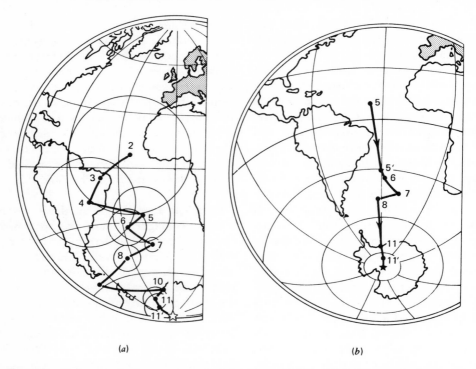

(a)   (b)

Figure 12-7. Polar wandering curve revised by Creer (1967). (a) Polar wandering curve (south pole) for Europe from Cambrian onward based on Irving's (1964) data. (Compare 12-4(a) and 12-6(a).) Standard error circles are plotted. (b) Polar wandering curve (south pole) for Europe and Russia; this is Creer's revised version of the curve in Figure (a). Key: 2, Cambrian; 3, Ordovician; 4, Silurian; 5, Devonian; 5′, Devonian, presumed Carboniferous magnetic age; 6, Carboniferous; 7, Permian; 8, Triassic; 9, Jurassic; 10, Cretaceous; 11, Tertiary; 11′, Upper Tertiary. (Reproduced by permission of John Wiley and Sons.)

red-beds. According to the remagnetization hypothesis one of these represents the true Devonian geomagnetic field, and the other the Carboniferous or permo-Carboniferous field. The revised polar wandering curve thus derived is shown in Figure 12-7b.

Creer then proceeded with paleogeographic reconstructions. The apparent polar wandering curves for five continents—South America, Africa, North America, Eurasia, and Australia—indicated similar polar movements of about 50° during the Upper Paleozoic. By taking a set of spherical shells, on each of which is drawn a particular continent and its polar wandering curve, Creer was able to superimpose the Upper Paleozoic portions of the curves with the Carboniferous poles

coincident. This brought the continents into close proximity as shown for four of them in Figure 12-8. He therefore concluded that during the Upper Paleozoic the paleomagnetic pole had moved about 50° relative to the *fixed* distribution of continents shown in Figure 12-8. This reconstruction differs from Wegener's (Figure 11-1a) and the computer fit (Figure 11-3) mainly because there is a gap between Laurasia and Gondwanaland of about 1500 km at its narrowest, between Africa and the eastern United States. Thus according to the paleomagnetic method of continental reconstruction a broad Tethys Sea completely separated the supercontinents of Laurasia and Gondwanaland during the Upper Paleozoic. Study of the divergence of the Mesozoic curves for the individual continents away from the superimposed master Paleozoic curve of Figure 12-8 shows, in principle, the sequence and direction of breakup of the continents. According to the scheme in Figure 12-8 the initial movements apparently occurred between the Carboniferous and the Permian when North America, Europe, and Australia were displaced from South America and Africa.

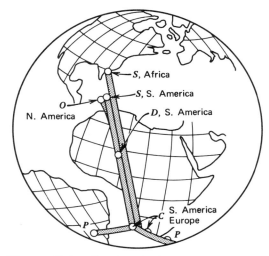

Figure 12-8. Sketch of photograph of polar wandering curves (south poles) for Europe, North America, Africa, and South America. The curves were drawn on spherical shells, and shifted over a globe until the wandering curves coincided (after Creer, 1967). Superposition of the Upper Paleozoic portions of the curves brings the continents into close proximity. Compare Figures 11-1(a) and 11-3. Key: *S*, Silurian; *O*, Ordovician; *D*, Devonian; *C*, Carboniferous; *P*, Permian. See paper by McElhinny and Luck (1970) for reconstruction of Gondwanaland from a common polar-wandering path for Lower Paleozoic data from southern continents. (Reproduced by permission of John Wiley and Sons.)

These stimulating experiments in paleogeography offer tantalizing glimpses of what paleomagnetic methods may provide in the future when more data are available and their interpretation is better understood.

Irving stated in 1964 that many decades of effort were yet required, and Rezanov would certainly maintain that Creer had over-extended the data. Creer himself emphasized two of the problems which limit interpretation of paleomagnetic data: the fact that the magnetic age of many sediments is younger than the fossiliferous age, and the difficulty in establishing equality of geological age for rock formations such as unfossiliferous red-beds in widely separated continents. A few quotations from Rezanov's translated paper will serve to illustrate his contention that even the most recent paleomagnetic data are too unreliable to be of any use in connection with the hypothesis of continental drift.

"We assembled only the most reliable determinations of the poles.

"So far as we could see, availabilities of large numbers of paleomagnetic measurements lead almost invariably to a prodigious scattering of paleomagnetic poles, and it becomes evident that no opinion at all may be formed on the continental drift on the basis of one single determination of the pole. But this is precisely the case with South America, in which the Silurian, Devonian, Carboniferous, and Permian paleomagnetic poles were found, each on the basis of one single determination. For Africa, two Carboniferous and two Permian poles are available, both for the same district; it develops that they are 5000 and 7000 km apart, respectively.

"The example with the 'migration' of the poles of the Nizhnyaya Tunguska area is a clear illustration of some kind of major error in the paleomagnetic determinations of the pole. Three regions are indicated, 6000–9000 km from each other, and the pole 'jumps' therein from region to region.

"It should be essential to investigate the causes responsible for inconsistencies of the paleomagnetic data for one and the same continent prior to drawing such responsible conclusions (tentative as they may be) in regard to such tremendous horizontal shifts and rotations within the continents.

"The width of the Atlantic Ocean is smaller than the range of reproducibility of the paleomagnetic measurements.

"But even this accuracy is unattainable for geological periods older than the Carboniferous. The scattering is as broad as 10,000 kms, i.e. the maximum possible discrepancy when magnetic axes of the earth are perpendicular to each other, in the Rhipean, Cambrian, Ordovician, and Silurian."

Many of the uncertainties arise from measurements of sedimentary rocks. A 1967 paper by C. S. Grommé, R. T. Merrill, and J. Verhoogen is an encouraging sign that we can expect better interpretations as more measurements are obtained for igneous rocks dated radiometrically. These authors determined the directions of NRM for two igneous intrusions in California with potassium-argon apparent ages of 136 m.y. and 142 to 129 m.y., and they obtained a mean paleomagnetic pole to which they assigned an averaged age of 138 m.y. In Figure 12-9 the new determination was compared with the five other Cretaceous paleomagnetic poles available for radiometrically dated igneous rocks in North America. The 95% confidence circles or ovals overlap and none of the six pole positions is significantly different from any other. This indicates that from about 138 to 84 m.y. ago either polar wandering relative to North America did not occur or its rate was too slow to be detected. Apparent polar wandering did occur earlier than 138 m.y. according to the only two other Mesozoic paleomagnetic poles available from radiometrically dated igneous rocks. These are shown in Figure 12-9 as *WM*, the White Mountain series dated at 180 m.y., and as *NG*, the Newark Group dated at about 202 m.y. The directions for the Newark Group are shown in Figure 12-2.

These data also provide new evidence that the Earth's field was a geocentric dipole farther back in time than the Cenozoic Period. The Cretaceous paleomagnetic poles from widely spaced localities in North America are coincident, and the paleomeridians have a common intersection.

Figure 12-9 compares the North American poles with 20 reliable late Mesozoic poles from other continents. The authors took the seemingly drastic step of omitting sedimentary rocks, but they pointed out that of the total of about 60 independent Jurassic and Cretaceous poles listed by Irving in 1964 only three from sedimentary rocks are reliable and are accompanied by confidence intervals. The paleomagnetic poles representing large and overlapping segments of Mesozoic time are so closely grouped for North America, Africa, and Australia that differences between individual poles do not indicate polar wandering. On the other hand the three groups of poles are distinctly separated suggesting that continental drift has occurred among them at some time between the late Mesozoic and the present.

The dated results plotted in Figure 12-9 show no apparent polar wandering for North America between 138 and 84 m.y. ago, nor for Africa between 209 and 109 m.y. ago, nor for Australia between 178 and 93 m.y. ago. Apparent polar wandering occurred for North America between 202 and 138 m.y. ago. These results suggest that effectively no true polar wandering occurred during the Jurassic and much of the Cretaceous between about 200 and 100 m.y. ago. The apparent polar wandering for North America between 202 and 138 m.y. ago the authors therefore interpret as continental drift.

They extended these conclusions by considering other published evidence that Australia showed very little apparent polar wandering from late Carboniferous to very early Tertiary (about 300 to 60 m.y. ago), and that considerable apparent polar wandering for Africa occurred between early Permian and late Triassic times (about 275 to 200 m.y. ago) and again since about the middle of the Cretaceous (about 100 m.y. ago). Thus the periods of apparent polar wandering for North America and Africa did not coincide in time within the Mesozoic

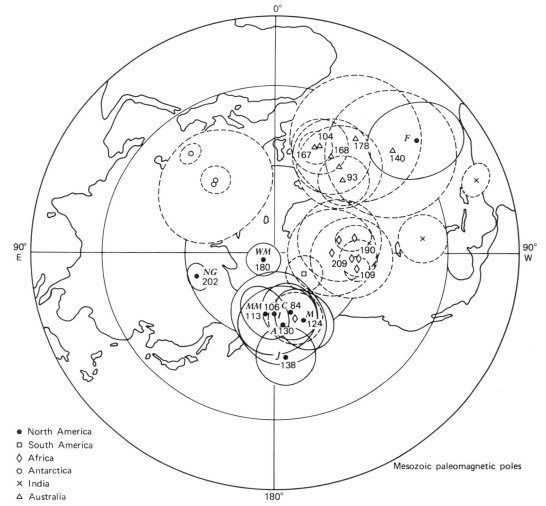

Figure 12-9. Equal-area map of northern hemisphere showing selected Mesozoic paleomagnetic pole positions for six continents (after Grommé et al., 1967); 95% confidence ovals or circles are shown as solid lines for North America poles and as dashed lines for others. Radiometric ages (in millions of years) are given where available. *J* is new determination by authors. (From Jour. Geophys. Res., **72**, 5661, 1967, with permission.)

Era. From the Australian evidence they concluded that true polar wandering during the Mesozoic was either absent or greatly subordinate. Therefore the apparent polar wandering for Africa and North America they interpreted in terms of continental drift.

The results shown in Figure 12-9 for North America suggest a period of continental drift followed by a long period of essentially no drift. Considering also the other results referred to above there is apparently a pattern for continents to experience long periods of essentially no drift preceded and followed by shorter periods of relatively rapid drift. The periods of drift for different continents do not necessarily coincide. Apparent polar wandering, and hence by implication continental drift, is an intermittent process.

Paleomagnetism and the Earth's Radius

The expanding Earth theory was outlined in Chapter 9. The usual model involves an originally continuous layer of sial broken into continental slabs as the Earth expands. The continents remain constant in area as they are moved radially outward, and as expanding ocean-basin floor is produced between them. This was B. C. Heezen's interpretation of the cross-section through the Atlantic Ocean shown in Figure 9-9. As the radius of the Earth increases the geocentric angle between two fixed points on a stable continental mass will decrease. Therefore paleomagnetic methods potentially permit calculation of the Earth's radius at different times in the past if the field remained dipolar.

Figure 12-10 following Irving (1964) summarizes three estimates (by Egyed, Carey, and Hilgenberg) of the rate of increase of the Earth's paleoradius expressed as fractions of the present radius which is set at 1.0. These rates were inferred independently of paleomagnetism.

Several methods have been used to estimate the paleoradius of the Earth using paleomagnetic methods. The simplest uses the equation

$$r_p = d/[\cot^{-1}(\tfrac{1}{2}\tan I_1) - \cot^{-1}(\tfrac{1}{2}\tan I_2)] \tag{12-1}$$

where I_1 and I_2 are paleomagnetic inclinations for two sites of the same age on the same stable continental mass and d is the distance measured on the surface between the two geomagnetic latitude circles. Cox and Doell averaged 80 values calculated for the Permian and obtained the value 6310 km, with the standard deviation of the mean being 230 km. This result is shown on Figure 12-10. The method is subject to uncertainties involving the average direction of magnetization of the sample and its time of magnetization compared to its time of formation and to uncertainties about matching the presumed or apparent ages of widely spaced rocks. It is doubtful that radius changes of less than

20% can be detected. Estimation of radius changes involves another assumption, that the area of the continental mass has remained unchanged by tectonic activity.

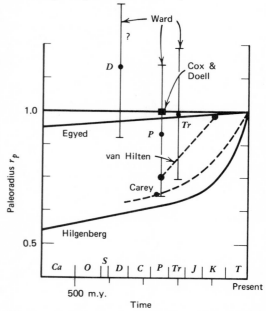

Figure 12-10. Changes in the Earth's radius since the beginning of the Cambrian (modified after Irving, 1964, Figure 10.31). Geological time is set out horizontally, and the paleoradius vertically; $r_p = 1.0$ is the present radius. The hypotheses of expansion are: Egyed 0.05 cm/year; Hilgenberg (1962); Carey (1958) $r_p = 0.75$ for late Paleozoic and the dashed line being surmise from his text. The values determined paleomagnetically by Ward (1963) are D, Devonian; P, Permian; Tr, Triassic, the limits being estimates of the standard deviations of the means. Points P, K connected by dashed line are from van Hilten (1968); Cox and Doell (1961) gave value for Permian of 6310 km with standard deviation 230 km. (Reproduced by permission of John Wiley and Sons.)

More generalized procedures include that of M. A. Ward who used the criterion that the most probable paleoradius is that for which the dispersion of paleomagnetic poles for a continent is a minimum. Three results determined for the Devonian, Permian, and Triassic have standard deviations of 20%;

there is no systematic trend of the mean values with time. D. van Hilten modified this method to take into account possible deformations of the continent and found it suitable only for the paleomagnetic data available in the Permian and Cretaceous Periods. The results are shown in Figure 12-10 without error bars and connected tentatively by the dashed line.

The paleomagnetic results are inconsistent with the hypotheses of Carey and Hilgenberg, but not sufficiently accurate to test the slow expansion rates proposed by Egyed. It appears therefore that the expanding Earth theory alone is inadequate to explain continental drift.

Magnetic Reversals in Igneous Rocks

The discovery by B. Brunhes in 1906 that some rocks are magnetized in a direction opposite to that of the Earth's present magnetic field received very little attention for many years. In the 1950's there developed a controversy over the origin of the normal (negative) polarities and the reversed (positive) polarities. It was established that the total numbers of reversed and normal rocks are about equal since the Precambrian. Many reversals of polarity were found in stratigraphically dated sequences of Mesozoic and Tertiary rocks, apparently occurring at intervals of a few thousand years. This suggested that the change of polarity was caused by a reversal of the Earth's magnetic field but the rocks were not accurately dated, and it could not be proved that the reversals occurred simultaneously in different locations. The only alternative solution appeared to be that certain rocks were capable of selfreversal due to some chemical or mineralogical peculiarity. Evidence against this is that recent lavas always have normal polarity. It is known, however, that self-reversal can occur. T. Nagata reported in 1952 that when he cooled a dacite lava from a temperature above its Curie point it became magnetized in a direction opposite

to the ambient field. At about the same time it was shown theoretically that there are several possible mechanisms by which a rock could acquire reverse magnetization especially in rocks containing two magnetic minerals. The complex relationships among the titanomagnetites and ilmenite-hematite series of minerals in igneous rocks suggested that selfreversals might have occurred. The concept that the Earth's magnetic field has reversed at intervals was a difficult one for many to accept.

Many laboratory and geological tests were devised to settle the controversy, and one of these involved study of the magnetization of igneous rocks and their associated thermally metamorphosed rocks. This is a particularly useful test because the country rock and igneous rock are usually of very different materials. According to the hypothesis of field reversals the metamorphic rock on cooling should acquire the same polarity as the igneous rock. If the hypothesis of self-reversals is to account for the observed frequency of normal and reversed polarities, then in roughly half of the cases studied the polarities of the metamorphosed rocks should be reversed compared to that of the igneous body. Results obtained by 1964 from this test and others were reviewed in Irving's book. A comparison of polarities observed in igneous rocks and their baked contacts from all continents and all geological periods except the lower Paleozoic produced 85 examples where the polarities of igneous rock and contact agreed, and only two where they disagreed.

An example illustrating the effects on several different rock types in the vicinity of an extensive igneous intrusive complex is illustrated in Figure 12-11. A 5000 ft. section exposed in the dry valleys of South Victoria Land, Antarctica, consists of a basement complex intruded by dikes overlain by Beacon sandstone. The whole sequence is intruded by the Jurassic Ferrar dolerite with three thick sheets making up almost one-half the total section. The schematic section shown in Figure 12-11 is through the horizontal component of magnetization. The arrows show the directions of magnetization observed in each unit based on detailed collections from over 100 sites. The remanence directions after magnetic cleaning are uniform throughout. Petrological estimates suggest that the temperature of the sediments exceeded 160°C when the dolerites were emplaced, and it appears that the sandstones and basement rocks were remagnetized at this time. In none of the country rocks was reversed polarity discovered.

By the time Irving published his book in 1964 the evidence was very strong that most reversed polarities were caused by reversals of the Earth's magnetic field, but the con-

clusive test that polarities should be constant in all rock units of the same age on all continents had not been made and occasional selfreversal for individual rocks was not precluded. If reversed polarities were produced by selfreversals then the contrasting polarities should be randomly distributed in space and time. Irving illustrated the two available examples of reversals dated radiometrically by independent groups of workers, and his figures are combined in Figure 12-12. This shows encouraging agreement; a pattern of alternating polarities with periodicity varying from 0.2 to 1.5 m.y. Irving summarized the status of results for late 1963 and included the following statements:

1. Recent and late Pleistocene rocks are magnetized in the same sense as the present field (negative polarity).

2. In early Pleistocene and older rocks frequent reversals of polarity occur in a manner unrelated to the rock type with a time scale of 10^6 years.

3. All late Carboniferous and Permian rocks (except the Upper Tartarian) are positively polarized indicating a time scale of about 50×10^6 years.

4. In rock formations with both polarities only a few percent of samples show intermediate directions suggesting that the transition period is short compared to the time for which constant polarity is maintained.

These first two studies of radiometrically dated lavas appeared to confirm previous conclusions from stratigraphic studies that polarity intervals were of about the same duration, and these were termed polarity epochs. Extension of lava studies with carefully chosen material and precisely determined polarities and ages, however, revealed the existence of polarity intervals with shorter durations of about 10^5 years. These were termed polarity events. Figure 12-13 shows how the successive discovery of polarity events changed the apparent distribution and duration of polarity intervals. By 1969 there were available 150 radiometric

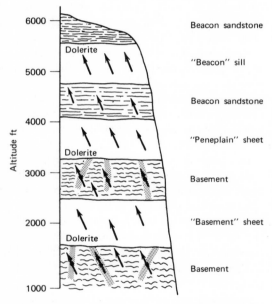

Figure 12-11. Schematic geological section in South Victoria Land (after Irving, 1964, Figure 7.26). Basement dikes, black, Admiralty Granites, wavy lines. Vertical section through horizontal component of magnetization which is approximately N 250°E. The arrows give, in a schematic fashion, the directions observed in each unit. The result is based on collections from over a hundred collecting sites by Bull, Irving, and Willis (1962). (Reproduced by permission of John Wiley and Sons.)

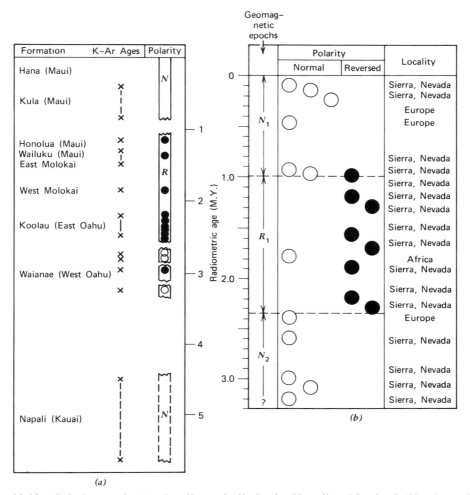

Figure 12-12. Polarity epochs dated radiometrically in the Hawaiian Islands, California, and elsewhere (from comparison by Irving, 1964, Figures 7.3 and 7.4). (*a*) Hawaiian Islands. The levels studied paleomagnetically and radiometrically are marked by dots (positive) and circles (negative) in the third column and crosses in the second column. (*b*) California and elsewhere. (Reproduced by permission of John Wiley and Sons.)

ages and polarity determinations meeting reasonable standards of reliability and precision in rocks less than 4.5×10^6 years old. In order to resolve the fine structure of reversals indicated by Figure 12-13 at least 450 determinations would be required from lavas whose ages are uniformly spaced at intervals no greater than 10^4 years. One difficulty here is in the nature of lava flows. The interval between successive flows may be

a few months or 10^5 years, and the task of locating volcanic formations with the required ages to give suitable coverage is a formidable one.

Figure 12-14 confirms that the polarity epochs and events are synchronous in widely spaced parts of the Earth, and no reasonable doubt remains that these events are produced by rather rapid switching of the Earth's magnetic dipole. The figure also

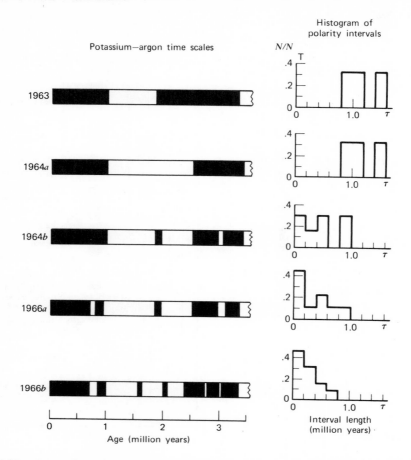

Figure 12-13. Successive versions of the radiometric time scale for reversals, showing how the discovery of polarity events changed the apparent distribution of polarity intervals. In the corresponding histograms, N_T is the total number of polarity intervals and N is the number in each class interval of the histogram. Compare Figures 12-14 and 12-21. (With permission; from A. Cox, Science, **163**, 237–245, 1969. Copyright 1969 by the American Association for the Advancement of Science.)

illustrates the problem of resolving the events caused by the uncertainties of K-Ar age determinations. The polarities may be intermixed for about 10^5 years on either side of a boundary between normal and reversed magnetization; the epochs are clearly demarcated but the events with durations of only 1 or 2×10^5 years are not. The transitions that bound epochs and events are even briefer time intervals as indicated by Figure 12-19.

Additional events have been discovered since the compilation of data in Figure 12-14,

from deep-sea cores as well as from lavas, and the geomagnetic reversal time scale as far back as 4.5×10^6 years including all reliable data published to the end of 1968 is shown in Figure 12-21. Attempts to extend the time scale further back in time have been made, but errors in K-Ar age determinations become too large for accurate work. Statistical analysis of the magnetic polarity of rocks as a function of their K-Ar ages has shown that the dating precision of rocks about 2.5 m.y. old is 3.6%, which agrees with independent estimates of precision

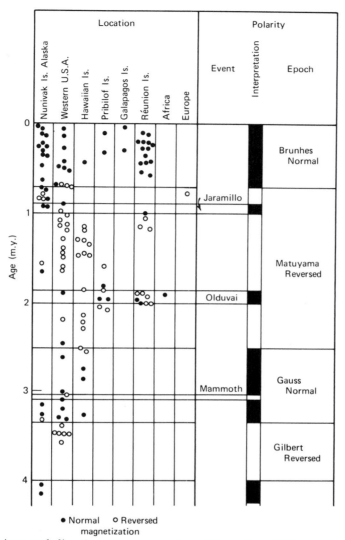

Figure 12-14. Ages and direction of magnetization of lavas from different continents, with corresponding polarity epochs and events. Compare Figures 12-13 and 12-21. (Adapted after Opdyke and Foster, 1967, by Bullard, 1968. Reproduced with permission.)

based on known sources of analytical error. For older rocks the dating error becomes large relative to the lengths of polarity epochs and much larger than the average event. For rocks 10 m.y. old dating precision of 3% gives a dating error of 3×10^5 years.

The difficulty in working with older rocks was illustrated by G. B. Dalrymple *et al.* in Figure 12-15. This shows the known time scale for geomagnetic polarity, extended in a hypothetical pattern from 3.5 to 12 m.y. to provide the basis for developing a probability model. Using the probability model they calculated the curve for the percentage of rock samples that would have a normal polarity as a function of their K-Ar ages assuming a dating precision of 3%. Figure

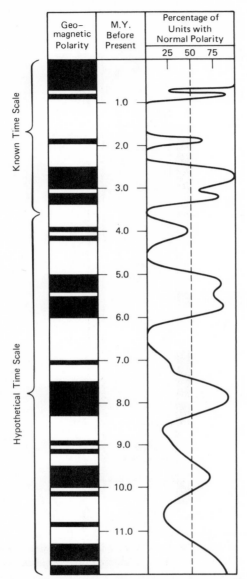

Geo–magnetic Polarity	M.Y. Before Present	Percentage of Units with Normal Polarity		
		25	50	75

Figure 12-15. Percentage of samples that would have normal polarity as a function of their K-Ar ages, assuming a dating precision of ± 3.0% (standard deviation) and an infinite sample density. The left-hand column shows the assumed polarity of the geomagnetic field. The assumed polarities prior to 3.6 m.y. are entirely hypothetical and are intended only to show the loss of resolving power of K-Ar dating for the earlier part of the reversal time scale (after Dalrymple et al., 1967, by permission of North-Holland Publishing Company, Amsterdam).

12-15 shows the loss of resolving power of K-Ar ages, and indicates that the radiometric time scale probably cannot be extended in detail much beyond 5 or 6 m.y. The definition of distinctive polarity transitions or longer periods of uniform polarity, however, is possible (Figure 12-16).

Figure 12-16 shows the type of results that have been obtained in the study of lavas up to about 20 m.y. old. Results of combined paleomagnetic-geochronometric studies of 44 volcanic units from the western United States are plotted in Figure 12-16*a*. No obvious pattern of reversals is apparent from the limited data points, but four epoch or event boundaries can be distinguished, three of them solely on the basis of the stratigraphic superposition of normal and reversed rocks. Partial compensation for the lack of resolving power of K-Ar dating in rocks of Pliocene age may thus be obtained by the use of well-defined stratigraphic successions.

This approach is shown in Figure 12-16*b*, which gives the results of a cooperative effort in Iceland by a team of 10 authors studying a predominantly basaltic succession ranging in age from oldest Tertiary to young Quaternary. Cores were collected from 21 overlapping lava profiles, designated *A* to *V*. The chronological sequence within each profile is known by superposition, and the relationship among the profiles was determined from stratigraphical correlations. The number of lava flows or flow units intersected was 1140 but, with allowances for overlap of one profile with another, the total succession comprises 900 separate lava flows or flow units totalling 8.8 km in thickness. Measurements were made on 2200 oriented samples from 1070 flows. They recorded 551 normal flows, 406 reversed flows, 73 anomalous flows, and they rejected 40. Most anomalous flows occurred between normal and reversed flows, and these are considered to give information about the intermediate field during the transition interval. Icelandic basalts are poor material for K-Ar dating, but 10 samples from six flows have been dated indicating a maximum age

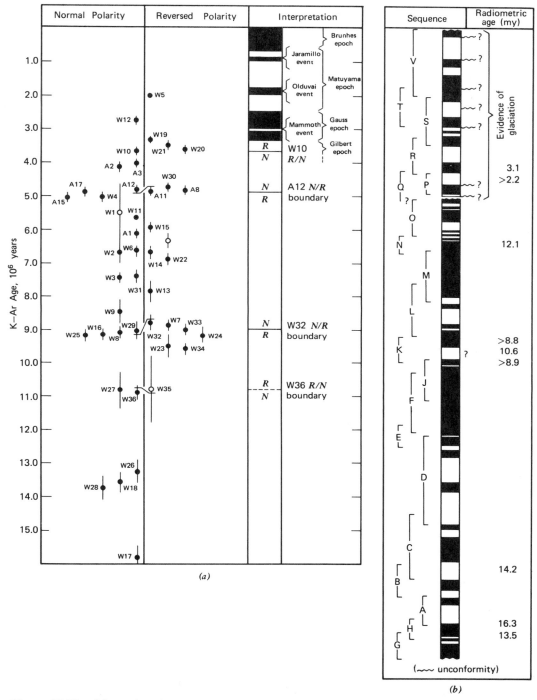

Figure 12-16. Magnetic polarity and K-Ar age determinations of volcanic rocks. (*a*) Upper Cenozoic rocks from the western United States (after Dalrymple et al., 1967). Shaded circles = primary data, open circles = secondary data, *x* = previously published data for the Gilbert epoch. The vertical bars are the estimated precision of the K-Ar ages at the 68% confidence level. The horizontal bars indicate the superposition relationships between flows of opposite polarity. The previously determined polarity time scale for the interval from 0 to 3.6 m.y. ago is shown (compare Figures 12-13 and 12-14). (*b*) Icelandic lavas. Generalized succession of polarity zones in overlapping lava profiles, designated *A*, *B*, etc. (after Dagley and nine co-authors, 1967). (Reproduced by permission of North-Holland Publishing Company, Amsterdam, and of *Nature*).

of about 20 m.y. Figure 12-16*b* gives the complete sequence of polarity reversals recognized continuous from the oldest lavas accessible with a small break between successions *O* and *P*. The normal polarity interval at the top of profile *V* probably represents the Jamarillo or Olduvai event (Figure 12-16*a*). The times of the reversals cannot be defined, but the analysis indicates at least 61 polarity intervals or 60 complete changes of polarity in this succession which gives an average rate of at least 3.0 inversions/ 10^6 years. Two of the normal polarity zones are noteworthy for their stratigraphic length and by implication for their duration. Profiles *E*, *F*, and *J* contain about 76 non-overlapping lavas, and profiles *L*, *M*, and *N* contain 101 nonoverlapping lavas of normal polarity. The authors suggested that this might provide a pair of marker horizons for dating purposes near 10^7 years.

Magnetic Reversals in Deep-Sea Sediments

The paleomagnetic study of sediment cores from the oceans has proceeded sporadically since 1938, but it was not until 1964 that reversely magnetized sediments were discovered. C. G. A. Harrison and B. M. Funnel observed five reversals in a short core from the Pacific Ocean. Sedimentation rates in deep oceans range from 1 to 10 mm in 10^3 years, and the depth of sediment corresponding to the boundary between the Gauss and the Gilbert epochs would thus be somewhere between 3 and 35 m. The continuous nature of deep-sea sedimentation may provide a more complete record of geomagnetic reversals than that available from volcanic rocks. Data from the volcanic rocks provide the absolute time scale needed to calibrate the cores. Then from core depth and time scale rates of sedimentation can be estimated. Because of its close connection with the evolution of man the Pleistocene has been a battleground for scientists since it was named more than 100 years ago, and the incomplete record of the Pleistocene on the continents kept the battle going. Now the prospect of dating the continuous sedimentary and faunal records in deep-sea cores provides a means for dating stratigraphic levels through the Pleistocene, and correlating these with biological and climatological changes. (Ericson and Wollin, 1968.)

The paleomagnetic study of long piston cores at Lamont Geological Observatory picked up momentum in 1966 when it was observed that cores from high latitudes around Antarctica were strongly magnetized and stable. The stratigraphy of seven of these cores is shown in Figure 12-17, with zones based on the appearance and disappearance of radiolarian species and the record of normal (negative) and reversed (positive) magnetization of closely spaced samples from the cores. The paleomagnetic stratigraphy and the radiolarian zones have the same time dependence throughout the area indicated on the location map. The sequence of normally and reversely magnetized sections in the left-hand core duplicates exactly the magnetic stratigraphy for lava flows shown in Figure 12-14. The 1966 version of the radiometric time scale for geomagnetic reversals is correlated in Figure 12-17 with the cores. The Jaramillo event (Figure 12-14) had not been established at that time but the short segment of negative polarity is recorded in four of the cores. The correlation from one core to another is striking. The zone thicknesses vary from one core to another reflecting different sedimentation rates in different locations, but the relative lengths of the magnetic zones in each core are effectively the same as the relative durations of the polarity epochs and events determined by K-Ar dating of lavas.

In Figure 12-18 the depths of polarity reversals in three cores from the north Pacific have been plotted against the geomagnetic reversal time scale. The lines through these points give the average sedimentation rates. For two of the cores rates of 0.75 cm and 1.3 cm/10^3 years are obtained,

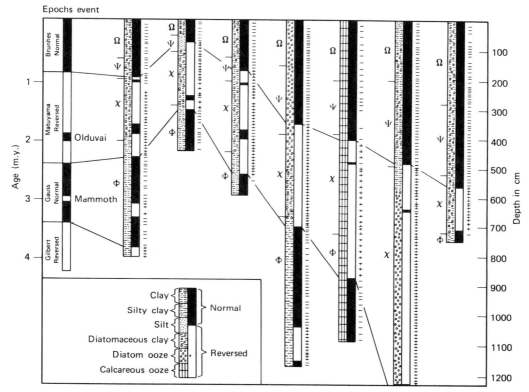

Figure 12-17. Magnetic stratigraphy in deep-sea cores from the Antarctic, with stratigraphic correlation and radiolarian faunal zones (greek letters). Normally and reversely magnetized levels are correlated across the Antarctic and with the dated polarity epochs (after Opdyke et al., 1966). See Foster and Opdyke (1970) for extension of the magnetic stratigraphy and its correlation with the sea-floor spreading anomaly sequence to 9 m.y. or more (Figures 13-13 and 13-15). (Reproduced with permission from Science, **154**, 349–357, 1966. Copyright 1966 by the American Association for the Advancement of Science.)

and the third core gives an average rate of 0.36 in the upper part and 0.80 in the lower. A well marked volcanic ash layer occurs in all three cores at different depths, and the correlation is confirmed because for all three cores Figure 12-18 gives an age of about 1.2 m.y. for this layer. These results indicate that average sedimentation rates in deep-sea environments remain constant through long periods although there are probably short-term fluctuations that average out. Assuming a constant rate of sedimentation and using diagrams such as Figure 12-18 it becomes possible to estimate the durations of the polarity events which are too short to be

measured accurately by dating lavas. The Jaramillo event appears to have lasted from 0.95 to 0.89 m.y. ago, and the Olduvai event extended from 1.95 to 1.79 m.y. ago. Given an average rate of sedimentation for a core it is now possible to estimate the date of anything that is recorded in the core, provided that the cores are amenable to paleomagnetic study. Unfortunately not all of them are.

These deep-sea cores with a geomagnetic time scale not only provide information about stratigraphy, paleontology, and sedimentation rates but also about the history of the Earth's magnetic field. Figure 12-19 shows the results of a detailed study of a portion of

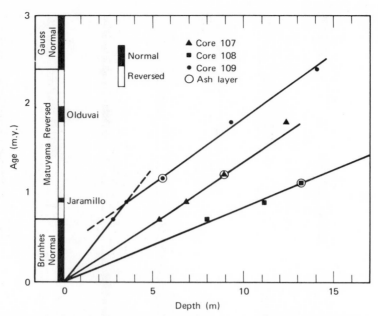

Figure 12-18. Rate of sedimentation for three cores from the North Pacific, given by plot of depths of polarity reversals against the geomagnetic polarity time scale (after Ninkovich et al., 1966). (Reproduced by permission of North-Holland Publishing Co., Amsterdam.)

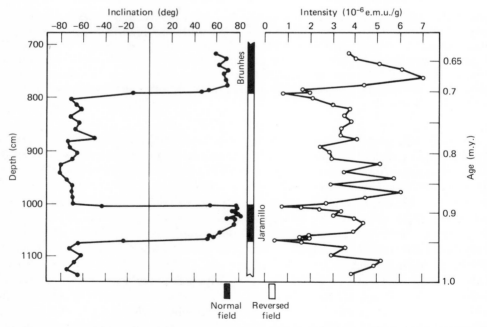

Figure 12-19. Magnetic inclination and intensity of magnetization of a portion of a core from the North Pacific (after Ninkovich et al., 1966). (Reproduced by permission of North-Holland Publishing Co., Amsterdam.)

a core from the north Pacific. The core depth scale is shown on the left and the polarity epochs and events in the center. The corresponding ages are given on the right. In an attempt to examine the nature of the Earth's magnetic field during a transition interval the cores were closely sampled across the reversals. The measured values of inclination and intensity of magnetization for samples taken at 1 cm intervals are shown in the figure. In a uniform sediment the intensity of magnetization provides a measure of the magnetizing field. The results indicate that the intensity of the Earth's field decreases through about 10,000 years by 60 to 80% before there is any change in dip, and then the field reverses during an interval of 1000 or 2000 years, and the field intensity then builds up again for another 10,000 years. This behavior is compatible with a reversal of the Earth's dipole field accompanied by the usual fluctuating nondipole field.

Measurements of the Earth's dipole moment show that the dipole field decreased between 1835 and 1965 at a uniform rate from 8.5 to 8.0×10^{25} gauss cm^3. If the rate remains constant the moment will pass through a zero point about 2000 years from now and then reverse its polarity. There is evidence, however, that this change is a part of a well-defined dipole fluctuation with cycle duration of about 10^4 years, and the probability that a geomagnetic reversal will result from the decrease in dipole moment currently in progress is 5% (Cox, 1969).

The continued paleomagnetic study of sedimentary cores has provided valuable information about the duration of longer events and has helped to extend the time scale back into the Gilbert reversed epoch, but it has not resolved the frequency of very short events. There are a number of inconsistencies and discrepancies in the detail of these events. The known Olduvai and Jaramillo events, for example, are not recorded in all cores as can be seen in the results of the first detailed study in Figure 12-17. The Gilsa

event (Figure 12-21) is represented in some cores and not in others. Some of the discrepancies may be due to disturbances in the coring process which appears to be something of an art; soft bands may be squeezed out or disturbed sediment sucked in. Slumping or turbidity currents could cause removal or addition of material at some sites on the deep-ocean floor and this becomes more likely approaching the continental margins.

The problem of short period polarity events in deep-sea sedimentary cores was reviewed by N. D. Watkins in 1968. He selected seven cores from the Southern Ocean and sampled them at 10 cm intervals reduced to 2 cm in the regions of polarity changes. All results are shown in Figure 12-20a. The normal and reversed polarities are correlated with the standard polarity time scale on the right. Figure 12-20b is a log of the inclination of the remanent magnetization for the core at the right of Figure 12-20a. Repeated measurements and resampling at 2 cm intervals prove that there is no question about the reality of the short polarity events shown in Figure 12-20. Whether or not they really reflect the ancient geomagnetic polarity behavior is regarded by Watkins as "much more problematical." He is convinced that there must exist a natural source or sources creating inconsistencies in the results for the upper Matuyama paleomagnetic stratigraphy, and his review led him to conclude that the main difficulties result from pene-contemporaneous organic redeposition of sediment. Watkins suggested that benthonic faunal activity may create a zone of redeposition up to 50 cm deep, and he discussed the possible effects of such activity. These include modification of the true position of a polarity boundary, reduction of the thickness of a given polarity zone, creation of new boundaries or "split" events, and the complete loss of short events. Reliable delineation of the fine structure of the geomagnetic history of deep-sea cores can only be expected for cores from areas of high sedimentation rates and no significant biological redeposition.

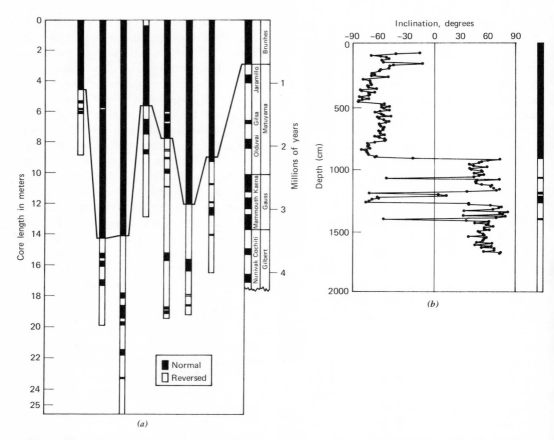

(a)

(b)

Figure 12-20. Short period polarity events in deep-sea sediments (after Watkins, 1968). (*a*) Polarity of remanent magnetism in seven selected deep-sea sedimentary cores from the Southern Ocean. Polarity time scale at right due to Cox and Dalrymple (1967). First column shows polarity, either normal (black) or reversed (clear); names of geomagnetic events in second column; names of geomagnetic epochs in third column. (*b*) Inclination of remanent magnetism in specimens from core on right of Figure (*a*), following demagnetization at 150 oersteds. Inclination is negative when normal polarity. Polarity log at right: black is normal, clear is reversed. (Reproduced by permission of North-Holland Publishing Co., Amsterdam.)

Figure 12-21 published in January 1969, compared with Figure 12-14 published in 1967, shows the discovery of several short events in only two years, and it is predicted and expected that others remain to be discovered. The time scale for Figure 12-21 is based on K-Ar dating, and each short horizontal line shows the measured age and the magnetic polarity of one volcanic cooling unit. The duration of events is based in part on paleomagnetic data from sediments and in part on interpretation of the profiles through the linear magnetic anomalies over mid-oceanic ridges. Interpretation of these anomalies depends upon the theory of sea-floor spreading. Therefore let us turn next to this theory which has proved to be the central theme for the revolution of the 1960's.

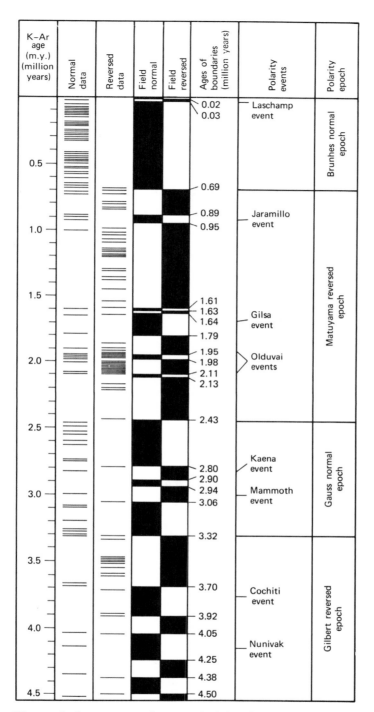

Figure 12-21. Time scale for geomagnetic reversals (after Cox, 1969). Each short horizontal line shows the age as determined by potassium-argon dating and the magnetic polarity (normal or reversed) of one volcanic cooling unit. Included are all published data which meet reasonable standards of reliability and precision. Normal-polarity intervals are shown by the solid portions of the "field normal" column and reversed-polarity intervals, by the solid portions of the "field reversed" column. The duration of events is based in part on paleomagnetic data from sediments and magnetic profiles. Compare with Figure 12-13. (From A. Cox, Science, **163**, 237–245, with permission. Copyright 1969 by the American Association for the Advancement of Science.)

THE 1960'S: SPREADING SEA-FLOOR CONCEPT

The Bandwagon Began to Roll in 1960

In the controversy between the mobilists and the fixists, alternatively known as horizontalists and verticalists, the stalemate in the 1940's was followed by a revival of interest in the 1950's arising from the paleomagnetic interpretations and the increasing amount of information becoming available about the physiography and physical properties of the ocean basins and continental margins. The reluctance of many geologists and geophysicists to reopen the discussion is indicated by the contents of the 1959 book *Physics and Geology* by J. A. Jacobs, R. D. Russell, and J. T. Wilson. They were not impressed by the evidence cited for continental drift, and they noted that no attempt had been made to reconcile continental drift with modern observations of the ocean. Although they concluded that the paleomagnetic evidence for polar wandering was rather convincing, they added that polar wandering could hardly be considered a major cause of orogenesis; it was possibly a by-product. They noted that convection current theories were widely held at this time, but they felt that arguments in favor of convection were inconclusive, and none of the theories had been developed into a specific or convincing explanation of the formation of any of the details observed in mountains or continents. By 1960 the time was ripe for new syntheses of the data, and it is notable that the attention of most investigators was directed toward the ocean basins rather than toward possibly drifting continents. A short time after the publication of this volume J. T. Wilson had become one of the most active contributors to the mobilist thesis and a spokesman for the revolution in the earth sciences. After half a century of indecisive arguments the theory of continental drift supported by the concept of sea-floor spreading generated such enthusiasm in the early 1960's that it took on the character of a bandwagon.

S. K. Runcorn edited a book, *Continental Drift*, in 1962, fifty years after Wegener published his first review on continental drift, with the hope that the volume would stimulate a serious interest in this subject formerly considered by many earth scientists as already closed. In this book B. C. Heezen noted specifically the ideas generated by the observation of high heat flow from the mid-oceanic ridges. Many scientists realized that such high heat flow could be accounted for by rising convection cells in the Earth's mantle, and these could also explain the tensional character of the ridges. According to Heezen among those who had expressed such views by 1960 or 1961, either in public lectures or in print, are M. Ewing, E. Bullard, H. Hess, R. Revelle, M. Menard, and R. Dietz. Of these and doubtless many others who were in a state of ferment over the heady excitement of rapidly accumulating data it was Hess and Dietz who developed the convection current scheme into the concept of sea-floor spreading.

The Contributions of H. H. Hess and R. S. Dietz

Voluminous literature on sea-floor spreading as a cause of continental drift has appeared since 1961, when Dietz introduced the term with the first formal publication on the topic "Continent and Ocean Basin Evolution by Spreading of the Ocean Floor." Hess is generally given priority for originating the theory, however (and Dietz acknowledges this), although his paper on the "History of the Ocean Basins" was not published until 1962. Hess wrote that although Holmes and others had suggested convection currents in the mantle to account for deformation of the Earth's crust, mantle convection was still considered a radical hypothesis not widely accepted by geologists and geophysicists. By 1960, however, scientists were beginning to realize that acceptance of mantle convection would permit the construction of a reasonable story to describe the evolution of ocean basins with whole realms of previously unrelated facts falling into a regular pattern. Hess's theory proposed that the sea floor is essentially the out-cropping of the peridotite mantle partly hydrated to form serpentinite. This is covered by a thin veneer of sediments and volcanic rocks. The major structures of the sea floor are direct expressions of the convection process, with the midoceanic ridges marking the sites of rising limbs of mantle-convection cells and the oceanic trenches being associated with convergences or descending limbs of convection cells. The continents are carried passively on the convecting mantle, and they do not plow through the oceanic crust as proposed by earlier hypotheses of continental drift. The leading edges of continents are strongly deformed where they impinge upon the downward moving limbs of convecting mantle, and the cover of oceanic sediments and volcanic rocks may also ride down into what Hess called the "jaw crusher" of the descending limb to be metamorphosed and eventually welded onto the continents.

Hess's concept of the rising convection cell at the midocean ridges is illustrated in Figure 12-22. Hess was impressed by the uniform thickness, 4.7 ± 0.7 km, of the main crustal layer (layer 3) beneath the oceans. He argued that basalt flows could not conceivably be so uniform in thickness over such a large area, and that it was more likely that the bottom of this layer represented a present or past isotherm, the temperature at which peridotite was converted into serpentinite in the presence of water. Figure 10-2 shows the intersection of a geotherm with the serpentinization reaction. This occurs at a temperature near 500°C, and in Figure 12-22a Hess has adopted the 500°C isotherm as the level at which serpentinization occurs. Water migrating upward with the rising limb of a mantle convection cell coexists with unaltered peridotite until the temperature falls to 500°C at the isotherm, and here the water

reacts to produce serpentinite. Divergence of the rising limb at the surface carries the layer of serpentinized mantle across the ocean floor, and the 500°C isotherm migrates to deeper levels as shown in Figure 12-22a. But in the absence of water the boundary between unaltered and partly serpentinized mantle remains at the depth corresponding to the isotherm beneath the midocean ridge. This explanation for the uniform thickness of the main crustal layer requires that the temperature distribution beneath the midocean ridge remains uniform for periods of time of the order of 10^8 years, which presents some problems of its own.

For most of the ocean basin the Moho is defined as the boundary between crustal material with P-wave velocity of 6.7 km/sec (range from 6.0 to 6.9) and mantle material with P-wave velocity of 8.1 km/sec, as shown in Figure 12-22b. This boundary is not found beneath the ocean ridge crest where the seismic wave velocities are considerably lower. Hess suggested that the anomalous seismic velocities could be explained by the higher temperature of the rising material, together with fracturing where the convective flow changes direction from vertical to horizontal. The fractures are healed as the partly serpentinized peridotite and the underlying peridotite move together to the flanks of the ridge. A new crustal layer is thus generated from mantle material at the midoceanic ridges, and this layer spreads across the ocean basin floor to a descending limb of the convection cell. Here the process is reversed and the descending material is dehydrated when its temperature reaches 500°C releasing water upward to the sea. It has been shown experimentally that the strength of serpentinite drops appreciably near its dehydration temperature, and this may have significant tectonic implications if it occurs beneath ocean trenches assumed by Hess to be the sites of descending limbs of the convection cells.

According to the theory of sea-floor spreading if the rate of movement for convection is

1 or 2 cm per year, the floors of the ocean basins must be completely renewed every 200 or 300 million years. The sea floors and the shapes of the ocean basins are thus comparatively young features compared to the ancient continental blocks. This could account for the relatively small thickness of sediments on the ocean floor and for the apparent absence of sediments older than the Jurassic period. Figure 12-22*b* shows how younger sediments should show progressive overlap on a midocean ridge if the mantle does move laterally away from the ridge crest. This prediction has since been con-

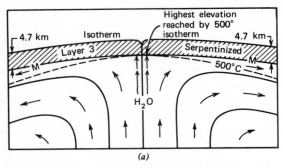

(a)

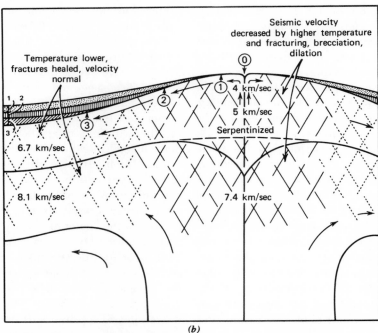

(b)

Figure 12-22. Representation of sea-floor spreading by Hess (1962). (*a*) Diagram to portray highest elevation that 500°C isotherm can reach over the rising limb of a mantle convection cell, and expulsion of water from mantle which produces serpentinization above the isotherm. (*b*) Diagram to represent (1) apparent progressive overlap of ocean sediments on a mid-ocean ridge which would actually be the effect of the mantle moving laterally away from ridge crest, and (2) the postulated fracturing where convective flow changes direction from vertical to horizontal. Fracturing and higher temperature could account for the lower seismic velocities on ridge crests, and cooling and healing of the fractures with time, the return to normal velocities on the flanks. (Reproduced by permission of The Geological Society of America.)

firmed in striking fashion by the results of JOIDES, the deep sea drilling program, in the Atlantic (Figure 14-7).

If new sea floor is generated at the mid-ocean ridges the older, displaced sea floor must be carried downward at the sites of converging convection cells. Continental masses are initially moved along in conveyor-belt fashion until they attain a position of dynamic balance overlying a convergence. There the continents come to rest, but the mantle continues to shear under and descend beneath them possibly forming an ocean trench. If a new region of divergence develops beneath a continental mass the mantle convection currents will tend to rift the continent apart. Presumably the Atlantic Ocean marks an ancient rift which separated North and South America from Europe and Africa. The Indian Ocean Rise may extend through the Red Sea into the African Rift Valleys, tending to fragment the continent. Downward movement of convection cells beneath continents or at continental margins places the continents under compression, which accounts for orogenesis and alpine folding. If the continental block is being carried along passively by the convecting mantle the margin is tectonically stable, but if the downward convection limb is uncoupled from the continent marginal mountain ranges tend to be formed. It is anticipated that the sites of divergence and convergence may change with time causing new tectonic patterns.

Holmes's early model of mantle convection and continental drift is remarkably similar to the model of sea-floor spreading developed by Hess and Dietz, but there is a significant difference which sets the scheme of Hess and Dietz apart not only from Holmes's model, but also from the many similar convection models that were published in the early 1960's. In 1931 it was generally believed that sialic patches, and even a thin layer of sial, existed above the thick basaltic layer forming the oceanic crust. Holmes therefore depicted his convection cells operating beneath the

oceanic crust which was stretched, thinned, and in part disrupted as the continental blocks were carried apart (Figure 11-2). He did refer to the addition of new basaltic material from the mantle to the crust. Hess and Dietz on the other hand proposed that there is no crust in the usual sense beneath the oceans. The thin veneer distinguished by seismic measurements is actually the exposed mantle surface, the upper part of an active convection cell, whose properties have been modified by serpentinization where the cell approaches the surface at a midoceanic ridge (Figure 12-22*a*).

Some Other Convection Models

In a detailed review, "Sea Floor Relief and Mantle Convection," H. W. Menard in 1965 concluded that neither the facts of marine geophysics nor the hypothesis of some kind of convection in the mantle required acceptance of the bolder hypothesis of sea-floor spreading as developed by Hess and Dietz. Menard was not convinced that the sea floor is periodically swept clean by convection, nor that the oceanic crust is easily created or destroyed. He noted that because the oceanic crust does exist the hypothesis of sea-floor spreading is meaningless unless this crust has the same composition as the mantle. A phase change from basalt to eclogite is not acceptable (Chapter 10), and Menard argued against Hess's serpentinization hypothesis of Figure 12-22. He showed that the 500°C isotherm was never likely to have been at the right depth to produce a uniform layer of serpentinized peridotite with the thickness of the present crust, and that the 500°C isotherm beneath the crust of the midocean ridges could rise to within 3 km of the ocean floor. Even if the 500°C isotherm did not rise above the 5 km depth at the ridge crest (Figure 12-22) Menard doubted that in such a thermally unstable environment the proposed limitation of crustal thickness by the isotherm could produce the uniform thickness required. He also considered it implausible that the crust could be transported from the ridge crests for several thousand kilometers without change in thickness.

Menard did agree, however, that the characteristic features of ocean rises and ridges are readily explained by transient convection cells in the mantle beneath the crust as illustrated in Figure 12-23. The sea-floor crust is stretched and thinned and crustal blocks are moved about differentially along faults. Menard's examination of the distribution of ocean rises on polar projection maps led him to conclude that 87% of the rises lie on circles centered on continental shields with most of the circles having about the same radius. About one-half the length of the rises, including of course the midoceanic ridges, is centered relative to the ocean basins. The circular pattern of ocean rises and ridges suggested to Menard that perhaps the primary control of convection is the upward separation of light from dense fractions of the mantle under continents, with the denser material sinking and spreading out

from below the continents. At some distance from the center, the material moves upward to complete the cell and to form the oceanic rises. The evidence from marine geophysics suggests that material rises and spreads laterally under the crust, but it does not indicate anything about the depth of or even the existence of a deep return flow in an organized convection cell.

According to Hess and Dietz mantle convection cells reach the ocean floor (Figure 12-22), but Menard pictures them moving beneath a thin ocean crust (Figure 12-23). J. T. Wilson discussed the same problem in 1963, and placed the convection cells deeper within the mantle beneath a lithosphere about 100 km thick (Figure 12-24a). The lithosphere is a tectonic unit including the crust and a rigid layer of the upper mantle. Horizontal flow is restricted to the weaker asthenosphere layer which is equated with the seismic low-velocity zone, and both upward and downward flow are similarly restricted to narrow zones within the mantle.

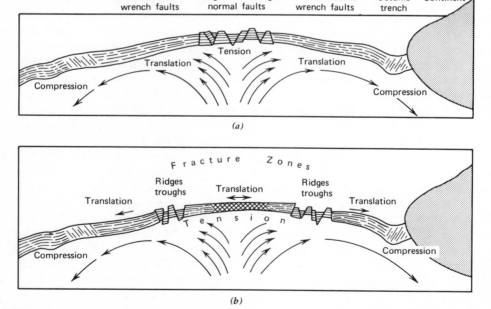

Figure 12-23. Convection current hypothesis; topographic and tectonic effects resulting from upwelling beneath the oceanic rise system. (From "Marine Geology of the Pacific" by H. W. Menard. Copyright 1964, McGraw-Hill Book Company. Used with permission.)

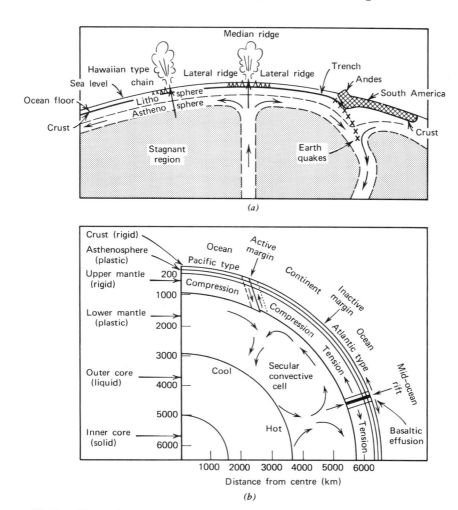

Figure 12-24. Alternative mantle convection models. (*a*) Diagrammatic section showing a type of convection which might explain the origin of pairs of lateral ridges, median ridges, Hawaiian-type chains and mountain building (after Wilson, 1963). (*b*) Convection currents confined to depths below the Benioff seismic zones (Figure 2-6) (after Bernal, 1961). Note that only shallow-focus earthquakes are associated with the mid-oceanic ridge system (Figures 2-5 and 14-3). (Reproduced by permission, from *Nature.*)

Downward flow beneath continental margins such as the Andes occurs along the inclined zones of deep-focus earthquakes (Figure 2-6). The convection current carries the lithosphere over the asthenosphere and drags it downward beneath the oceanic trench and marginal mountains. See Figure 14-5.

Figure 12-24*a* also illustrates a possible explanation for linear chains of volcanoes in the ocean basins. Active volcanoes are formed near the crest of the ridge and the horizontal currents carry the volcanic piles successively off the ridge, detaching them from their source. This process could explain the aseismic lateral ridges extending from median ridges. Similarly a fixed source of lava rising from stagnant mantle beneath the asthenosphere could produce a Hawaiian-type chain of volcanoes. If the lava source were in the asthenosphere then a single active volcano would be formed, and this would migrate as the asthenosphere source migrated carrying the lithosphere and super-incumbent volcano with it.

J. D. Bernal inferred that the upper mantle was rigid to a depth of 900 km because of the existence of deep earthquake zones (Figure 2-6), and he proposed that convection cells are restricted to the lower mantle as illustrated in Figure 12-24b. This model includes the asthenosphere shown in Figure 12-24a, but it plays no role in the convective scheme. The convection cells place the upper mantle and crust in tension beneath the midoceanic ridges, drag the ocean floor and western hemisphere continents to the west, and create

the Pacific trenches by overthrusting along the inclined plane defined by intermediate and deep-focus earthquakes. He also indicated the existence of a rift zone extending to 900 km in depth beneath the midoceanic ridges, but this is unlikely because all earthquakes in these regions are of shallow-focus types.

S. K. Runcorn noted in 1962 that the scale of continental drift required mantle-wide convection cells. He suggested that gradual growth of the core would cause the convection cells to decrease in size and increase in number in a series of rather abrupt changes in convection patterns. Each change in convection pattern could produce drifting and repositioning of the continents.

In 1968 J. C. Maxwell suggested that

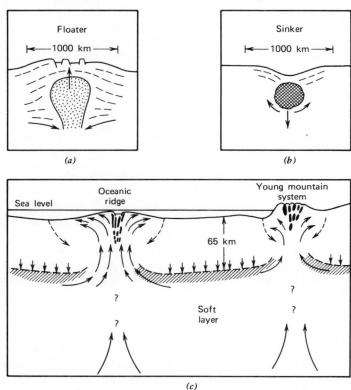

Figure 12-25. Vertical movements as alternative to closed convection cells (after **Maxwell**, 1968). (a) Sinker (Figures 9-8(f) and 10-9). (b) Floater (Figure 10-8). (c) Postulated association of oceanic ridges and young mountain systems with diapiric mantle material. Black blobs represent mantle rocks brought up into the crust (Figure 6-6). (Reprinted by permission, *American Scientist*, journal of The Society of the Sigma Xi.)

closed convection cells were not required to explain the features of ocean basins and continental margins. Figures 12-25a and b show the initiation of convection or advection by the uprise or sinking of a portion of the mantle that has become slightly less dense or more dense respectively than its surroundings. These two types, the floater and the sinker, may accomplish the transport of large volumes of material but they may occur independently of each other, and therefore they are not necessarily components of a convecting system. Maxwell proposed the scheme illustrated in Figure 12-25c. Both midoceanic ridges and young mountain systems are loci of long-continued high thermal energy and hot mantle material probably rising from the low-velocity zone. Diapiric uprise of this material is driven by the sinking of cool areas of normal crust making up the continents and ocean basins. Maxwell accepts sea-floor spreading because he states that the rising hot mantle material must spread laterally from the oceanic ridges if the process is to continue, but this system seems to be incapable of bringing about continental drift.

The enthusiasm with which various convection models were developed in the early 1960's to explain the observations of marine geophysics and to support continental drift met with opposition. For example G. J. F. MacDonald published a series of papers using evidence from gravity and heat flow measurements to show that continents have deep structures, and that chemical differences between oceanic and continental regions must extend downward to several hundred kilometers (Chapter 7). This implies that if convection cells exist within the Earth, they cannot extend upward to near-crustal levels. This would only permit schemes of the type shown in Figure 12-24b. MacDonald's comparison, however, of the figure of the Earth derived from satellite gravity data with the ideal ellipsoid of revolution assuming hydrostatic equilibrium, demonstrates that stress differences exist in the mantle, and that an average viscosity of 10^{26} poises is required. This value is incompatible with a convecting mantle. Despite these arguments, evaluation of possible mantle convection cells has continued, and we shall return to this topic in Chapter 14.

13. *Magnetic Anomalies in the Ocean Basin*

INTRODUCTION

Neither paleomagnetism nor magnetic anomaly is listed in the index of the symposium on the *Crust of the Earth* edited by A. Poldervaart, and published in 1955. In their contribution reviewing the geophysical contrasts between continents and ocean basins M. Ewing and F. Press stated that:

"Magnetic measurements are as yet meager, but those available indicate that over large oceanic areas the magnetic field is unusually smooth. As data accumulate, valuable conclusions may be drawn from edge effects of continents and from correlation with submarine topography."

The increasing value of magnetic data was confirmed by the fact that in a 1966 symposium on "The History of the Earth's Crust" at the Goddard Institute for Space Studies in New York, edited in 1968 by R. A. Phinney, no fewer than six of 16 contributed papers were concerned with geomagnetism or magnetic anomalies of the ocean basins, and revolutionary conclusions were indeed being drawn. The interpretations and correlations were not related merely to local features of submarine topography, but to worldwide oceanic processes.

During the interval 1955 to 1966 there had been three major developments related to geomagnetism. Two of these have been reviewed in Chapter 12: the location of paleomagnetic pole positions with respect to continental masses at specific times during geological history and the recognition and dating of reversals of the Earth's magnetic dipole. The third was the discovery of linear magnetic anomalies over the ocean basins and their correlation with the polarity reversal time scale.

Several papers suggesting some relationship between the magnetic anomalies and sea-floor spreading had appeared by 1966, but the parts of the theory did not fall into place until the Goddard conference. The editor, Phinney, introduced the symposium volume with a historical sketch of recent developments. He wrote that

"In July 1966, the results of Vine and Pitman and Heirtzler were known to only a few colleagues. Word got around during the fall, by preprint, and at the Goddard conference, slightly preceding formal publication in *Science*. By January 1967 the impact was such that nearly 70 papers on sea-floor spreading had been submitted for the April meeting of the American Geophysical Union."

The linear anomalies in what Ewing and Press described as an unusually smooth magnetic field thus deserve our careful attention.

Oceanographic cruises from the Lamont Geological Observatory and the Scripps Institution of Oceanography initiated the intensive collection of magnetic data. In

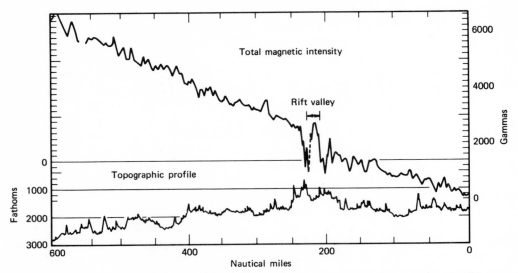

Figure 13-1. Profile of total magnetic intensity and topography, mid-Atlantic ridge (after Heezen et al., 1959). Magnetic values in gammas relative to an arbitrary zero. (Reproduced by permission of The Geological Society of America.)

their special paper on "The Floors of the Oceans. I, North Atlantic" B. C. Heezen, M. Tharp, and M. Ewing of Lamont reported that the first magnetic data for the Atlantic Ocean from a ship-towed magnetometer were obtained in 1948, and that by 1957 nearly 20 crossings had been made of the crest of the mid-Atlantic ridge. These revealed a characteristic pattern of a large positive anomaly of more than 500 γ over the rift valley with negative anomalies of

300 to 500 γ over the adjoining rift mountains. This is illustrated by Figure 13-1 which gives a topographic profile of the mid-Atlantic ridge and the profile of total magnetic intensity. They also reported rough fields with 5- to 15-mile wide 100 to 200 γ anomalies over oceanic ridges and rises. It is now known that a large part of the ocean basins is covered by a magnetic pattern in the form of stripes, but these were not reported until 1958.

1958–1968: LINEAR MAGNETIC ANOMALIES

1958: Discovery in the Pacific by R. G. Mason

Research ships from the Scripps Institution had been towing total field magnetometers behind them since 1952, but the isolated, long profiles provided no basis for quantitative interpretation. In 1955 the U.S. Coast and Geodetic Survey Ship appropriately named *Pioneer* began surveys with closely spaced lines, and an extensive area of the Pacific was

mapped with the detail and accuracy associated with airborne magnetometer surveys on land. The first magnetic results were reported by R. G. Mason in 1958. The survey was extended to an area 250–300 miles wide off the foot of the continental slope between latitudes 32 and 52°N as shown in Figures 13-2 and 13-3. The detail for a portion of this survey is related to the geomorphic provinces and principal fault zones in Figure 13-3. Further cruises were undertaken by Scripps ships in 1958,1959,and 1960 in order to extend the survey to the west on both sides of the Pioneer and Mendocino faults.

The magnetic maps were constructed by first plotting the measured field value along the ship's track and contouring the result with an interval of 50 γ. The standard deviation of a single observation is about 15 γ, but the relative error between any two points on the same traverse is negligible. It is unlikely that normal diurnal variations, whose effects were not removed, have any significant effect on the positions and magnitudes of the more important anomalies. The regional magnetic field was determined by a smoothing process for mean observed fields in 20 mile squares, and this was subtracted from the observed field to obtain the magnetic anomaly maps shown in Figures 13-2 and 13-3. Figure 13-2 shows the anomaly pattern in terms of positive (black) and negative (white), and Figure 13-3 shows contour intervals of 250 γ.

The magnetic survey revealed a narrow pattern of remarkably straight anomalies of about 400 γ magnitude and about 30 km width, trending north-south for about 1000 km. The largest anomaly is less than 2% of the Earth's field. The more prominent anomalies are remarkably regular throughout their lengths, and the pattern is clearly offset by the Murray, Pioneer, and Mendocino fracture zones. The offset suggests for the Murray fault zone a right-lateral displacement of about 155 km in the eastern section, and more than 640 km on the western section; a left-lateral slip of about 265 km across the Pioneer fault; and about 1185 km across the Mendocino fault zone. These displacements are of the same magnitude as those required by continental drift, but the fact that the magnetic pattern is preserved in the upper part of the oceanic crust indicates that this too is rigid and not a sea of sima through which the sialic continents can sail.

The remarkable regularity of the magnetic pattern points to a simple cause but no satisfactory explanation was forthcoming at this stage. It was clearly established that the pattern showed little correlation with topography over most of the area. The anomalies could be produced by flat, slab-like structures approximately underlying the areas of positive anomaly, and there are two factors that restrict the possible depths and thicknesses of such slabs. Because of the sharpness of the anomalies their sources must lie at shallow depths beneath the ocean bottom. The anomalies require magnetization contrasts corresponding to susceptibilities in the range 0.005 to 0.015, which is appropriate for the contrast between magnetic basalts and relatively nonmagnetic sediments. The upper limit to the polarization contrast that can be assumed gives a minimum thickness for the production of anomalies with the observed magnitude.

Figure 13-4 shows the profile of a positive magnetic anomaly along the line *A-A* just above the Mendocino fault in Figure 13-3. Also shown are three schematic crustal sections through infinite north-south structures with the shapes of the slabs adjusted by trial and error so that the computed theoretical anomaly for each structure (not shown in the figure) fits the observed anomalies. The first example represents an isolated body of magnetic basalt within the second layer of the crust, the second an elevated fault block of the main basaltic crustal layer, and the third a vertical slab of the main crustal layer plus the second layer with magnetic contrasts compared with the adjacent crust possibly caused by intrusion of highly magnetic material from the mantle. Any one of these structures would explain the observed anomalies in two-dimensional profiles at least, but no topographic or seismic expression of any of these possible structures has been established in the surveyed area.

The various explanations proposed to account for the anomalies include the following:

1. The excess magnetic material represents the topography of the upper surface

of the more magnetic crust, which has been covered for the most part by a blanket of sediments.

2. Lava flows have spread out over the floors of the ocean filling pre-existing troughs or depressions due to block faulting or of unknown origin.

3. More strongly magnetic material has been injected into old lines of weakness in the ocean floor possibly connected with the

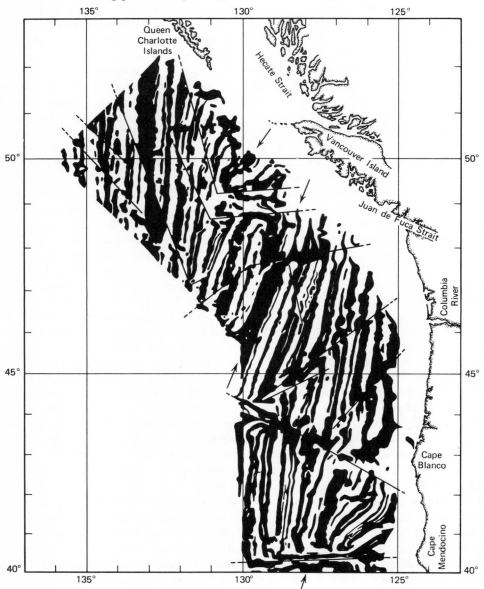

Figure 13-2. Summary diagram of total magnetic-field anomalies southwest of Vancouver Island. Areas of positive anomaly are shown in black. Straight lines indicate faults offsetting the anomaly pattern; arrows, the axes of the three short ridge lengths within this area—from north to south, Explorer, Juan de Fuca, and Gorda ridges (after Raff and Mason, 1961). (Reproduced by permission of The Geological Society of America.)

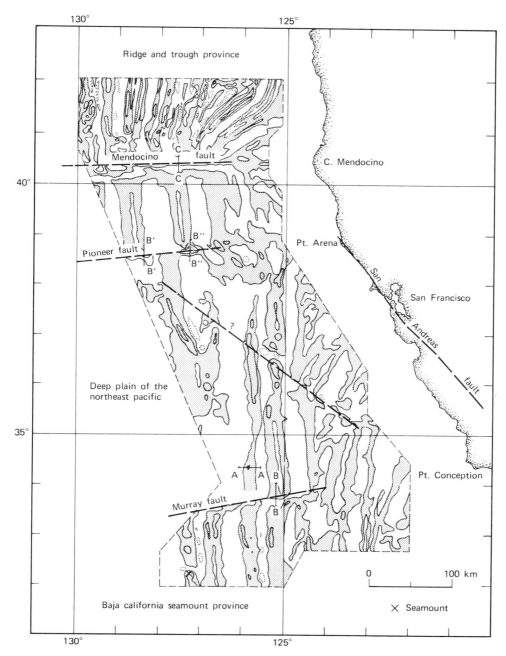

Figure 13-3. Skeleton magnetic map, showing geomorphic provinces, principal known faults, and location of profiles. The contour interval is 250 gammas. Areas of positive anomaly are stippled. Negative contour lines are broken (after Mason and Raff, 1961). (Reproduced by permission of The Geological Society of America.)

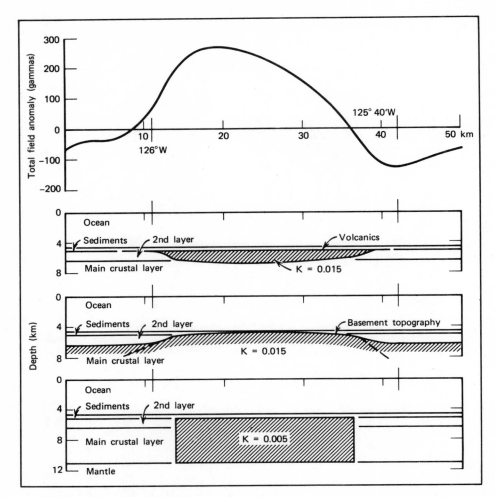

Figure 13-4. Possible interpretations of magnetic profile *a-a* in Figure 13-3 (after Mason and Raff, 1961). The theoretical anomaly computed for each of the three infinite north-south striking structures fits the observed anomaly almost exactly. (Reproduced by permission of The Geological Society of America.)

Earth's spin or mechanisms involving Earth tides.

4. Among other suggestions A. D. Raff in 1961 noted that the magnetic anomalies in the area of the *Pioneer* survey and in some other parts of the Pacific run generally parallel to the ridges there, and that "it looks as though the two are related."

5. The lineation may be the record of a regenerative process of the oceanic crust along oceanic rises, either by global expan-

sion or by the rise of mantle material along oceanic ridges, and bands of lineations should then run parallel to the ridges.

6. The occurrence of linear anomalies in bands parallel to the oceanic rises could be explained by the hydration of olivine of the peridotite mantle into serpentine and magnetite according to the sea-floor spreading hypothesis (Figure 12-22).

7. F. J. Vine, D. H. Matthews, and L. W. Morley proposed a specific model in 1963

correlating the anomalies with sea-floor spreading and reversals of the Earth's magnetic field, and this has subsequently been generally accepted.

1963: Explanation by F. J. Vine, D. H. Matthews, and L. W. Morley

By 1963 many profiles showing bathymetry and the associated total magnetic field, similar to that given in Figure 13-1, had established the existence of a central magnetic anomaly over the oceanic ridges in the North Atlantic, the Antarctic, and the Indian Oceans, and a consistent pattern of anomalies on the ridge flanks. Largely on the basis of measurements made from *H.M.S. Owen* in 1962 during the International Indian Ocean Expedition Vine and Matthews suggested in *Nature* in September, 1963, that the pattern of linear magnetic anomalies discovered in the Pacific (Figure 13-2) was due to strips of the sea floor being magnetized in opposite directions. They explained the existence of such strips by proposing that as new oceanic crust is formed over a convective upcurrent in the mantle beneath an oceanic ridge, according to the idea of sea-floor spreading, it becomes magnetized in the direction of the Earth's magnetic field when it cools below the Curie temperature. If the Earth's field reverses periodically as sea floor spreading occurs, then successive strips of oceanic crust paralleling the crest of the ridge will be alternately normally and reversely magnetized. The successive strips will then cause an increase or decrease in the total intensity of the ambient magnetic field, thus producing the series of linear anomalies.

In an article published in 1964 Morley and Larochelle referred to the presentation of essentially the same hypothesis by L. W. Morley at the Annual Meeting of the Royal Society of Canada in Quebec City in June 1963. Apparently similar conclusions were reached independently. Morley suggested that a nearly unbroken record of the reversals of the Earth's magnetic field may exist in the permanent magnetization of the rocks on the ocean floors, and he suggested that determination of the rate of mantle convection might give the history of the reversals of the Earth's field.

The idea was developed in detail by Vine and Matthews. They found that the magnetic effects of some individual topographic features of igneous origin on the crest of the Carlsberg Ridge in the Indian Ocean could be distinguished. Two volcano-like features were studied. One had negative anomalies as expected for normal magnetization in this low magnetic latitude, but the other had a pronounced positive anomaly suggesting that it was reversely magnetized. The estimated effective susceptibilities of the material forming the two features was ± 0.0133. Computed magnetic profiles across the Ridge assuming normal magnetization were found to bear little resemblance to the observed profiles. This suggested to Vine and Matthews that whole blocks of the survey area might be reversely magnetized. Computed profiles for models with blocks of the crust alternately normally and reversely magnetized did fit reasonably well with the observed profiles, and this led to formulation of the hypothesis outlined above.

The models shown in Figure 13-5 were developed to show how the reversed magnetization hypothesis could account for the shapes of the anomaly profiles encountered in the Pacific (Figures 13-2, 13-3, and 13-4), and the pronounced central anomaly which occurs over the oceanic ridges. The crust is divided into strips 20 km wide which are alternately normally and reversely magnetized. The thickness of the magnetic crust is limited by the depth to the Curie-point isotherm, which was assumed to be 20 km below sea level for the deep ocean, and 11 km beneath the center of the ridges where heat flow is higher. The Curie-point isotherm is probably not far below the 500°C isotherm, which Hess estimated to be at somewhat higher levels (compare Figure 12-22a). The effective susceptibility adopted for the magnetic crust

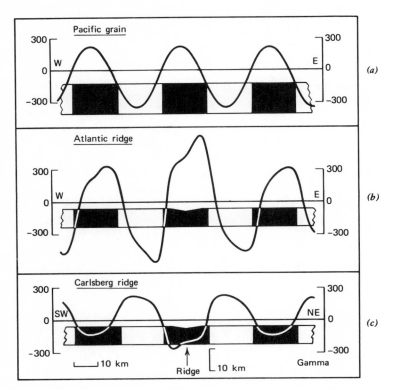

Figure 13-5. Magnetic profiles computed for various crustal models (after Vine and Matthews, 1963). Crustal blocks marked as normally or reversely magnetized. Effective susceptibility of blocks, 0.0027, except for the block under the median valley in profiles b and c, 0.0053. (a) Pacific grain. Total field strength, $T = 0.5$ oersted; inclination, $I = 60°$; magnetic bearing of profile, $\theta = 073°$. (b) Mid-Atlantic Ridge. $T = 0.48$ oersted; $I = 65°$; $\theta = 120°$. (c) Carlsberg Ridge, $T = 0.376$ oersted; $I = -6°$; $\theta = 044°$. (By permission of *Nature*.)

is 0.0027, considerably less than that derived for the isolated features described above. For the block beneath the median valley in the two Ridge profiles, the effective susceptibility was doubled, to 0.0053, because this most recent block is the only one which has a uniformly directed magnetic vector. The effective magnetization of adjacent blocks is reduced by the intrusion and extrusion of new material in a wide zone extending beyond the limits of the central zone. The computed profiles shown in the figure for each crustal section illustrate the essential features of the observed profiles, and confirm that these can be produced by alternating directions of

magnetization, without recourse to major inhomogeneities of rock type within the main crustal layer, or to unusually strongly magnetized rocks.

Let us note that at this time, not one of the three basic assumptions of the hypothesis was generally accepted: (a) reversal of the Earth's magnetic field, (b) the contribution of remanent magnetism to oceanic magnetic anomalies, and (c) sea-floor spreading. In addition three observations and inferences were not explained by the hypothesis:

1. The best known examples of linear anomalies were those in the northeast Pacific (Figure 4-33), but these did not

appear to be parallel to any existing or pre-existing oceanic ridges.

2. In surveys of known ridges elsewhere, the only anomaly that could be correlated from one profile to another was the central anomaly, and the existence of linear anomalies paralleling the central anomaly was not established.

3. The hypothesis did not explain the observed variations in the amplitude and wavelength of anomalies on either side of the ridge.

The hypothesis was at first received with some scepticism, but rather suddenly between 1965 and 1967 the bandwagon which had been rolling since 1960 gathered such momentum that it carried most scientists along with it. The developments during this interval have had, and will continue to have, such far-reaching implications for the geophysical sciences that we must examine in some detail the successive stages. I think that there are many lessons for students to learn in following the development of the ideas feeding this revolution; for it *has* to be accepted as a revolution. With evidence and interpretations being exchanged at meetings, with so many papers being published within a few months of each other, and with different journals having publication delays varying from 2 to 12 months or more after receipt of the manuscript, it is not easy to decipher the succession. As examples of the ways in which scientific theories develop I have selected two trends which appear quite clear. Magnetic survey work proceeded apace over many oceans, and Vine tested all new data against the sea-floor spreading hypothesis claiming proof by 1966. J. R. Heirtzler and his associates at Lamont, who were actively engaged in gathering new data, felt that some other interpretation was required, but they were unable to produce one. Then some time during 1966 they were persuaded that sea-floor spreading did provide the solution, and by 1968 they had out-spread Vine, one of the original spreaders.

1965-1966: Confirmation by F. J. Vine and J. T. Wilson

By 1965 two of the basic assumptions of the Vine-Matthews hypothesis, field reversals and the contribution of remanence to the magnetic anomalies, had become accepted, and the periodicity of the field reversals was being defined (Figure 12-13). Then J. T. Wilson showed that the apparent absence of oceanic ridges in the northeast Pacific is due to the complications caused by the large horizontal fault zones (Figures 2-3, 13-2, and 13-3) whose effect was not considered in the original hypothesis. In his 1964 book, *Marine Geology of the Pacific*, H. W. Menard had previously identified a belt of ridges and troughs extending north of the Mendocino fracture zone and vanishing, apparently at another fault, at about 50°N.

In order to explain two puzzling features about these lines of large horizontal shear Wilson developed the concept of transform faults. The two features are that many of the dislocations terminate abruptly, and recent seismic activity is often confined to short parts of the fracture zones. The horizontal movements had conventionally been interpreted as due to transcurrent faults, but this tacitly assumed that the faulted medium is continuous and conserved. If new crust is being generated on ridges, then other kinds of faults have to be envisaged. Figures 13-6*c* and *d* show how a sinistral transcurrent fault would offset a ridge. Figure 13-6*a* shows a dextral ridge-ridge type transform fault connecting two expanding ridges. In 13-6*b* the fault is shown after a period of ridge expansion, the arrows with solid heads indicating sea-floor spreading from the ridges. Motion has not changed the apparent offset of the ridge, and the relative motion is limited to the fracture between the ridge crests; note that the motion along the shear between the ridge crests is in the reverse direction to that expected if the fault were regarded as transcurrent. D. C. Krause (1966, p. 425) independently developed a

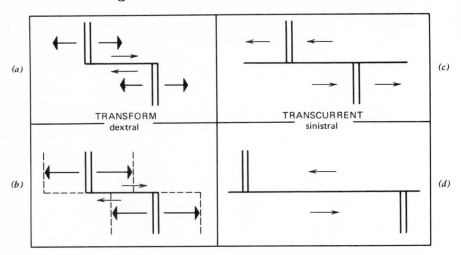

Figure 13-6. Motion on transform faults compared with transcurrent faults (after Wilson, 1965). (*a*) Dextral ridge-ridge type transform fault connecting two expanding ridges. (*b*) Fault shown in (*a*) after a period of movement. Note that motion has not changed the apparent offset. (*c*) Sinistral transcurrent fault offsetting a ridge, with offset in the same sense, but motion in the opposite sense to the transform fault in (*a*). (*d*) Fault shown in (*c*) after a period of motion. Note that the offset has increased. Open-headed arrows indicate components of shearing motion. Solid-headed arrows indicate ocean floor spreading from the ridge axis. (From Science, **150**, 482–485, 1965, with permission. Copyright 1965 by the American Association for the Advancement of Science.)

similar fault concept for an expanding ocean basin.

Application of the concept of transform faults by Wilson to the northeast Pacific in 1965 suggested that the Juan de Fuca Ridge, southwest of Vancouver Island, was linked to the East Pacific Rise by the San Andreas transform fault. Figure 13-2 includes the region of the Juan de Fuca Ridge and any interpretation of the Ridge should obviously be compatible with the magnetic observations. Wilson identified a crustal block with intense parallel anomalies striking N20°E, instead of the usual north-south, and located an axis about which the anomalies are symmetrically arranged. This axis coincides with the physiographic expression of the Ridge crest. This confirmed that at least some of the linear magnetic anomalies in the Pacific should be related to oceanic ridges, and also that these were parallel to a central anomaly over the crest of a ridge.

In a companion paper to Wilson's contribution Vine and Wilson reexamined the

Vine-Matthews hypothesis with a polarity reversal time scale to guide them. The three models shown in Figure 13-7 are based directly on the ridge models illustrated in Figures 13-5*b* and 13-5*c*, with the magnetized blocks of crust extending from 3 to 11 km below sea level. Each model in Figure 13-7 represents a 200 km section symmetrically disposed about a ridge crest at point *O*. According to the sea-floor spreading hypothesis the age of the crust at any distance ffom the ridge crest is a function of the spreading rate, and for the assumed spreading rates of 1 and 2 cm per year per limb in Figures 13-7*a* and 13-7*b*, the age of the crust at any distance from the ridge crest *O* is readily calculated. The polarity reversal time scale shown as 1964*b* in Figure 12-13 has major reversals at 1, 2.5, and 3.4 m.y., and short events at about 1.9 and possibly 3 m.y. These ages are shown below the crustal sections at appropriate distances from the ridge center, and the corresponding alternately magnetized blocks of the crust are fitted

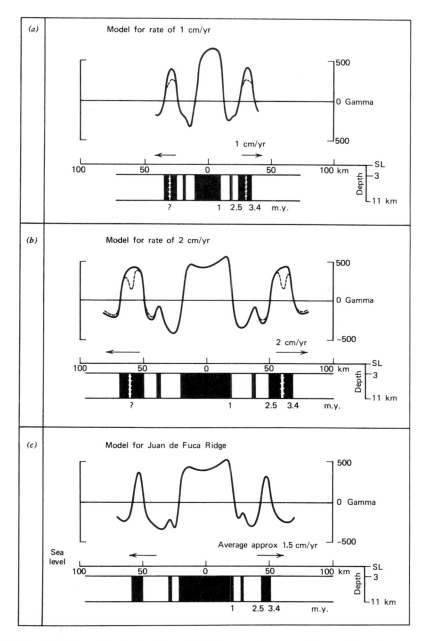

Figure 13-7. Models and calculated total field magnetic anomalies resulting from a combination of polarity reversals for the Earth's magnetic field and ocean-floor spreading (after Vine and Wilson, 1965). Normally and reversely magnetized blocks of crust are distinguished . (*a*) and (*b*) assume uniform rates of spreading. (*c*) Deduced from the gradients on the map of observed anomalies. The dashed parts of the computed profiles show the effect of including what was in 1965 considered to be a possible reversal at 3 m.y. (compare polarity sequence with Figure 12-13). (From Science, **150**, 485–489, 1965, with permission. Copyright 1965 by the American Association for the Advancement of Science.)

to these positions with dashed lines for the possible 3 m.y. reversal. For the distribution of magnetized blocks so deduced in Figures 13-7a and 13-7b and for the conditions stipulated in the figure legend, the computed magnetic profiles are as shown. The dashed parts of the profiles show the effect of including the possible reversal at 3 m.y. These computed profiles indicate that with a faster spreading rate, the profile gives more information about the distribution of magnetized blocks within the crust. The shape and amplitude of the central anomaly also change with spreading rate.

Figure 13-7c is a computed model for the Juan de Fuca Ridge (see also Figure 13-9). Vine and Wilson selected the steepest gradients of observed anomaly profiles on the magnetic intensity map, assumed that these delineated the boundaries between normally and reversely magnetized crustal blocks, and concluded on this basis that the crust was magnetized as shown in Figure 13-7c. They related the alternately magnetized blocks to

the polarity reversal time scale and added appropriate dates. The rate of spreading appeared to be rather erratic (dashed line in Figure 13-10a), with an average spreading rate of 1.5 cm per year per limb of the cell. The anomaly profile computed for this deduced model in Figure 13-7c shows reasonable agreement with the observed anomalies shown in Figure 13-8a. Unfortunately at this time the Jaramillo event had not been distinguished and in Figure 13-7 this was identified as the Olduvai event (Figure 12-14), and the calculated spreading rate was therefore too low (Figure 13-10).

The generalized model of Vine and Matthews used in Figures 13-5 and 13-7 differs from Hess's concept of sea-floor spreading, which requires that the crust is composed of serpentinized mantle peridotite, covered by a layer of only 1 or 2 km basalt (Figure 12-22). Vine and Wilson therefore worked out a second model for the Juan de Fuca Ridge, in which the magnetic material is confined entirely to layer 2, between 3.3

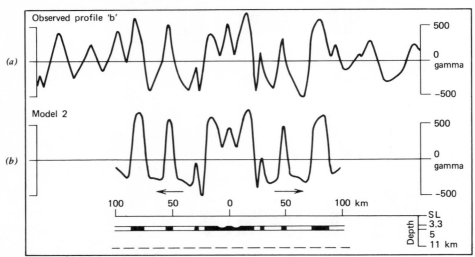

Figure 13-8. Interpretation of magnetic anomalies associated with the Juan de Fuca Ridge (after Vine and Wilson, 1965). (a) Observed profile 'b' across the Juan de Fuca Ridge; note its symmetry. (b) Model of magnetized crust and calculated anomaly assuming a strongly magnetized basalt layer only. Normal or reverse magnetization is with respect to an axial dipole vector; axial dipole dip taken as $+65°$. Effective susceptibility taken as $±0.01$, except for the central block, $+0.02$. (From Science, **150**, 485–489, 1965, with permission. Copyright 1965 by the American Association for the Advancement of Science.)

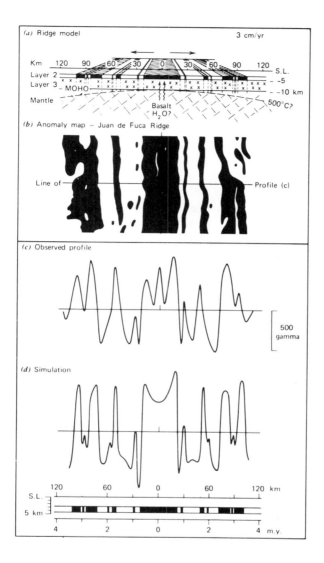

Figure 13-9. The tape-recorder, conveyor belt scheme for sea floor spreading summarized by Vine (1968). (*a*) A schematic representation of the crustal model, applied to the Juan de Fuca Ridge, southwest of Vancouver Island (Figures 13-2 and 13-8). Shaded material in layer 2, normally magnetized; unshaded, reversely magnetized. (*b*) Part of the summary map of magnetic anomalies recorded over the Juan de Fuca Ridge (Figure 13-2). Black, areas of positive anomalies; white, areas of negative anomalies. (*c*) A total-field magnetic anomaly profile along the line indicated in *b*. (*d*) A computed profile assuming the model and reversal time scale. Intensity and dip of the Earth's magnetic field taken as 54,000 gamma and +66°; magnetic bearing of profile 087°. (1 gamma = 10^{-5} oersted.) (S.L. = sea level.) Note: Throughout, observed and computed profiles have been drawn in the same proportion: 10 km horizontally is equivalent to 100 gamma vertically. Normal or reverse magnetization is with respect to an axial dipole vector, and the effective susceptibility assumed is ±0.01 except for the central block at a ridge crest (+0.02). (From F. J. Vine in "The History of the Earth's Crust", ed. R. A. Phinney, Figure 1, p. 75. Copyright 1968, Princeton University Press. Used with permission of the author.)

and 5 km below sea level as illustrated in Figures 13-8b, 13-9a, and 13-9d. The underlying serpentinite would contribute little to the observed anomalies. Figure 13-9 is a schematic representation of this crustal model, published later by Vine, with yet another revision of the polarity reversal time scale. Figures 13-9c and 13-9d are revised versions of Figures 13-8a and 13-8b. In this model the effective susceptibility is assumed to be considerably higher than in the generalized model. Figure 13-8b was derived in precisely the same way as Figure 13-7c, and comparison of the two computed profiles with the observed magnetic profile in Figure 13-8a suggests that the specific model after Hess is an improvement over the original generalized model. These are possibly the two extremes because a main crustal layer of serpentinite would surely be riddled with basic intrusions. Vine and Wilson concluded that the success of both models confirms the essential feature of the Vine-Matthews hypothesis that steep magnetic gradients are due to boundaries between normally and reversely magnetized crust.

At the Goddard conference in November 1966 (published 1968) and in a *Science* article of December 1966, Vine reviewed new evidence from several magnetic surveys which provided convincing confirmation that linear magnetic anomalies can be correlated from one profile to another, that these do parallel ridge crests, and that for many latitudes and orientations, the anomalies are symmetrical about the axis of the ridge (Figures 13-8a, 13-12b). This evidence he regarded as virtual proof of sea-floor spreading and its implications.

The successive revision of the polarity reversal time scale shown in Figure 12-13 has complicated the interpretation and dating of the anomaly profiles, as shown by Figures 13-8, 13-9, and 13-10 for the Juan de Fuca Ridge. These figures used the polarity time scales labelled in Figure 12-13 respectively as 1964b, 1966a, and 1966b. Since the publication of Figures 13-7 and 13-8, recognition

of the Jaramillo event in igneous rocks produced the reversal scale of 1966a in Figure 12-13 (Figure 12-14). Vine noted that he and Wilson could have used the magnetic profile shown in Figure 13-8 to predict the then-unknown Jaramillo event; the distribution of alternately magnetized blocks inferred from the observed profile in Figure 13-8 does not agree closely with the reversal scale of Figure 12-14 (excluding the short Mammoth event). This raises the possibility that if spreading rates remain reasonably constant, which might be anticipated for inertial reasons, then the spacing of the linear anomalies could be used to check the reversal time scale, to date more accurately the short events, and to extrapolate the time scale beyond the 4 m.y. limit of the K-Ar dating method.

An estimate of spreading rate is obtained by following the procedure described for Figure 13-7c. Normal-reverse crustal boundaries inferred from the magnetic profiles are correlated with the polarity epochs and events. In Figure 13-10a the distances of the inferred boundaries from the ridge crests are plotted against the 1966a time scale for the reversal sequence (Figures 12-13 and 12-14). Each point shows the distance from a ridge crest of a specific normal-reverse boundary. The horizontal time scale thus gives the apparent ages, or the duration of spreading, for the successive boundaries in the crustal sections. There are small departures from linearity for each of the three ridges illustrated, and the best straight line through each set of points gives the average spreading rate listed. The dashed line shows the earlier interpretation of Vine and Wilson for the Juan de Fuca Ridge (Figures 13-7 and 13-8), which was based on identification of the then-unknown Jaramillo event as the Olduvai event; this implied a more erratic spreading and slower average rate.

The deviations from linearity for the East Pacific Rise and the Juan de Fuca Ridge are exactly analogous, except that the magnetic profile for the South Pacific provides evidence that the Mammoth event near 3 m.y. is multiple (and it is so plotted in Figure 13-11). Figure 13-10*b* gives straight lines for the average spreading rates of these two ridges, and the distances of the inferred normal-reverse boundaries have been re-plotted on these lines; this causes slight changes in the apparent ages of the boundaries. Equivalent normal-reverse boundaries on each ridge remain very similar in apparent age, as shown by the dashed lines, and Vine suggested that the average ages of these equivalent pairs might be considered, tentatively, as revised values for the polarity reversal time scale. The slightly revised scale is shown in Figure 13-10*b*, and compared with the K-Ar scale at the top of the diagram which has been transferred from Figure 13-10*a*.

The polarity reversal time scales in Figure 13-10 are correlated with anomaly profiles extending out to 150 km from the crest of the East Pacific Rise, but anomalies have been recorded for much greater distances. If we assume that spreading from the East Pacific Rise proceeded at a constant rate of 4.4 cm per year for a period of 11.5 m.y., we can plot the positions of normal-reverse boundaries inferred from the anomalies measured at distances out to 500 km from the Rise crest as shown in Figure 13-11. These points, located on the spreading rate line in terms of distance, provide the sequence of normal and reversed polarities shown on the horizontal time scale. The time scale out to 5.5 m.y. is based on the plot for the East Pacific Rise together with the similar plot for the Juan de Fuca Rise using the average apparent ages for equivalent boundaries as in Figure 13-10*b*; but beyond 5.5 m.y. the time scale is based only on the East Pacific boundaries. The extrapolation depends upon the assumption of constant spreading rate, and this is therefore a reversal time scale relative to the East Pacific Rise. This pattern of longer and shorter events can be correlated with scales obtained from other ridge systems, and with scales obtained by other techniques,

such as the uncalibrated sequence of polarity zones in Figure 12-16.

Comparison of the anomaly profiles for the East Pacific Rise and the Juan de Fuca Ridge indicates that changes in spreading rate do occur. Figure 13-10*a* shows that the rate of spreading from the Juan de Fuca Ridge remained fairly constant at 2.9 cm per year for a period of 4 m.y. However, the inferred boundaries for distances greater than about 150 km from the Juan de Fuca Ridge crest depart from linearity when plotted against the time scale relative to the East Pacific Rise as shown in Figure 13-11. One interpretation of Figure 13-11 is that spreading from the Juan de Fuca Ridge decelerated about 5.5 m.y. ago from a rate of about 4 cm per year to its present rate of 2.9 cm per year. Alternatively, if we assumed that spreading from the Juan de Fuca Ridge had remained constant, we would obtain a different reversal time scale, and the points for the East Pacific Rise when plotted against this scale would have departed from the straight line. This would imply an acceleration of spreading from the East Pacific Rise about 5.5 m.y. ago.

From his review of the available data on magnetic anomalies in the Pacific Ocean Vine concluded that the north-south anomalies of the northeast Pacific (Figure 13-2) are related to a former crest of the East Pacific Rise, presumably overridden by the North American continent and possibly now located beneath the area of the Colorado Plateau uplift. He therefore constructed a profile just north of the Mendocino fracture zone, extending from the Gorda Ridge at $127\frac{1}{2}°$W out to the boundary of the north-south anomalies at 168°W (Figures 13-3 and 13-16 for the geography). If a constant spreading rate of 4.5 cm per year is assumed for this 3500 km section of the Pacific crust, the profile can be calibrated beyond 11 m.y. by using the same procedure described for Figure 13-11. This extends the time scale for reversals out to more than 80 m.y. which indicates that spreading was initiated from the

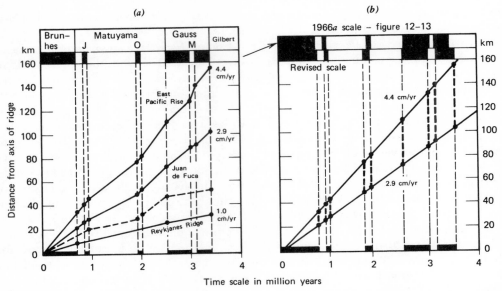

Figure 13-10. Magnetic anomalies and polarity reversals (after Vine, 1966). (*a*) Inferred normal-reverse boundaries within the crust plotted against the reversal time scale (Figure 12-13, 1966*a* scale). The dashed line represents a similar plot for the Juan de Fuca Ridge if one assumes an earlier time scale, as did Vine and Wilson in Figures 13-7 and 13-8. Note the similar deviations from linearity for the East Pacific Rise and Juan de Fuca Ridge. The average spreading rates are given for each ridge, and two of these are illustrated in Figure *b*. (*b*) Distances of polarity reversal boundaries from ridges plotted on the straight lines giving average spreading rates from Figure *a*. This suggests that the reversal time scale requires slight revision (assuming that there was a constant spreading rate for the past 4 m.y.). The revised sequence of reversals on the time scale is based on the dashed lines through the plotted points, and compared with the previous reversal time sequence from *a* at the top of the figure. (From Science, **154**, 1405–1415, 1966, with permission. Copyright 1966 by the American Association for the Advancement of Science.)

East Pacific Rise in the late Cretaceous. Vine emphasized the speculative nature of this time scale and stressed the necessity for using it to predict and correlate anomalies in other oceanic areas. This procedure has since been used extensively (Figure 13-13).

1965-1966: Scepticism and Conversion of J. R. Heirtzler and Lamont Associates

The marine magnetics program at the Lamont Geological Observatory was initiated by M. Ewing. The early magnetic measurements in the Atlantic Ocean (Figure 13-1) were subsequently extended to the Pacific, Antarctic, and Indian Oceans. At a symposium on "The World Rift System" in

1965, M. Talwani, X. Le Pichon, and J. R. Heirtzler reviewed the patterns of magnetic anomalies over the Midocean Ridge Rise system, dividing them into "axial anomalies" and "flank anomalies." Figure 13-12*b* shows axial anomalies lying within the lines *BB*, and the rather abrupt change to more irregular, longer-wavelength anomalies on the flanks; in some locations, the amplitudes become higher too. Their review ended with the following comments:

"We are not sure of the ultimate origin of the ridge magnetic anomalies. However we do believe that the symmetry of the magnetic pattern and its parallelism to the strike of the ridge requires that the magnetic anomalies owe their existence to the formation of the

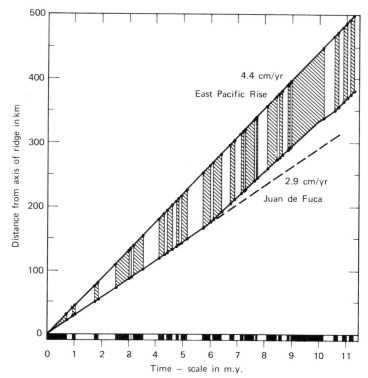

Figure 13-11. Extrapolation of the reversal time scale (after Vine, 1966). Magnetic boundaries across the East Pacific Rise, deduced out to 500 km from the crest on the basis of magnetic anomalies, and plotted on a line representing a constant spreading rate of 4.4 cm/year (extrapolation of line in Figure 13-10b). Similar boundaries for the Juan de Fuca Ridge are plotted out to 150 km assuming a constant spreading rate of 2.9 km/year. The reversal time scale out to 5.5 m.y. is based on both plots. Beyond that time, the time scale is based on the East Pacific Rise boundaries, and the distances of the boundaries for the Juan de Fuca Ridge then depart from the line of constant spreading rate; beyond 5.5 m.y., the Juan de Fuca points trace a line indicating faster spreading rates (assuming constant East Pacific rates). (From Science, **154**, 1045–1415, 1966, with permission. Copyright 1966 by the American Association for the Advancement of Science.)

ridge, which in turn is caused by a change of density in the upper mantle . . . However we feel that the variation in amplitude of the axial anomalies as well as the completely different character of the flank anomalies argues against the Vine and Matthews hypothesis.'' (Irvine, 1966, p. 346.)

The Reykjanes Ridge southwest of Iceland was selected in 1961 for magnetic survey because the crest was known to have a large anomaly, and there were indications that it had a linear character. Comparison with the long, regular anomalies of the northeast

Pacific appeared worthwhile. A proposal was submitted in 1962 and the survey was carried out during October and November 1963 as a cooperative project by Lamont Geological Observatory and the U.S. Naval Oceanographic Office. As shown in Figure 13-12a, an area about 350 km square was surveyed, located about 350 km southwest of Keflavik in Iceland. Aeromagnetic survey aircraft operating from Keflavik flew 58 lines approximately perpendicular to the ridge axis, at 0.46 km altitude, with track spacing of 5–10 km. This survey had more accurate

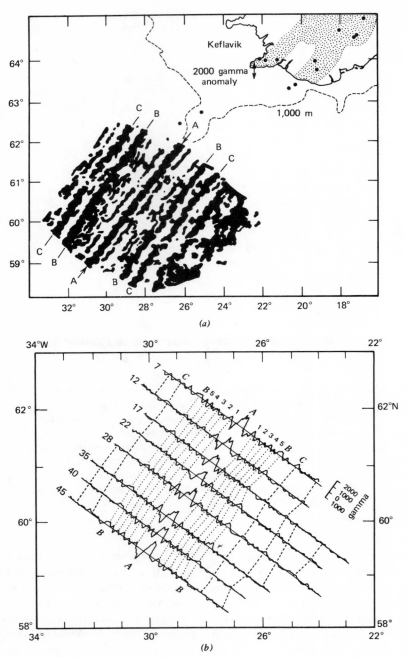

Figure 13-12. Magnetic anomalies over Reykjanes Ridge south of Iceland (after Heirtzler et al., 1966). (*a*) Black striped area indicates positive magnetic anomalies; the central anomaly over the ridge is marked by the arrows. (*b*) Eight of the 58 profiles of total magnetic field across the Ridge, projected on lines at right angles to the Ridge. Note the continuity of the central anomalies along the ridge axis, and the symmetry about this axis of the profiles. (From *Deep-sea Research*, by permission of Pergamon Press.)

navigation than other extensive marine surveys completed by that time. The anomalies were determined from the total geomagnetic field intensity measurements by subtracting values for a regional field which was a least squares fit to a simple linear function of latitudes and longitudes. The result is shown in the anomaly map of Figure 13-12a, which was published in 1966. This was the first study to reveal in detail the magnetic pattern over a large area of the midoceanic ridge in a place where its identification as a ridge was beyond doubt.

Clear details of the anomalies are given in Figure 13-12b by the profiles projected on lines perpendicular to the ridge. The axial anomaly A over the ridge of the crest is an outstanding feature, with a width of 40 km and an amplitude of 3000 γ, peak-to-peak. The axial anomalies consist of six pairs of symmetrically disposed anomalies each about 15 km wide with 100 to 300 γ amplitudes. The larger wave-length anomalies of the flank provinces can be correlated from one profile to another through long distances (dotted lines in Figure 13-12b). In the axial area between the lines BB, the regional field is depressed by several hundred gammas (this does not show on the anomaly map and profiles). The authors suggested that this might be caused by elevation of the Curie isotherm (500°C) because of high heat flow in this axial region, with the result that the amount of magnetized material in the crust decreases progressively towards the axis. They concluded that except for the body causing the axial anomaly, the source of the magnetic anomalies in the axial zone lay entirely within the 2–4 km thick upper layer.

In their discussion of the origin of the anomalies Heirtzler *et al.* (p. 440) concluded that the Vine-Matthews hypothesis "in its present form does not explain the characteristic change in magnetic pattern from the axial zone to the flanks and the difference between the axial anomaly and the adjacent ones."

The change from axial anomalies to flank anomalies was previously listed as the third

difficulty facing the Vine-Matthews hypothesis, and by 1966 it appeared to be the main argument of the Lamont group against the hypothesis. In 1966 Vine suggested that the change might reflect an increase in frequency of reversals of the Earth's field together possibly with a decrease in its intensity. Because this type of change would be worldwide, the boundary should occur at different distances from ridge axes, according to the spreading rate in a region. Vine reported from a preliminary examination of many ridge profiles that such a change may have occurred approximately 25 m.y. ago. Before he published this tentative explanation, however, the Lamont group had been converted to the Vine-Matthews hypothesis, following Hess's model for sea-floor spreading (Figure 13-9 and not Figures 13-5 and 13-7).

Acceptance of the hypothesis must have come between the end of 1965 and the fall of 1966. The paper on the Reykjanes Ridge was submitted in November, 1965, and published in June, 1966. W. C. Pitman and J. R. Heirtzler submitted a paper to *Science* on September 12, 1966, which was published on December 2 just two weeks before Vine's paper in the same journal. The Lamont contribution described the magnetic anomalies over the Pacific-Antarctic Ridge, which were as linear and symmetrically disposed about the ridge as those illustrated in Figure 13-12 for the Reykjanes Ridge.

Impressed by this symmetry the authors applied the Vine-Matthews hypothesis to the ridge, using normally and reversely magnetized blocks with upper surface set by the average bathymetry and a thickness of 2 km. The crust sloped upward toward the ridge crest in this model, reflecting the sea-floor surface, rather than remaining horizontal as in Figure 13-9. The anomalies were correlated with the K-Ar reversal time scale back to the Gilbert epoch as in Figure 13-10a, which gave an average spreading rate of 4.5 cm per year. Normally and reversely magnetized blocks were then placed beneath the remaining observed anomalies to the edge of the profile. Assuming a constant spreading rate, Pitman and Heirtzler extended the polarity reversal scale back to 10 m,y., just as Vine did in Figure 13-11. This reversal time scale relative to the Pacific-Antarctic Ridge was then applied to the Reykjanes Ridge, using Vine's spreading rate of 1 cm per year (Figure 13-10*a*). The scale was contracted by a factor of 4.5 to allow for the difference in rates, and a magnetic profile for the Reykjanes Ridge was computed from the crustal model so obtained. The similarity of computed and observed anomaly profiles was considered good enough to indicate constant spreading rates from these two ridges at least for the past 10 m.y. This led the Lamont investigators to conclude that:

"We feel that these results strongly support the essential features of the Vine and Matthews hypothesis and of ocean-floor spreading as postulated by Dietz and Hess . . . permits one (using a constant spreading rate) to date reversals of the geomagnetic field back to 10 million years ago."

At the Goddard symposium on "The History of the Earth's Crust" held at essentially the same time, November 1966, Heirtzler stated that: "several studies of magnetic anomalies nearing completion at Lamont not only support axial spreading but indicate that this process extends completely from the ridge axis to the continental shelf in many areas." (Phinney, 1968, p. 90.) He noted that the following criteria must be satisfied before a series of magnetic anomalies could be associated with ocean-floor spreading: (a) the anomalies must be linear in a direction parallel or nearly parallel to ridge axis; (b) the anomalies must be in accord with the history of field reversals found in other oceans, modified by the presence of fracture zones. Since 1966 the efforts of many scientists at the Lamont Geological Observatory have been directed toward extrapolation and testing of the hypothesis in diverse ways.

1968: Extrapolation by J. R. Heirtzler, G. O. Dickson, E. M. Herron, W. C. Pitman, III, and X. Le Pichon

A series of four contiguous papers in the March 1968 issue of the *Journal of Geophysical Research* showed that a magnetic anomaly pattern, parallel to and bilaterally symmetrical about the midoceanic ridge system, exists over extensive regions of the North Pacific, South Pacific, South Atlantic, and Indian Oceans. The selected results compared in Figure 13-13 demonstrate further that the pattern is the same in each of these oceanic areas, and that the pattern may be simulated in each region by the same sequence of crustal blocks. The blocks comprise a series of normally and reversely magnetized material 2 km thick. The observed anomaly pattern and the general configuration of the crustal model conform to the prediction of Vine and Matthews (Figure 13-9 modification of Figure 13-5) and this is regarded by the Lamont investigators as strong support for the concept of sea-floor spreading. The authors did note, however, that if the Vine-Matthews theory is basically in error, then the conclusions of their analysis do not apply.

Figure 13-13 shows observed magnetic profiles for each of the oceans; the ship tracks are given in the original papers. Each profile has been projected perpendicular to the ridge axis. Some of the key anomalies have been numbered for reference purposes, and the dashed lines show the correlation of these anomalies from one profile to another. The crustal models that could explain the observed anomalies according to the Vine-Matthews theory are shown in the usual way by the strips of alternately black (normally magnetized) and white (reversely magnetized) units. Simulated profiles computed for the crustal models are shown between the observed profiles and the models. The ridge crests of each profile have been plotted directly above each other, and their positions indicated by 0 km. A distance scale is given

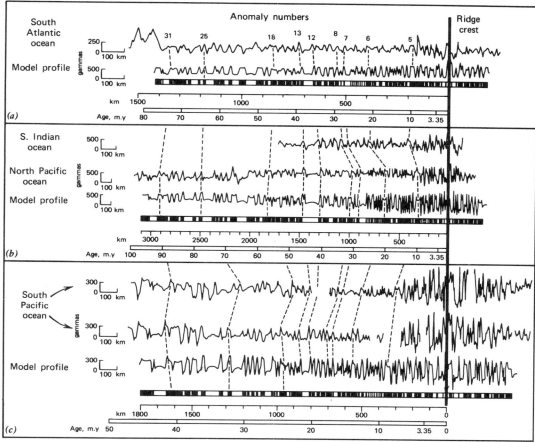

Figure 13-13. Sample magnetic profiles from various oceans (after Heirtzler et al., 1968; reviewing data in three companion papers by Dickson et al., 1968, Le Pichon and Heirtzler, 1968, and Pitman et al., 1968). Beneath each observed profile is a theoretical profile for comparison, calculated from the assumed sequence of crustal blocks normally magnetized (black) and reversely magnetized (white). Each block is taken to be 2 km thick. With each model is a time scale constructed arbitrarily by assuming an age of 3.35 m.y. for the end of the Gilbert reversed epoch (see Figures 12-14 and 12-21), and assuming a constant spreading rate for 50 to 100 million years beyond that time. Dashed lines relate similarly shaped anomalies identified by the numbers at the top of Figure (*a*). Note that specific anomalies in different profiles do not have the same "ages" on the three time scales. (From Jour. Geophys. Res., **73**, 2119–2136, 1968, with permission.)

below each crustal model showing the distances of anomalies from the ridge crests. Notice that there are three different distance scales in the figure. An arbitrary age scale is given with each model, based on the date of 3.35 m.y. for the beginning of the Gauss normal polarity epoch (Figures 12-14 and 12-21), and assuming a constant rate of spreading. These are not true age scales, as

shown by the fact that specific anomalies in different profiles do not have the same "age" on the scales. The main objective of the fourth paper in the series was to derive a time scale for the sequence of geomagnetic field reversals predicted by the crustal models.

We have already discussed the procedure adopted by Vine, Pitman, and Heirtzler for extending the polarity reversal time scale relative to a particular ridge back to 10 m.y., by assuming a constant spreading rate from that ridge during this period. Using the same procedure Vine extrapolated the scale back to 80 m.y. relative to the East Pacific Rise. Figures 13-10 and 13-11 illustrate the procedure, but Figure 13-11 also demonstrates that the rates of spreading of one ridge relative to another may change with time. Assumptions of constant spreading rate must therefore be made only with due caution and after comparison with relative reversal time scales for other ridges.

Figure 13-14 shows the relative rates of spreading from ridges in different oceans; its construction is essentially the same as Figure 13-11. The vertical axis gives the recorded distances of anomalies from a ridge crest. The horizontal axis is a relative time scale based on the assumption that the South Atlantic ocean floor has been spreading at a constant rate of 1.9 cm per year for the past 80 m.y. This rate was calculated from the distance covered within the 3.35 m.y. since the beginning of the Gauss normal polarity epoch. The ages of the normal and reversed magnetic intervals listed for the South Atlantic crustal model in Figure 13-13 were obtained by dividing the distances to the model magnetized bodies by the spreading rate of 1.9 cm per year. Both distances and ages are shown on the axis of Figure 13-14, and these may be compared with the crustal model polarity sequence in Figure 13-13. The ages of the specific anomaly numbers at the bottom of the diagram are located with respect to this time scale by their distance from the crest of the South Atlantic Ridge.

The time scale extrapolated from the South Atlantic profile was selected as a standard because:

1. The anomaly pattern for the South Indian Ocean is not sufficiently long;

2. The North Pacific profile is distorted in the ridge axis;

3. The spreading in the South Pacific appears to have changed relative to both the

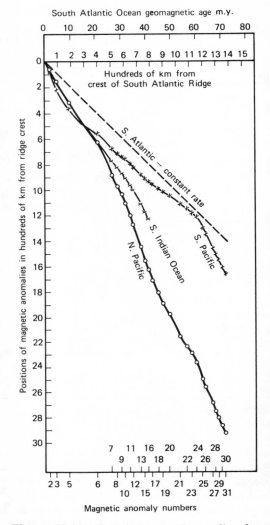

Figure 13-14. Relative rates of spreading from ridges in different oceans (after Heirtzler et al., 1968). The distance of a given anomaly in the South Atlantic from the ridge crest, plotted against the distance of the same anomaly in the South Indian, North Pacific, and South Pacific Oceans (data from Figure 13-13). Anomaly numbers are given, facilitating comparison with Figure 13-13. The South Atlantic anomaly distance scale is used to provide a relative time scale, assuming constant spreading rate (compare Figure 13-11). (From Jour. Geophys. Res., **73**, 2119–2136, 1968, with permission.)

North Pacific and the South Atlantic in the vicinity of anomalies 5 and 24, suggesting that these are real changes in rate;

4. There is evidence that the spreading rate from both the North and South Atlantic Ridge has been constant for the past 10 m.y.

The authors emphasized the possible error inherent in the extrapolation and this warning should be heeded. The numbers listed as "ages" in Figures 13-13 and 13-14 do not represent the passage of the stated number of years. All "ages" are relative to an assumed constant rate of spreading. In Figure 13-13 each age scale is derived independently of the others for each specific ridge. In Figure 13-14 all ages are relative to the assumed constant spreading rate for the South Atlantic. Each anomaly has the same age in Figure 13-14 but not according to the "age scales" in Figure 13-13. The age of 80 m.y. just beyond anomaly 32 compares with a 73-m.y. date given by Vine for the same point on a magnetic profile.

In Figure 13-14 the distances to model magnetized bodies in the other oceans have been plotted against the South Atlantic relative time scale. The relative spreading rate curves illustrated are continuous, but nonlinear. The straight dashed line shows where the spreading rate curve for the South Atlantic would plot. Time scales for the North Pacific and the South Pacific crustal models relative to the South Atlantic standard could be taken directly from the horizontal axis, and adjusted to fit the model blocks for the crustal sections in Figure 13-13. The nonlinear age scales so determined would differ from the scales given in Figure 13-13; each specific anomaly would then be located at the same age on the three scales.

Figure 13-15 shows a geomagnetic reversal time scale, relative to the South Atlantic spreading, derived from Figures 13-13 and 13-14. The anomaly numbers are plotted against their ages. It is not legitimate to compare this hypothetical polarity time scale with the fine detail of marine core paleo-magnetic data (Figures 12-17 and 12-20a), but comparison with polarity scales such as those in Figure 12-16 may prove to be useful. It now appears to be generally agreed that the linear magnetic anomalies of the ocean basins do provide a record of the geomagnetic polarity reversals. Even events of quite short duration have been inferred from profiles in regions of rapid spreading, but it is unlikely that all short events have been distinguished. The greatest uncertainty is in the assignment of ages. Despite the assumptions and extrapolations involved in the provisional age scale of Figure 13-15, scientists directly involved in this research consider it unlikely that the ages can be wrong by as much as a factor of two.

Figure 13-14 demonstrates considerable variations in spreading rate, but there are no systematic gaps in the anomaly patterns, which suggests that any cessation of spreading occurred simultaneously in all oceans. Some variations in spreading rate were apparently rather abrupt. For example the ratio between spreading rates in the North and South Pacific was constant (at about one) between anomalies 31 and 24, increasing between anomalies 24 and 18 (to more than three), and after anomaly 5 it decreased again to about 0.7. There have been several suggestions that sea-floor spreading has been episodic with worldwide discontinuities. This is consistent with paleomagnetic evidence that drifting of continents has been intermittent, and other evidence comes from the distribution of sediments.

In 1962 ships from the Lamont Observatory began a reconnaissance survey of sediment thickness in the Pacific Ocean using a continuous seismic profiler. Results published in 1968 are shown in Figure 13-16. The upper layer of sediments (acoustically transparent) ranges in thickness from 20 m to 1 km. Beneath the transparent layer over most of the western North Pacific, but not in the eastern Pacific, is an acoustically opaque layer with thickness usually between 20 and 300 m. Evidence from cores indicates that the

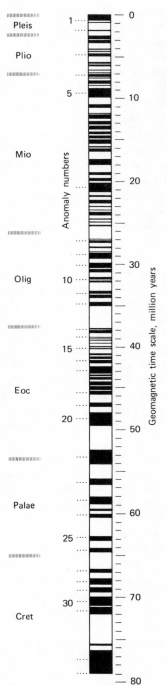

transparent layer consists of Cenozoic sediments, and that the pre-Cenozoic sediments represented by the opaque layer are limited to the western part of the Pacific basin. The limited sediment core data are consistent with the concept of a spreading ocean floor. The present pattern of biological productivity produces a high sedimentation rate in the equatorial belt, caused by upwelling along the equator. Figure 13-16 indicates that this pattern has continued throughout the Cenozoic with no indication of a change in the position of the equator during this time. The uniform thickness of the transparent layer along the axis of the equatorial belt is sharply reduced across the East Pacific Rise. It shows a marked decrease in thickness from 99° to 107°W, and less pronounced thinning between 150° and 168°W.

Similar patterns have been detected in many traverses across the midoceanic ridge system. J. Ewing and M. Ewing reported in 1967 that at the axes of ridges sediments are too thin to be resolved in the profiler records. At a distance of 100 to 400 km from the axis there is an abrupt increase in thickness of the sediments. Beyond the discontinuity the flank sediments are of almost constant thickness. Ages for the discontinuities, estimated by dividing the distance from the ridge crest by the spreading rate calculated for the region from the magnetic anomaly pattern, are about 10 m.y. in all regions. If uniform spreading rates are assumed it is very difficult to explain the discontinuity. It requires unlikely changes in the rates of sediment accumulation on a worldwide basis. Intermittent sea-floor spreading is the preferred interpretation.

These studies of sediment distribution led the authors to propose three main episodes of sea-floor spreading:

Figure 13-15 continued

Figure 13-15. The geomagnetic time scale (after Heirtzler et al., 1968), relative to the South Atlantic spreading (Figure 13-14). From left to right: Phanerozoic time scale for geologic eras, numbers assigned to bodies and magnetic anomalies, geomagnetic field polarity with normal polarity periods black (Figure 4-8). (From Jour. Geophys. Res., **73**, 2119–2136, 1968, with permission.)

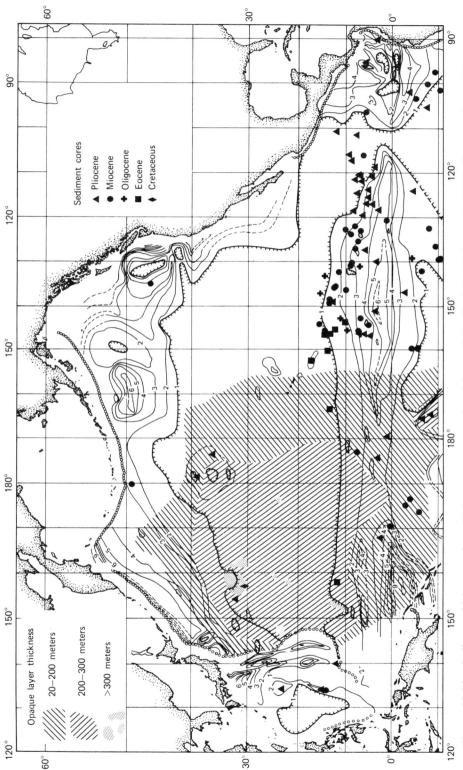

Figure 13-16. Sediment thickness and pre-Pleistocene core samples of the north Pacific (after Ewing et al., 1968). Contours show thickness of upper transparent layer in tenths of seconds of reflection time; each contour represents approximately 100 meters. Thicknesses of opaque and lower transparent layer are represented by hatching. Ages of cores generally correspond to a depth of about 10 meters in the sample. Open circles show axes of trenches. (J. Ewing *et al*., Geophysical Monograph, **12**, 147, 1968, with permission of Amer. Geophys. Union.)

1. Mesozoic when the ocean basins were formed.

2. Early Cenozoic during which most of the midocean ridge area was created. The pre-Cenozoic sediments in the Pacific Ocean suggest that this cycle began when the region now at 145°W was at the axis of spreading. Then followed a long period of quiescence during which most of the Cenozoic sediments were deposited; this may have exceeded 15 m.y., possibly reaching 30 or 40 m.y.

3. Latest Cenozoic beginning 10 m.y. ago during which the crestal regions were generated.

Identification of the two latest major readjustments of spreading is based on changes in sediment thickness at crest-flank and ridge-basin boundaries, which coincide closely with anomalies 5 and 32. Anomaly 5 is the outer limit of the axial magnetic pattern which is shown as the line *BB* in Figure 13-12*b*. Anomaly 32 coincides with the ridge-basin boundaries in the Atlantic and the opaque layer boundary in the Pacific (Figure 13-16), which is dated as Mesozoic. Therefore Le Pichon (1968) suggested that the provisional scale in Figure 13-15 should be modified by dating anomaly 32 at 60 m.y. (early Paleocene) instead of 77 m.y., and by placing an interruption of spreading about 10 m.y. long at anomaly 5, preceded by a general slowing down of the movement. This adjusted time scale is consistent with the available data.

According to the Vine-Matthews hypothesis, mapping the magnetic anomalies of the ocean basins is tantamount to mapping isochrons for the ocean basin floor. The locus of a specific anomaly is a line of constant age, where the age corresponds to the time of magnetization, which is interpreted as the time that this part of the ocean floor was brought to the surface at an active ocean ridge. The areas in which the sequence of anomalies shown in Figure 13-15 had been recognized by 1967 are summarized in Figure 13-17. Instead of showing magnetic anomalies as in Figures 13-2 and 13-12, the map shows lines corresponding to provisional ages at 10 m.y. intervals, taken from Figure 13-15. These lines may be considered as 10 m.y. "growth lines" from the ridges which are indicated by the zero isochron. The dotted lines show the fracture zones that have displaced the isochrons. In view of the preceding remarks about intermittent spreading and the uncertainty of the age scale in Figure 13-15 the dates assigned to the isochrons must be regarded as provisional. However, the ability to contour the ocean basins with isochrons, even if they are of uncertain ages, provides us with the prospect of unravelling the history of the ocean basins, and the resultant movement of the continents, with precision of detail inconceivable during the debate about continental drift that occupied the first half of this century.

PROBLEMS AND INTERPRETATIONS SINCE 1968

At the end of his comprehensive review on "Reversals of the Earth's Magnetic Field" in 1968 Sir Edward Bullard wrote:

"The lecture on which this paper is based was given in June 1967; it was a well chosen time, the threefold story of the reversals of the field had just become clear and could be easily and elegantly set out. In the few months needed to write the paper there has been an avalanche of new results which has revealed

many discrepancies and many matters needing elucidation; the usual chaos of the Earth sciences is clearly about to be reestablished at a higher level of understanding. We can now see many particular facts in the light of a global theory and realize that they are anomalous, whereas previously they appeared merely as isolated facts. The worldwide nature of the reversals can tie together the spreading of the ocean floor and the volcanic and

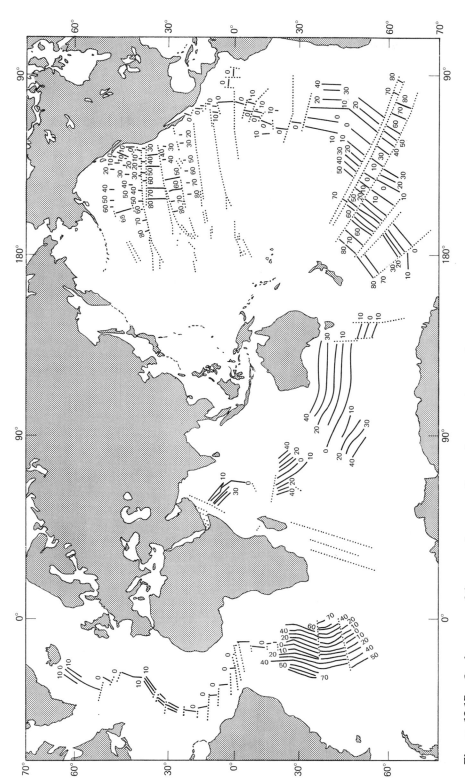

Figure 13-17. Isochron map of the ocean floor according to the magnetic anomaly pattern using time scale in Figure 13-15. Numbers on isochron lines represent age in millions of years. Dotted lines represent fracture zones (after Heirtzler et al., 1968, from Jour. Geophys. Res., **73**, 2119–2136, with permission).

sedimentary history of the last 50 m.y. with a detail never before approached. It is not surprising that ambiguities, difficulties and contradictions are emerging.''

We have examined the development of the threefold story of the reversals of the Earth's field and its implications for global tectonics, as they appeared in 1968. We should now review the discrepancies referred to by Bullard as well as more recent evidence. Unfortunately, if this book is to cover the range of material indicated by its title and yet remain within the length limits of economic feasibility, there is simply no space for an adequate discussion. I can only introduce selected topics and suggest a few references for reading and seminars.

Magnetization of Basalt

Bullard reviewed evidence that the titanomagnetite minerals in reversely magnetized lavas are more highly oxidized than in normally magnetized lavas and stated that the undoubted statistical correlation of oxidation and magnetic polarity must be considered one of the major unsolved problems of Earth science. In a detailed study of over 550 specimens from 14 lavas and seven dikes in Iceland N. D. Watkins and S. E. Haggerty in 1968 reported that the oxidation spectra in each body, and from baked sub-basaltic sediments, exclude the existence of self-reversals of magnetic polarity. The results show a strong correlation between the percentage of reversed magnetic polarity specimens and higher oxidation in the lavas but no correlation in the dikes. There is at present no satisfactory explanation.

The state of magnetization of submarine basalts, and the causes of magnetic anomalies in oceanic crustal material, are not adequately understood. In a 1968 review N. D. Watkins listed the assumed properties of the mobile basaltic blocks involved in models for seafloor spreading such as Figure 4-40. These are:

1. A high intensity of magnetization, J.

2. An effectively uniform intensity of magnetization, in terms of instrument response at sea level.

3. A Q factor greater than unity. $Q = J/\chi H$, where χ is the susceptibility and H is the ambient geomagnetic field intensity. The Q factor is thus defined as the ratio of the permanent (or remanent) intensity of magnetization to the induced intensity of magnetization. Typical oceanic anomalies could not be caused by induced magnetization alone. If Q is not much greater than 1.0, however, then the induced component would strongly diminish any negative anomaly due to a reversely magnetized body.

4. A high magnetic stability, S. S describes the degree to which NRM is retained in a rock. The NRM developed in an original reversed magnetic field might be subsequently obliterated in an unstable rock.

Watkins reviewed results indicating that variations of J and S in lavas correlate with the oxidation state, as revealed by the titanomagnetite minerals. The Q factor in many basaltic bodies shows considerable variation with position in the body, which is probably attributable to initial cooling conditions. In Icelandic basaltic rocks Watkins reported that Q values of less than 1.0 were found in over 20% of lava specimens, and more than 30% of dike specimens, showing that bodies originally possessing reverse magnetization may not always produce a negative magnetic anomaly. Watkins concluded that single bodies of basaltic material may develop sufficient variation in J, χ, Q, and S to produce a series of anomalies parallel to the cooling surface of the body. A series of inclined and truncated bodies of this kind could thus give rise to a series of linear anomalies.

The few magnetic studies of basalts dredged from the ocean floor that have been completed suggest that they have unusual magnetic properties, compared to continental basalts. For example Ozima *et al.* (1968) reported that three out of eight samples from Pacific seamounts and the deep ocean crust showed selfreversal of TRM when heated to 300°C in air. When heated, all eight samples showed a much greater increase in saturation magnetization, I_s, than do continental basalts. The authors suggested that a "certain physical-chemical condition prevailing in the deep ocean bottom is responsible for their peculiar magnetic properties." They also pointed out that hot emanations or lavas rising along fractures would produce an increase in I_s, and might conceivably cause linear magnetic anomalies. A process of this type, however, is not likely to produce magnetic anomalies with widths of the order of tens of kilometers.

Direct measurement of magnetic polarity in basalts dredged from the Reykjanes Ridge at approximately 60°N (Figure 13-12) was reported by J. de Boer, J. Schilling, and D. C. Krause in 1969. The original orientations of the basalts on the ocean floor, the positions in which they were magnetized, were deduced from standard volcanic criteria such as the shape of pillows, the location of pillow necks and stems, and vesicle distribution. Cores were drilled from the oriented specimens and the average directions of magnetization for each specimen gave the directions of remanent magnetization. The measurements indicated that basalts from the axial zone (zone A in Figure 13-12) were magnetized in a field with normal polarity along a direction roughly parallel to that of the present field. A specimen dredged from an adjacent area with negative anomaly (between A and 1 in Figure 13-12) was magnetized with inclination $-20° \pm 5°$, in opposition to the $+74°$ inclination of the present field at the recovery site. This suggests that the basalt was magnetized in a reversed magnetic field supporting the Vine-Matthews theory.

Selected Interpretations

In order to examine anomalies the trend of the regional geomagnetic field must be removed from the observed magnetic field. Bullard (1967) deplored the proliferation of methods employed to remove the trend, and the problems thus introduced when results from different surveys were fitted together. The problems should be eased in the future if an October, 1968, recommendation is followed. The International Association of Geomagnetism and Aeronomy—International Union of Geodesy and Geophysics recommended that the International Geomagnetic Reference Field be used as the common survey datum. This field is described by a spherical harmonic analysis of degree and order 8 and it includes coefficients of annual change.

Before 1968 many surveys had employed methods where the trend was estimated from the survey data itself. This procedure invariably produces a series of anomalies of alternating sign, even if all of the variations in total field intensity are positive (or negative) with respect to the smoothed geomagnetic field derived from global observations. According to Watkins and Richardson (1968) it has yet to be satisfactorily established that all the linear anomalies of the ocean basins are truly of variable sign. This is germane to interpretation of the anomalies in view of Watkins's conclusions that linear variations in the regional field can be produced by variations in the magnetization of a single, cooling basaltic body.

The estimated rate of spreading from a ridge has been correlated with a number of physical features. For faster spreading there appear to be more earthquake epicenters on the transform faults compared to the ridges. The East Pacific Rise, with a fast spreading rate, exhibits smoother topography with no central rift valley than the rugged and more slowly spreading mid-Atlantic ridge. According to Menard (1967) the topographic relief is proportional to the thickness of layer 2 and inversely proportional to the spreading rate. He calculated the discharge of lava from the crests of ridges necessary to form layer 2 by multiplying spreading rate times the thickness times a unit length. The estimated discharge on each flank is 25 to 45 km³/

10^6 yr/km, which gives 5 to 6 km^3/year for the whole ridge system. A rate of 1.8 km^3/ year for the age of the Earth would be sufficient to produce the total volume of the continents.

"Some Remaining Problems in Sea-Floor Spreading" were discussed at the 1966 Goddard symposium by Menard (Phinney, 1968), and he drew attention to features of the ridges and transverse fracture zones that were more complicated than the simple spreading theory implies. He suggested that the East Pacific Rise was once straight, and that portions of the crest have migrated with spreading from each short segment now being caused by a separate convection cell. Menard showed in Figure 13-18 how one-sided convection, with the ridge crest migrating at the same rate that the sea floor appears to

spread, could produce a series of symmetrical anomalies indistinguishable from the model with uniform spreading from a fixed crest (Figure 13-9).

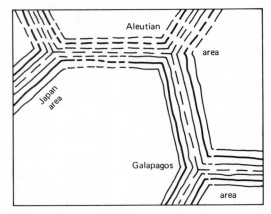

Figure 13-19. Schematic map of magnetic anomalies as they would appear if they were a paleomagnetic record produced by very large polygonal sections of the earth's crust receding from each other. Only parts of this scheme are confirmed by data (after Raff, 1968, Jour. Geophys. Res., **73**, 3699–3705, with permission).

A. D. Raff in 1968 related the schematic map of anomalies in Figure 13-19 with the approximate configuration of three radial rifting lines existing in the Galapagos area, and the sharply bending anomalies in the Aleutian area. He suggested that this pattern could be produced by upwelling of mantle material at a center with radial spreading giving the effect of large polygonal plates receding from each other. The development of the "magnetic bight" near the Aleutians has been the subject of much discussion.

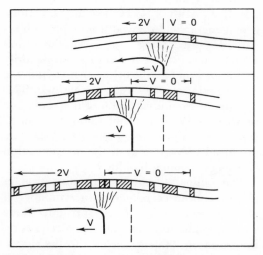

Figure 13-18. A means of producing symmetrical magnetic anomalies without moving one flank of a rise. The underlying convection cell migrates at the apparent spreading rate. One side spreads at twice the apparent rate and the other is fixed. This is one extreme possibility, the other extreme is that the convection is fixed and the two flanks spread at the same rate. (From H. W. Menard in "The History of the Earth's Crust", ed. R. A. Phinney, Figure 3 on p. 113. Copyright 1968 by Princeton University Press. Reprinted by permission of Princeton University Press.)

Detailed analyses of the magnetic anomaly pattern and fracture zones of the northeast Pacific (Figures 13-2 and 13-3) have led to conclusions that the process of sea-floor spreading was episodic (see also discussion of Figure 13-16). Peter *et al.* (1970) suggested two changes in the direction and rate of spreading; one between anomalies 22 and 24 and the second between anomalies 5 and 7 (see also Bassinger *et al.*, 1969). Menard and

Atwater (1968) assumed five events of reorganization. Vine and Hess (1970) reviewed evidence that the geometry of spreading changed within the past 5 m.y. in the Pacific and also in the North Atlantic, and they suggested that seismically inactive structures may well have been active at some time during the Tertiary. It can be anticipated that if there is a change in the spreading pattern, the change from one crustal geometry to another will be rapidly accomplished (Menard and Atwater, 1968).

It will be some time before agreement is reached about interpretation of magnetic anomaly patterns in terms of changing rates of movement, and of regional or global periods of cessation of spreading. In December 1969 D. A. Emilia and D. F. Heinrichs reviewed 15 published magnetic profiles from the Atlantic, Pacific, and Indian Oceans, using diagrams similar to Figures 13-10 and 13-11. The diagrams could be interpreted in two ways:

1. If the postulated relations between magnetic anomalies and the paleomagnetic time scale (as in Figures 13-13 and 13-15) are assumed correct, then the results indicate variable worldwide increases and decreases in the rates of spreading. This differs from episodic spreading, where all movement stops completely for a period.

2. If the spreading rates have remained nearly constant for the past 3.32 m.y., some of the accepted relations between the anomalies and the paleomagnetic time scale are incorrect.

They preferred the second interpretation, because it does not require complication of the basic spreading hypothesis. They concluded that the anomaly previously correlated with the Olduvai event is probably caused by the Gilsa event (Figure 12-21), and that the Olduvai event is represented by a minor anomaly previously unassigned to the Earth's polarity history. No particular magnetic anomaly had previously been assigned to the Gilsa event. J. D. Phillips and B. P. Luyendyk concluded in 1970 that the deep-sea drilling results

(Chapter 14) and oceanic magnetic anomalies together suggested that the rate of motion between the North American and African plates (Figure 14-3) had been relatively constant through the Cenozoic. D. P. McKenzie and W. J. Morgan showed in 1969 that a complex series of tectonic events can be produced by the evolution of triple junctions (Figure 13-9) without changes in the direction or magnitude of the relative motion between plates (Chapter 14).

During the next few years we can anticipate that interpretations of linear anomalies in terms of crustal structure, faulting, and relative movements will undergo successive modifications. For example in 1969 G. Peter and R. Lattimore reinterpreted the magnetic structural map of the Juan de Fuca and Gorda Ridges which we reviewed in Figures 13-2, 13-3, and 13-8 to 13-11. The 1970 review of magnetic data bearing on sea-floor spreading north of Iceland by P. R. Vogt, N. A. Ostenso, and G. L. Johnson illustrates well some of the problems and the choices that have to be made between alternative interpretations.

The theoretical model for interpretation of the linear anomalies is most elegant, but the complications introduced by real rocks and real geological structures inevitably lead to some speculative correlations and applications. Recent results suggest that the layer of oceanic rocks responsible for the magnetic anomalies may be much thinner than indicated in the models reported in this chapter. As Bullard noted, the acquisition of new data is likely to reestablish the usual chaos in the Earth sciences, but we do now have a universal model as a point of departure.

A Note of Caution from N. D. Watkins and A. Richardson

Watkins stated in 1968 that we know so little about the magnetic properties of submarine basalts and other oceanic crustal material that this permits a considerable degree of circular reasoning in interpretations of linear anomalies. Any model of blocks based on the geomagnetic polarity reversal scale can be fitted to the observed anomalies by suitable variations of block depths, widths (spreading rate variations), and intensity of magnetization; and Watkins added that magnetic details may optionally be interpreted as spurious, minor or multiple events, or possibly inherent magnetic features of finite bodies. Watkins and Richardson (1968) therefore concluded that to invoke variable spreading rates seems to be forcing data to fit what is still a hypothesis and should be discouraged until the geomagnetic polarity time scale beyond 4.5 m.y. is established.

The significance of highly symmetrical anomalies across some ridges and the repeat of a standard pattern fitting the radiometric polarity time scale in different oceans were not denied by Watkins and Richardson; but they made the point that if areas of active crustal spreading are to be delineated accurately, it is necessary to consider alternative causes of linear magnetic anomalies when the classic pattern is not clear. They emphasized that when geological information about the structure and rock types was available, this can not be ignored in the interpretation of the anomalies.

Most of our knowledge of the midoceanic ridges is based on individual traverses, and there have been few detailed studies of specific areas. Data collected on cruises from Woods Hole Oceanographic Institution in 1964 and from the Scripps Institution in 1965 permitted van Andel and Bowin to present a detailed geological and geophysical study of the crest and upper western flank of the mid-Atlantic ridge between 22° and 23°N. They prepared a schematic structural diagram for a 25 km cross section including the median valley and, on the basis of this and the known bathymetry, Watkins and Richardson presented the hypothetical geological cross section shown in Figure 13-20b. The picture is one of basalts overlying their low grade metamorphic equivalents, greenstones and greenschists. Following metamorphism,

uplift of up to 3 km occurred along faults dipping 30 to 45° toward the median valley; the overburden was displaced by sliding away from the crest. Eruption of basalts into the valleys followed the rifting. The vertical relief of up to 2 km completes the schematic section shown in the figure.

Figure 13-20d is the observed total magnetic profile at 22.5°N, and model a shows the conventional reversed polarity block model used to interpret the profile d. Figure 13-20a shows also the bathymetry of the sea floor. Notice the volume of rock above the magnetized strip which is ignored with respect to the development of anomalies.

Watkins and Richardson stated that "it is virtually impossible to see how the essential assumption that vertically-sided finite blocks of alternating polarity and high J and S values can be applied to the bathymetry depicted in" Figure 13-20b. Also the vertical movement required by the regional tectonic analysis argues against the idealized uniform crustal spreading blocks depicted in Figure 13-20a. They therefore sought an alternative interpretation based on the geological features.

Figure 13-20c is a version of model b showing the assumed distribution of basalts with normal (black) and reversed (white) polarity. Magnetic properties derived from local dredged rocks were used. The greenstone and greenschist have negligible magnetization. The lower depth of 5.7 km below sea level for model c is a standard value (model a). The computed profile for model c shown in Figure 13-20e demonstrates that the observed profile in d can be explained by a series of faulted geological features with the basic stratigraphy illustrated by van Andel and Bowin (model b). An exact fit could be obtained by suitable variation of the subjectively selected parameters, and the amplitude of computed profile e would be reduced to the same as observed profile d by halving the J and Q values of the basalt.

The magnetic anomalies *can* be correlated with the standard reversed polarity block

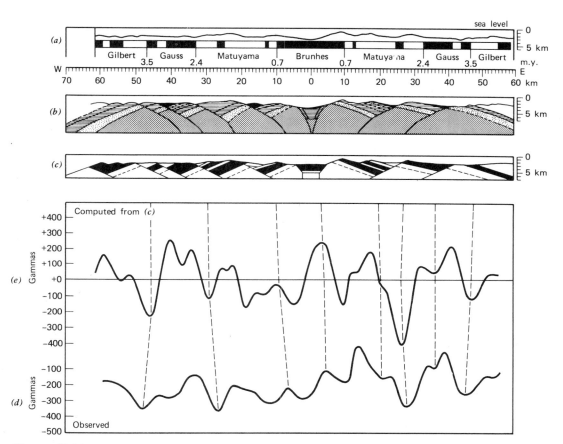

Figure 13-20. Interpretation by Watkins and Richardson (1968). Observed (*d*) and computed (*e*) west-east total field magnetic sea-level profiles across the mid-Atlantic ridge at latitude 22.5° North. Horizontal and vertical scales for models *a*, *b* and *c* are the same, given in km at the bottom and right hand side of the diagram. Model *a* shows the conventional crustal spreading finite block model used to interpret the magnetic profile when averaged with other profiles (Figure 13-13). The additional horizontal scale in model *a* is in millions of years. Note the difference between the bathymetry and the limits of the blocks. Model *b* employs the known bathymetry, and is a geological model which is hypothetical, although the central segment between 5 km west and 15 km east of the central part of the median valley (0 km) is essentially from Figure 7 of van Andel and Bowin (1968). Geological symbols used in model *b* are: fine dots = greenschist; coarser dots = greenstone; horizontal shading = basalt; black = young basalt. Model *c* shows the geometry and polarity of the blocks employed for the computation of curve *e*: black = normal polarity; white = reversed polarity (except for the greenschist and greenstone which is magnetically negligible and therefore without polarity). Curve *e* employs magnetic parameter averages obtained from local dredged material as follows: 'young basalt' has $J = 9.7 \times 10^{-3}$ emu/cm³, $\chi =$ (susceptibility) $= 0.7 \times 10^{-3}$ c.g.s. and $Q = 28.4$; 'basalt' has $J = 4.3 \times 10^{-3}$ emu/cm, $\chi = 2.4 \times 10^{-3}$ c.g.s., $Q = 3.6$; both the greenschist and greenstone have effectively J, χ, and $Q = 0$. Vertical scales for *d* and *e* in gammas. (Reproduced by permission of North-Holland Publishing Company, Amsterdam.)

model shown in Figure 13-20a, but this neglects the known relief and volume of rocks above the magnetized blocks. A better approximation to actual conditions might be obtained by using the average depth to the ocean floor as the top of the magnetized blocks following the procedure of the Lamont group, but the total relief of 2 km is about the same as the thickness of the hypothetical blocks. The magnetic anomalies cannot be correlated with the topography alone. The geological model in Figure 13-20c, incorporating both topography and magnetized structural layers, does provide a good correlation; but this model too is artificial in the sense that the choice of normal and reversed polarities was arbitrary. The success of the geological model in this location where the geology can be reconstructed, however, does emphasize the need for more information about the structure and magnetic properties of the oceanic crust.

Near-Bottom Magnetic Results from the "Fish"

Since 1967 ships from the Scripps Institution of Oceanography have taken a closer look at the properties of the ocean floor by towing a submersible instrument package referred to as the "Fish" at heights above bottom between 35 and 180 m. The Fish records its height above the sea floor and the magnetic field intensity, while the towing ship makes conventional bathymetric and magnetic measurements at the surface. Figure 13-21 shows results from two locations. Figure 13-21a is from a traverse across the East Pacific Rise just south of the Gulf of California. Figure 13-21b is a traverse across anomaly number 10 in the northeast Pacific, at latitude 32°25′N, just east of longitude 126°W (Figures 13-15 and 13-3).

Each diagram shows a profile of the bottom topography, the path of the Fish relative to the bottom, the total magnetic field intensity as recorded along the Fish path, and the total magnetic intensity as measured at the

surface. The magnetic intensity measured at the surface is correlated in the usual way with uniformly magnetized blocks of crust. Figure 13-21a shows the Brunhes normal polarity interval extending across the Rise crest and the Jaramillo event in the Matuyama reversed epoch on the western flank of the crest. I have added to Figure 13-21b a schematic representation of the younger normal event of anomaly 10 (Figure 13-15).

The surface anomalies give a picture that can be interpreted in terms of uniformly magnetized blocks of crust, but the measurements at depth indicate that the blocks are interspersed with material of different magnetization producing anomalies of very large amplitude and narrow width compared to the surface anomalies. The smoother trace and lower amplitude of the surface magnetic field results from the greater distance of the instrument from the magnetic bodies.

The deep anomalies in Figure 13-21a show the same general character right across the portion of the East Pacific Rise traversed. Superimposed on the profile are three large offsets of about 2000γ each, occurring at the transitions between normally and reversely magnetized crustal material as inferred from the surface magnetic profile. Each offset occurs within a horizontal interval of about 140 m. From the estimated spreading rate of 3 cm per year this indicates that it takes no more than 4700 years for the Earth's field to reverse, and it implies a maximum limit of about 280 m for the width of the intrusion center at the ridge crest. This is surprisingly narrow: other estimates range up to 10 km for the intrusion center.

In both locations the topography is lineated parallel to the surface magnetic anomalies. Larson and Spiess concluded that the topography of the 200 m abyssal hills forming the relief on the East Pacific Rise is not responsible for the deep anomalies. They suggested that the anomalies are compatible with the hypothesis that the magnetic field has under-

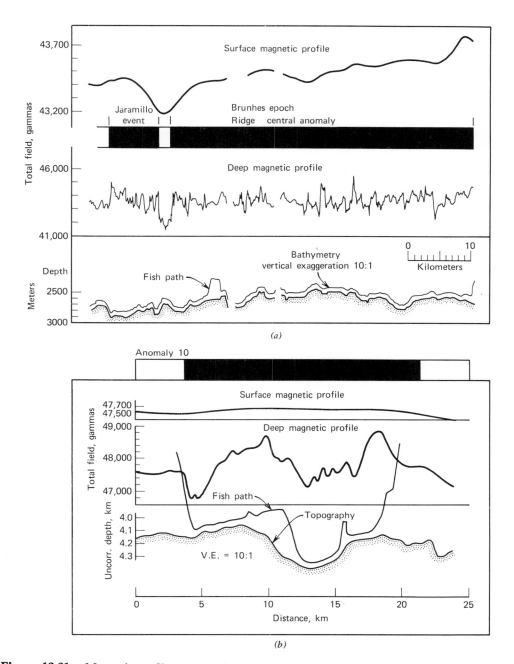

Figure 13-21. Magnetic profiles measured near the ocean floor by the "Fish". Each diagram shows bottom topography, total magnetic field intensity recorded along the Fish path, and measured at the surface. The polarity reversal sequences associated with the surface anomalies are shown. (*a*) Profile across the East Pacific Rise crest from west (at left) to east (after Larson and Spiess, 1969). Surface profiles are vertically exaggerated ten times with respect to the deep magnetic profiles. The trace above the bathymetry is the track of the instrument above the ocean floor. The sediment cover is illustrated as true thickness to the basement reflector. (*b*) Profiles across the younger normal event of anomaly 10 (Figure 13-15) in the north-east Pacific, latitude 32°25′ N, just east of longitude 126° W (Figures 13-15 and 13-13) (after Luyendyk et al., 1968). (Reproduced by permission. (*a*) Science, **163**, 68–71, 1969. Copyright 1969 by the American Association for the Advancement of Science. (*b*) Jour. Geophys. Res., **73**, 5951, 1968.)

gone many fluctuations of short period and small intensity during the past 2 m.y.

Figure 13-21*b* does show, however, an apparent correspondence of the deep magnetics with the topography. Low amplitude smaller width anomalies occur within the valley, while larger widths and amplitudes occur across the ridges on either side of the valley. Luyendyk *et al.* in 1968 suggested that the hills may represent the topographic expression of large dikes magnetized in the opposite sense to the crust containing them, and the low amplitude anomalies within the valley might represent hydrothermally altered material in fracture zones. Most other studies indicate no correspondence between magnetic profiles and bathymetry.

In 1969 Luyendyk applied magnetic model studies to the data and concluded that the anomalies are caused by slow, continuous changes in magnetization within layer 2 of the crust in a direction perpendicular to the lineations. The fluctuations are due to time variation either in the paleofield intensity or in the properties of injected magmas. This conclusion was questioned by G. Peter in 1970. His discussion and Luyendyk's reply demonstrate the uncertainties in the location of magnetized rocks and the causes of the magnetic anomalies.

Luyendyk *et al.* confirmed statistically that a true correlation does exist between the deep magnetic anomalies in successive profiles. Their reason for doing this is worth repeating:

"The literature of the past few years is replete with reports on the correlation of marine magnetic anomaly profiles by their supposed similarity over distances up to thousands of kilometers. Although we believe the majority of these correlations to be valid, we have noticed the widely different concepts of significant correlation, both in the literature and among our colleagues. We have therefore decided to subject our assertion of lineation of our data to the rigors of cross-correlation analysis. We would like to encourage other workers in marine magnetics to conduct similar statistical tests."

Figure 13-21 confirms that we have much to learn about the causes of magnetic anomalies in the ocean basins.

14. *Plate Tectonics*

INTRODUCTION

Whatever the detailed explanations may be for the development of the linear magnetic anomalies their correlation with reversals of the Earth's magnetic field imprinted upon new oceanic crust generated at the mid-oceanic ridges gave considerable momentum to the spreading floor theories of Hess and Dietz (Chapter 12). Earth scientists were quick to realize the significance of this correlation, and in 1967 and 1968 four major papers appeared introducing what has become known as plate tectonics. The increasing support for many aspects of the sea-floor spreading hypothesis led Vine and Hess to reformulate the concept in 1968 in terms of the new evidence which had become available. Their review was published in 1970.

Figure 13-17 shows a pattern of recent motion of the ocean floors away from three active ridges in the Atlantic, Pacific, and Indian Oceans. A fourth active ridge traverses the Arctic Ocean. According to the concept of sea-floor spreading the uprise of mantle material and its spreading from the oceanic ridges causes the ocean floor and coupled continental blocks to move away from the ridges. The ocean crust goes down into the mantle at the oceanic trenches, which form sinks for the system. Figure 14-1 is a provisional attempt by Vine and Hess to show the relationships between continental and oceanic crust. The ridge crests, fracture zones, and isochrons have been transferred from

Figure 13-17 and a 65 m.y. growth line ha., been added in the Pacific. Although there are many gaps in the isochron map there is no major gap in geographical coverage. This permits delineation of the shaded areas in Figure 14-1 as oceanic crust created within the past 65 m.y. during Cenozoic time.

Figure 14-1 shows that although there are oceanic trenches available for large segments of the spreading Pacific Ocean floor, other plates can not be accommodated in trenches. The general picture indicates that the continents on either side of the Atlantic are coupled to the blocks spreading from the mid-Atlantic ridge, and that the North American continent has over-ridden the northern part of the East Pacific Rise. The location of sinks between the ridges in the Atlantic and Indian Oceans is not obvious. Spreading rates are slower in the Atlantic than in the Pacific, and possibly this could be related to the inference that spreading from the mid-Atlantic ridge is driving large continental plates westward over the faster-spreading Pacific Ocean floor.

The concept of plate tectonics has developed rapidly since 1967, and it has major implications for all aspects of geology. Already the topic merits a whole book in itself. In the first section I attempt to show the stages of formulation of the concept by seven scientists, but after that I can only consider briefly a few selected fragments of the total picture.

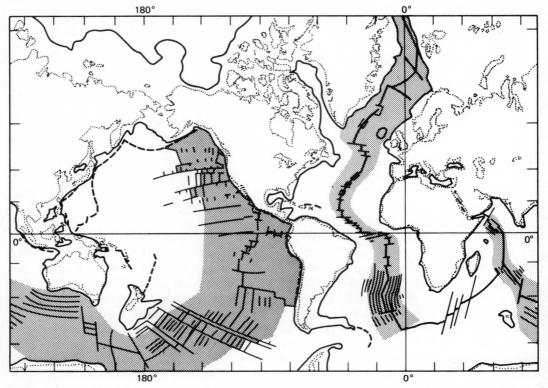

Figure 14-1. Relationships between continental and oceanic crust (after Vine and Hess, 1970). The shaded area is oceanic crust created within the past 65 m.y. (Cenozoic time); ridge crests, fracture zones, and isochrons transferred from Figure 13-17; dashed lines are trenches, see Figure 2-2. (With permission of John Wiley and Sons.)

THE CONCEPT OF TECTONICS ON A SPHERE

The concept of plate tectonics combines the satisfactory parts of the hypotheses of continental drift and sea-floor spreading. The Earth's surface is considered to be made up of a few rigid crustal plates or blocks which are in motion relative to each other. In the early 1960's there were many convection cell models published with ocean ridges representing sites of upwelling and ocean trenches the sites of convergence and descending limbs of cells (Chapter 12). Since 1967 more attention has been paid to deciphering the relative surface motions than to the energy source for the motion. The boundaries of the rigid plates are not necessarily related to convection cells in the mantle,

and the lack of a satisfactory driving mechanism remains a major problem.

Wilson's transform fault concept was extended to a spherical surface, as illustrated in Figure 14-2, by D. P. McKenzie and R. L. Parker in 1967 and by J. W. Morgan in 1968. Euler's theorem states that a layer on a sphere can be moved to any other conceivable orientation by a single rotation about a properly chosen axis through the center of the sphere. The relative motion of two layers can therefore be described by a rotation about some axis. This is the theorem used by Bullard and associates to fit together the continents on either side of the Atlantic (Figure 11-3).

Paving Stones of the North Pacific by D. P. McKenzie and R. L. Parker

In 1967 McKenzie and Parker outlined the key elements of plate tectonics, which they termed a paving stone theory, defining ridges and trenches respectively as lines along which crust is produced and destroyed. They tested the theory in the North Pacific with particular reference to focal mechanisms of earthquakes and to volcanism.

They determined the pole of relative rotation between the Pacific and the North American continent from the slip directions of two widely spaced faults, the San Andreas fault and the site of the 1964 Alaskan earthquake in the Kodiak Island region. Construction on a sphere gave a pole position of 50°N, 85°W. According to the paving stone theory, all slip vectors of the North Pacific earthquakes should lie on small circles around this pole. They found that about 80% of 80 published fault plane solutions for shallow earthquakes during and after 1957 had slip vectors in agreement with the prediction. Figure 14-4 shows the general consistency of the slip directions around the Pacific. The area that they examined amounts to about a quarter of the Earth's surface, and they expected it to be equally applicable to the other three-quarters. Subsequent global analyses confirmed their predictions.

McKenzie and Parker also noted that the distribution of trenches, active andesitic volcanoes, and intermediate- and deep-focus earthquakes is controlled by the trend of the faults which in turn influences the fault plane behavior for the shallow earthquakes. The shallow earthquakes associated with these features have overthrust fault solutions; they are not caused by faults of strike-slip transform nature.

Aseismic Crustal Blocks of the World by W. J. Morgan

When Morgan formalized the concept of plate tectonics in 1968, he divided the surface of the Earth into about twenty plates in each of which there is no distortion of any kind. The boundaries between plates are of three types which are shown in Figure 14-1. At the extensive ridge type new surface is generated and crustal surface is destroyed at the compressive type including trenches. The third type is represented by the great faults at which crustal surface is neither created nor destroyed. Figures 2-5 and 14-3 show that present earthquake activity is largely concentrated in narrow belts which are essentially continuous, and these belts form the boundaries to a series of plates which are essentially aseismic. Shallow-focus earthquake activity delineates the boundaries of the plates that are in relative motion across the surface of the Earth. Intermediate- and deep-focus earthquakes appear to be restricted to compressive boundaries such as trenches and young fold mountain chains as shown in Figure 14-3. Notice that although there is a circum-Pacific earthquake belt (Figure 2-5) the intermediate- and deep-focus earthquakes are not continuous around the Pacific.

Figure 14-2 shows Wilson's transform fault concept on the Earth's surface. Blocks 1 and 2 are crustal plates, the short double lines represent offset midoceanic ridges, and the series of curved lines crossing the ridges represent transform faults. If spreading occurs from the oceanic ridges it is convenient to assume that block 1 remains stationary, and to consider the movement of block 2 relative to block 1. According to Euler's theorem the relative motion can be described by rotation of block 2 about a specific axis with pole at A. This is defined by two parameters to locate the pole and a third to specify the angular velocity. For a given angular velocity the relative velocity between the blocks increases as the sine of the angular distance from the pole A. For relative movement to continue about pole A, the transform faults where blocks 1 and 2 are in contact must lie on small circles centered on A. This gives us two methods for calculating the positions of poles of relative motion, or poles of opening, from measurable features at the Earth's surface: (a) the first method involves location of the pole that fits the strikes of fracture zones, and (b) the second involves location of the pole best fitting the variation in spreading rate.

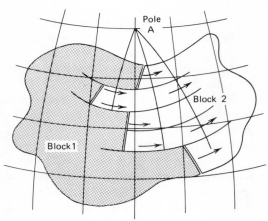

Figure 14-2. Transform faults on a spherical surface. On a sphere the motion of block 2 relative to block 1 must be a rotation about some pole. All faults on the boundary between 1 and 2 must be small circles concentric about the pole A. (From W. J. Morgan, Jour. Geophys. Res., **73**, 1959, 1968, with permission.)

Morgan examined the traces of the fault zones offsetting the mid-Atlantic ridge and the Pacific-Antarctic ridge, and the old and new fracture zones in the North Pacific Ocean. Each group of fracture zones satisfied the conditions required by Figure 14-2, and Morgan estimated the positions of poles of relative motion by drawing great circles perpendicular to the strikes of offsets of the ridges, or of fault segments, and determining the circles of intersection. The spreading velocities estimated in different parts of the mid-Atlantic ridge from the magnetic anomalies agreed roughly with the velocity pattern predicted for the pole located from the fracture zones.

Global Patterns of Surface Motion by X. Le Pichon

Le Pichon had access to more complete data than Morgan on the pattern of spreading (Chapter 12) and the locations and extents of some large fracture zones, and in 1968 he redetermined the three poles of rotation located by Morgan, with good agreement. He also located poles for the fracture zones in the Indian Ocean and the Arctic Ocean. The circles numbered 1 to 5 in Figure 14-3 are the positions of the five poles for the principal lines of opening in the Atlantic (1), North Pacific (2), South Pacific (3), Arctic (4), and Indian Oceans (5). Pole number 6 is a preliminary result for the S.W. Indian Ocean representing motion of Antarctica relative to Africa, which was deduced by assuming closure of Gondwanaland and summing angular velocities; Le Pichon's result differs significantly from that reported by Morgan. Angular velocities were derived for each pole from the available spreading rate data. The general picture is one of simplicity with the two major openings in the Atlantic and the Pacific occurring about approximately the same axis, not greatly inclined from the Earth's rotational axis (poles 1, 2, and 3), and with these openings being linked by two oblique openings, one in the Indian and one in the Arctic Ocean.

Le Pichon concluded that his results were detrimental to the hypothesis of an expanding Earth. Expansion of the Earth would be approximately the same along all radii, because the Earth has remained spherical. If the opening were due to expansion, however, the calculated rates of opening along great circles indicate differential expansion by as much as 500 km along different radii during the past 10 m.y.

Given a spherical Earth of constant radius and the derived poles and rotational vectors of the five major openings, Le Pichon attempted the first worldwide analysis for the movement of the major blocks relative to each other, predicting rates of crustal exten-

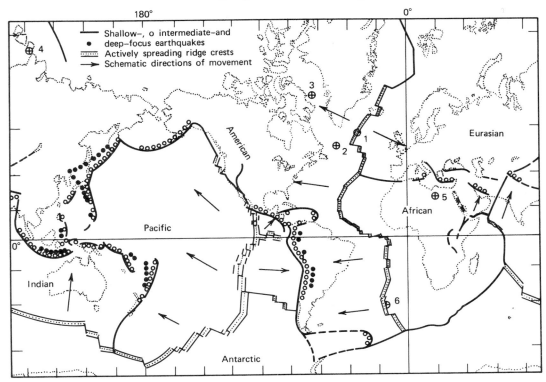

Figure 14-3. Summary of seismicity of the Earth (Figure 2-5) and aseismic plates (after Vine and Hess, 1970). Plates assumed by Le Pichon (1968) are named, and instantaneous centers of rotation deduced for plate pairs are plotted as numbered circles with internal crosses. Centers are: 1, for South Atlantic (America–Africa); 2, for North Pacific (America–Pacific); 3, for South Pacific (Antarctica–Pacific); 4, for Arctic Ocean (America–Eurasia); 5, for N.W. Indian Ocean (Africa–India); 6, for S.W. Indian Ocean (Antarctica–Africa). Spreading rates at ridge crests are indicated schematically and vary from 1 cm/year near Iceland to 6 cm/year in the equatorial Pacific. (With permission of John Wiley and Sons.)

sion, and rates of compression at the plate boundaries. Instead of the 20 crustal blocks considered by Morgan, Le Pichon chose six major blocks in such a way that the problem could be solved. Smaller plates were incorporated into the chosen six which are shown in Figure 14-3: these are the Pacific, American, African, Eurasian, Indian, and Antarctic blocks. Boundaries to the blocks are shown by:

1. Actively spreading ridge crests whose spreading rates are indicated by the separation of the pair of lines; they vary from 1 cm per year near Iceland to 6 cm per year in the equatorial Pacific.

2. Transform faults.

3. Oceanic trenches or Tertiary mountain belt systems.

All boundaries of the plates have shallow-focus earthquakes associated with them, and the compressive boundaries (3) also have associated intermediate- or deep-focus earthquakes. The directions of relative movement of these major blocks are shown by the arrows. Le Pichon concluded that this picture was in reasonable agreement with physiographic, seismic, and geological data and then proceeded to study and reconstruct surface movements since the Mesozoic. If the Antarctic block is considered as fixed relative to the Earth's rotational axis then all other plates and plate boundaries will move across the mantle.

J. D. Phillips and B. P. Luyendyk in 1970 considered both the deep-sea drilling results and magnetic anomalies. They located the relative motion pole for the North Atlantic somewhat further south and east than that given in Figure 14-3, and their angular rate of opening of the central North Atlantic differs markedly from rates offered by Morgan and Le Pichon. Their data show only one relative rotation pole during the late Cenozoic contrary to conclusions published by others in 1969. They favored the idea that the central North Atlantic and the South Atlantic represent different plate systems.

The New Global Tectonics by B. Isacks, J. Oliver, and L. R. Sykes

The third major paper of 1968, by Isacks, Oliver, and Sykes, dealt specifically with the strong support given by the observations of seismology to the concepts of sea-floor spreading and plate tectonics. The correlation of earthquakes with the boundaries of plates, and of intermediate and deep earthquakes with compressive boundaries, has already been noted. Figure 14-4 shows again the ridges, transform faults, and trenches bounding the plates given in Figure 14-3, and also the slip vectors determined from studies of shallow-focus earthquakes. Each arrow depicts the motion of the plate on which it is drawn relative to the

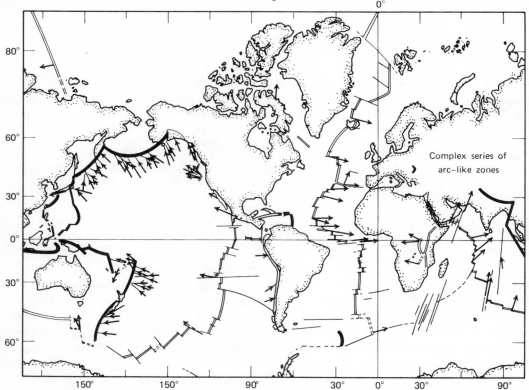

Figure 14-4. Summary map of slip vectors derived from earthquake mechanism studies (after Isacks et al., 1968). Arrows indicate horizontal component of direction of relative motion of block on which arrow is drawn to adjoining block. Crests of world rift system are denoted by double lines; island arcs, and arc-like features, by bold single lines; major transform faults, by thin single lines. (From Jour. Geophys. Res., **73**, 5855, 1968, with permission.)

plate on the other side of the tectonic feature. These directions correspond remarkably well with the movement directions of Figure 14-3 derived by Le Pichon from different evidence. In particular many of the vectors of differential movement calculated by Le Pichon for arcs, shown in his Figure 6, are very similar in direction to the slip vectors of Figure 14-4. McKenzie and Parker had previously reviewed the slip vectors around the North Pacific.

Figure 14-5 is a block diagram illustrating schematically the main features of sea-floor spreading, plate tectonics, or the "New Global Tectonics." The rigid lithosphere, including the crust and part of the upper mantle, is on the order of 100 km thick. The asthenosphere, a layer of no strength on the appropriate time scale, corresponds to the low-velocity zone extending down beneath the lithosphere to several hundred kilometers. The mesosphere makes up the greater part of the mantle and is relatively passive with strength. The lithosphere is continuous except at the boundaries of the moving plates. The diagram shows plates moving away from midoceanic ridges and down into the asthenosphere at island arcs with relative surface movement between plates occurring across transform faults. A return flow compensating for the movement of the lithosphere is shown in the asthenosphere with uprise at the opening ridge crests.

Seismology also provides evidence about movements at the ridges and trenches which supports this model. In connection with Figure 13-6 we noted that the relative motion across a transform fault is the reverse of that expected from a transcurrent fault, and that relative motion was restricted to the portion of the fault between the ridge crests. Shallow earthquakes along the oceanic ridges occur only along the ridge crest where they result from normal faulting (tension) and along the fracture zone between ridge crests. They do not extend beyond them as they would for transcurrent faults. Furthermore focal mechanism solutions indicate motions appropriate for transform faults.

Island arcs and ocean trenches have long been interpreted as sites of compression and downward movement of material, but

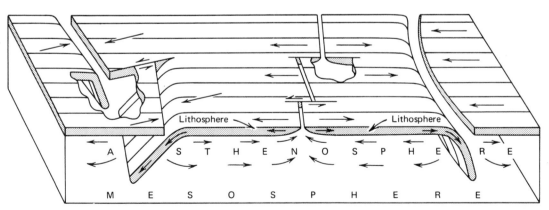

Figure 14-5. Relative movements of plates (after Isacks et al., 1968). Block diagram illustrating schematically the configurations and roles of the lithosphere, asthenosphere, and mesosphere in a version of the new global tectonics in which the lithosphere, a layer of strength, plays a key role. Arrows on lithosphere indicate relative movements of adjoining blocks. Arrows in asthenosphere represent possible compensating flow in response to downward movement of segments of lithosphere. One arc-to-arc transform fault appears at left between oppositely facing zones of convergence (island arcs), two ridge-to-ridge transform faults along ocean ridge at center, simple arc structure at right. See Richter (1969) for criticism and reply by Isacks et al. (1969). (From Jour. Geophys. Res., **73**, 5855, 1968, with permission.)

analyses of gravity data and focal mechanisms for earthquakes indicate tension at trenches. The seismic evidence reviewed by Isacks et al. gives strong support for the picture of a slab of relatively cool lithosphere being transported deep into the mantle as illustrated in Figure 14-5. Remarkable evidence is the discovery of anomalous zones about 100 km thick, with low attenuation of seismic waves and relatively high seismic velocities, which occupy the positions of the down-going lithosphere in Figure 14-5. The deep seismic zones beneath island arcs (Figure 2-6) are situated on the upper sides of the anomalous zones. The changing pattern of seismic activity with depth beneath the island arcs is also consistent with the model in Figure 14-5. Infrequent shallow earthquakes beneath the trench axis indicate extension and normal faulting probably caused by downward bending of the lithosphere (Figure 14-9). A thin zone of active seismicity beneath the inner margin of the trench is characterized by dip-slip mechanisms indicating underthrusting of the oceanic plate. This zone is continuous with the deep zone (Figure 2-6), for which the 1968 evidence favors maximum compressive stress parallel to the dip of the seismic zone which is the presumed direction of the slab. The maximum depths of earthquakes divided by estimated spreading rates indicate that the underthrusting of the lithosphere is about equivalent to the amount of spreading during the past 10 m.y. More recent work shows that for Benioff zones with no known deep earthquakes, or for those with a gap in seismicity between 300 and 500 km depth, the intermediate focus earthquakes indicate tension in the downdip direction. The deep-focus earthquakes, if they occur, have nearly vertical axes of compression. This information according to T. J. Fitch and P. Molnar in 1970 is consistent with models in which gravitational body forces are important factors in the dynamics of the lithosphere (Figure 14-8).

Figure 14-6 shows the relationships of continents and oceans to the lithosphere plates of Figure 14-5, as depicted schematically by J. F. Dewey and J. M. Bird in 1970 for some specific cross sections. They also presented a more detailed version incorporating more geological information (Figure 14-16). Figure 14-6 shows the continents as superficial passengers on rigid plates up to 150 km thick. Oceanic plates are being consumed in marginal trenches, against a continent in *a* and against an island arc in *b*. Small ocean basins are trapped within the complex trench-transform fault systems associated with island arcs in the Western Pacific as in *c* and *d*. Continents are too buoyant to be carried down into the mantle, and *g* represents the Indian plate partly underthrust beneath the Asian plate with the Himalayan uplift resulting.

The seismic evidence is not everywhere as clearly favorable for plate tectonics. I. S. Sacks in 1969 reported that preliminary results from earthquakes in the Peru-Brazil region contradict the pattern expected for the simple model of downgoing lithosphere shown in Figure 14-6*b*. There is a low-Q zone beneath the Andes, but the region has high-Q pathways through it and the difference in Q between them is probably more than an order of magnitude. The holes are not distributed in any systematic manner relative to the trench-mountain system.

C. F. Richter commented on the new global tectonics paper in 1969 congratulating Isacks, Oliver, and Sykes for their service in coordinating the expanding subject and noting some points which they had cautiously avoided or not fully considered. The most significant questions that he raised were in connection with the tectonics of Japan and the Alpide belt. In particular he stated that the tectonic complexity of Japan is too often obscured in global discussions by mapping on a very small scale, and he presented the major features which have to be explained by any general global scheme. In their reply the authors agreed that, although they believe the overall fit of seismological data with Le Pichon's plate model is highly significant,

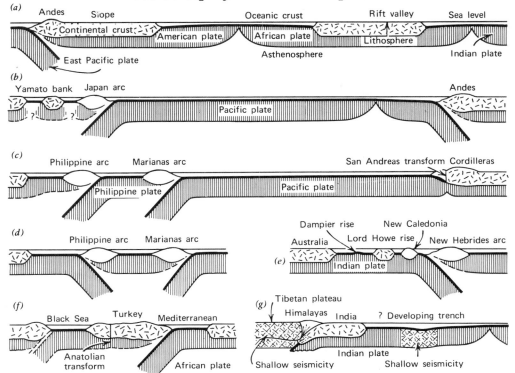

Figure 14-6. Schematic sections showing plate, ocean, continent, and island arc relationship (after J. F. Dewey and J. M. Bird, Jour. Geophys. Res., **75**, 2625, 1970, with permission).

"so significant that the hypothesis is effectively proved," the real Earth was certainly more complex on a finer scale and more detailed studies of selected regions should tend to increase the complexity of the model.

Small Crustal Plates by P. Molnar and L. R. Sykes

Morgan considered global tectonics in terms of 20 major plates, and Le Pichon managed with only six as shown in Figure 14-3. He chose six major blocks in such a way that he could solve specific problems. Figure 14-3 shows that in the region of the Caribbean and Middle America there are areas bounded by linear tectonic features which were treated as parts of the larger American and Antarctic plates. The Caribbean plate underlies the Caribbean Sea and is bounded by the Middle America arc,

the Cayman trough, the West Indies arc, and the seismic zone through northern South America. The Cocos plate is bordered by the East Pacific rise, the Galapagos rift zone, the north-trending Panama fracture zone near 82°W, and the Middle America arc.

From their study of the seismicity of the region Molnar and Sykes in 1969 concluded that the tectonics is dominated by the interaction of these two small plates which are moving relatively to the larger plates around them. The plates are nearly aseismic. Molnar and Sykes determined the focal mechanisms of 70 earthquakes and worked out the relative movements of the plates, the rates of motion, and the locations of underthrusting. This is one of the few parts of the world where an active oceanic ridge approaches an island arc, and elucidation of the tectonics in such regions is a necessary test for the global theories. Figure 14-6 indicates the existence of small plates associated with the island arc systems of the Pacific Ocean.

Evolution of Triple Junctions of Plates by D. P. McKenzie and W. J. Morgan

Plate tectonics as developed above is concerned with the relative motions of rigid plates presently occurring and not with changes that occurred through geological time. We saw in Chapter 13 that patterns of magnetic anomalies and fracture zones have been interpreted in terms of changes in the spreading pattern. The changes proposed include (a) changes in the directions of spreading, (b) changes in the relative rates of spreading, and (c) episodic spreading where all movement stops completely for a period. If plate tectonics is to provide a complete theory of global tectonics it must include geological time as a variable.

In 1969 McKenzie and Morgan discussed and classified the geometry and stability of all possible triple junctions, where three plates meet in a point (Figure 13-9). There are 16 possible combinations of ridges, trenches, and transform faults which, taken three at a time, can form triple junctions. All except two of these are stable in certain conditions. McKenzie and Morgan showed that evolution of these triple junctions can produce many of the changes in tectonic style which would otherwise appear to have been caused by a change in the direction or magnitude of the relative motion between plates. They reexamined the magnetic lineations and fracture zones in the northeast Pacific (Figures 13-2 and 13-3) in terms of the evolution of triple junctions and concluded that the main features of the geological history of this complex region could be explained with the assumption that the relative velocities of the major plates had remained unchanged during the Tertiary. This and other examples show how a complex series of events can be produced simply by the geometry of plates.

THE RESULTS OF DEEP SEA DRILLING

Hess's 1962 model for sea-floor spreading away from the midocean ridge is shown in Figure 12-22b. It predicts progressive overlap of ocean sediments onto the ridge; the greater the distance from the ridge the older the deepest sediments should be. The results from deep-sea drilling in the South Atlantic, on leg 3 of the *Glomar Challenger* cruise, have confirmed this prediction. This cruise, under the aegis of the Joint Oceanographic Institutes for Deep Earth Sampling (JOIDES) has as its primary aim the recovery of complete sedimentary sections of the ocean floor. Results were presented by A. E. Maxwell and seven coauthors in 1970.

Figure 14-7a shows eight sites from which long cores were obtained on a traverse across the mid-Atlantic ridge near 30°S. Basement composed of basalt was reached at seven sites. Each site was located within the well-defined magnetic anomaly pattern for this latitude (Figure 13-13) which gives a predicted age for the basaltic basement rocks if a constant rate of sea-floor spreading is assumed (Figures 13-14, 13-15, and 13-17). Abundant calcareous microfossils were used to assign ages to the sediments in the core. Figure 14-7a gives the magnetic age of the basement and the paleontological age of the deepest sediment just above the basement. The agreement between them is reasonably good.

The ridge has numerous offsets and changes of strike (Figure 14-4), and it is not always easy to locate a point from which a given site should have originated according to sea-floor spreading. The distance of each site from the ridge crest was estimated in two ways: (a) by measuring the linear distance to the nearest ridge axis, and (b) by rotating the site back

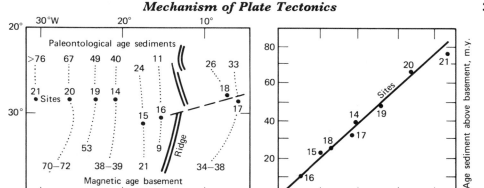

Figure 14-7. Results obtained from deep drilling in the South Atlantic Ocean. (*a*) Trend of the axis of the Mid-Atlantic Ridge in the South Atlantic as determined from geophysical evidence plus earthquake epicenter data. The drilling sites are shown relative to the ridge axis. For each site, the diagram shows the paleontological age of the deepest sediment just above basement, and the magnetic age of the basement according to the magnetic anomaly and the geomagnetic time scale (Figures 13-13 and 13-15). (*b*) The age of the sediment from *a* plotted as a function of distance from the ridge axis. The points lie close to a straight line, indicating a constant spreading half rate of 2 cm/year (compare Figure 13-14, where constant rate was assumed for the South Atlantic). After Maxwell *et al.*, Science, **168**, 1047–1059, 1970. Copyright 1970 by the American Association for the Advancement of Science.

to the ridge along a small circle about Morgan's rotation axis with pole at 62°N, 36°W (compare pole 1 in Figure 14-3). These distances are compared in Figure 14-7*b* with the paleontological age of the sediment immediately above the basalt basement. The relationship is clearly linear, corresponding to a spreading half-rate of 2 cm/year. The studies of magnetic anomalies had previously given the same rate assuming constant spreading in the South Atlantic (Figures 13-13 and 13-14). The agreement between the paleontological and magnetic estimates of sea-floor spreading is regarded as strong support for the hypothesis of sea-floor spreading and the concept of magnetic stratigraphy. It has yet to be established, however, that the basalt reached is definitely

basement rock. Deeper drilling may possibly reveal more sediments below basaltic layers.

Similar deep-drilling results in the Pacific Ocean, reported in 1970 by A. G. Fischer and nine coauthors, have yielded a wealth of useful data, but they have not yet proved as directly relevant to interpretation of sea-floor spreading. It has been tentatively inferred that the ocean crust becomes progressively older from the East Pacific Rise to Hawaii and further west, but chert in the older sediments prevented penetration into the basement below the sediments for a definitive test. The results of drilling suggest that the "opaque layer" of Figure 13-16 is an alternating sequence of pelagic ooze and chert.

MECHANISM OF PLATE TECTONICS

Plate tectonic theory has had considerable geometrical success in accounting for complex tectonic phenomena, but there remains the major problem of the energy source and the

mechanism which maintains the motion. Whatever drives the motions provides the energy for earthquakes and volcanism, and the radioactive decay of U, Th, and K is probably the only source of energy large enough. Some kind of thermal convection is required to convert this into heat. The problem relates directly to the geochemistry of the mantle and crust as well as to the mode of convection within the mantle.

Figure 14-8*a* shows a standard Rayleigh-Benard mantle convection model, with flow in the asthenosphere exerting viscous drag to the strongly coupled lithosphere and forcing it into the mantle where the cell turns downward (see also Figures 12-23 and 12-24). This produces tension in the surface slab and compression in the down-going slab.

Figure 14-5 depicts an alternate form of convective motion in which the lithosphere forms the upper part of a convection cell, as originally proposed by Hess (Figure 12-22). In connection with Figure 14-5 Isacks et al. suggested that the pattern of flow in the mantle may be controlled largely by the motions of the cold plates of lithosphere. In Figure 14-8*b* it is assumed that the cold slab of lithosphere is denser than the asthenosphere, and therefore it sinks, pulling behind it the surface litho-

sphere, under tension. This process would be aided by the conversion of crustal gabbro to eclogite in the sinking slab. D. P. McKenzie reviewed the consequences and causes of plate motions in 1969, and he refers to this mechanism as an extreme form of convection which takes place if viscosity is a rapidly varying function of temperature. He presented a rudimentary analysis of the process and concluded that, despite its appeal, this mechanism probably cannot maintain the motion of the large plates. In Figure 14-5 the arrows in the asthenosphere represent possible compensating flow in response to the downward motion of the lithosphere plates. McKenzie emphasized that the mantle and lithosphere are in thermal and mechanical contact, and that motions cannot occur in either without the other being affected. Thus in Figure 14-5 sinking of the cold lithosphere causes motion in the mantle, and in Figure 14-8*a* motion in the mantle moves the lithosphere. Probably all of the mechanisms illustrated schematically in Figure 14-8 are in operation, and the relative contributions of each will not be

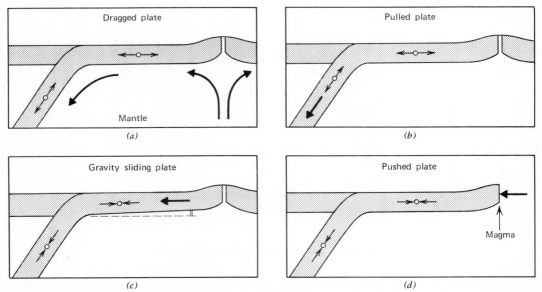

Figure 14-8. Diagrammatic representation of mechanisms proposed for moving plates of lithosphere (compare Figure 14-5). States of compression and tension in the plates are indicated. See text for discussion.

known until a detailed three-dimensional analysis has been completed on the problem. Apparently complete solution is beyond the capability of the present generation of computers.

In 1969 A. L. Hales suggested that gravitational sliding of the lithosphere on a partly melted asthenosphere (Figure 6-20) could play an important role in sea-floor spreading as illustrated in Figure 14-8c. The ocean floor is assumed to be horizontal, with the surface of the low velocity zone dipping gently away from the ridge. He calculated that a slope of 1/3000 for the surface of the low velocity layer would be sufficient for sliding at a rate of 4 cm/year, and that if a block 2000 km long and 100 km thick is tending to slide toward a continent in this way, then the stresses at the interface with the continent would be of the order 2×10^8 dyne/cm^2. The stresses are more than enough to cause failure. The potential energy per year is of order 10^{24} ergs over a front of 2000 km, and this is comparable with the energy released in earthquakes. Hales presented these numbers only to show that stresses and energies of the right order are possible. Similar effects might be produced if the surface of a lithosphere slab dips away from the ridge possibly because of distortion arising from convection as in Figure 14-8a.

L. Lliboutry proposed a fourth process in 1969 as shown in Figure 14-8d. He suggested that the lithosphere plates are pushed apart (upslope) by compressed magma which fills a deep crack connected with magma at the top of the asthenosphere. As the plates diverge the magma solidifies on the walls of the crack and erupts at the surface intermittently; but once initiated the crack never disappears. Neither currents in the asthenosphere nor sinking of the plates at one end are excluded as contributory forces to plate motion.

Lliboutry also discussed the way in which the horizontal lithosphere beneath the ocean floor could change its configuration so abruptly to a near-planar slab dipping steeply into the mantle. He pointed out that the usual picture of a plastic extension of the upper lithosphere and a plastic compression of the lower lithosphere would produce the curvature shown in Figure 14-9a and that this would be retained in the softer asthenosphere. As an alternative he described the mechanism illustrated in Figure 14-9b. The oceanic lithosphere impinging upon the continental lithosphere initially bends a little as the continental lithosphere exerts a resisting force dipping at about 45°. When the shear stress in a vertical plane reaches the shear strength value, a vertical fault is produced originating an earthquake. The oceanic lithosphere is thus sheared into vertical fragments, but the sequence of fault blocks

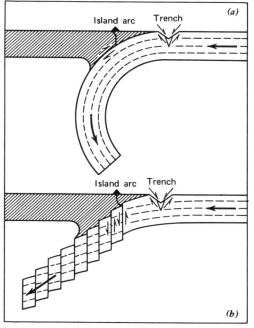

Figure 14-9. Lithosphere sinking in the mantle beneath island arcs. Mechanism at a sink. (a) Plastic bending of the lithosphere. (b) Discontinuous shearing by faulting along vertical planes. Only this process agrees with the plane form of the sunk lithosphere, and with the focal mechanism of shallow earthquakes. (From Lliboutry, L., Jour. Geophys. Res., **74**, 6525, 1969, with permission.)

as a whole takes on the shape of an inclined plate. In 1970 Å. Malahoff reviewed the mechanisms for gravity and thrust faults under the oceanic trenches and presented a picture that seems to be consistent with Figure 14-9*b*. He suggested that between 0 and 10 km below the trench failure was largely by gravity faults; between 10 and 40 km depth there were largely low dipping (20°) thrust faults; and between 40 and 600 km there were possibly step faults allowing the plate to move downward as shown in Figure 14-9*b*. He suggested that the gravity faulting beneath the trench explained why their sedimentary fill was usually undistorted.

In 1970 N. L. Carter and H. G. Ave'Lallemant described deformation experiments on dunite and peridotite which suggested that flow in the upper mantle may be governed primarily by a non-linear creep law. In a companion paper (Ave'Lallemant and Carter, 1970) they proposed that syntectonic recrystallization is an important or dominant mode of flow in the upper mantle. They examined their results with reference to the motions of spreading lithosphere plates, and presented schematic "flow-fields" corresponding to the hypotheses where:

1. The displacements are due to drag by creep of convecting material in the mantle (Figure 14-8*a*).

2. Where the motion of the lithosphere is opposed by drag due to creep in the mantle below (Figures 14-8*b*, 14-8*c*, and 14-8*d*).

They illustrated the planes of maximum shearing stress for these two models and the fabrics that would develop by syntectonic recrystallization during flow. The preferred crystal orientations influence the seismic wave velocities, and they compared the effect of the proposed tectonite fabrics with recorded seismic anistropy in the upper mantle. The problem remains inconclusive but the approach is promising.

H. R. Shaw suggested in 1970 that the dissipation of tidal energy in the solid Earth may play a crucial role in magma generation and in the mechanism of sea-floor spreading. Only about 3% of the heat flow from the surface results from tidal energy, but this is enough to produce about 30 km^3 of magma a year. The tidal energy is strongly focused by dissipation mechanisms and can locally dominate other energy sources. Shaw concluded that dissipation is maximized in a partly molten region with a liquid fraction of 10 to 30%. Magmatic injection in the vicinities of oceanic ridges (Figure 14-11) transfers heat both from the stored heat in the mantle and from tidal sources. The possible tidal magma production alone can account for much of the average spreading rate. A compensating mass transfer from deeper regions is required to maintain a quasi-steady state. Thus a tidal-magmatic mechanism can act as a trigger to convective circulation. He also discussed the heat flux from continents and oceans and concluded that globally constant heat flux is consistent with equality of heat fluxes due to mass transfer and radioactive heat production in continental and oceanic sections explained in terms of the tidal-magmatic model.

MIDOCEANIC RIDGES AND MANTLE CONVECTION

Interest in sea-floor spreading was originally concerned mainly with the formation of new oceanic crust at the ridge crests, and the magnetic lineations have apparently left a record of the spreading history. The topography and crustal structure of the ridges is known in some detail as well as the gravity and heat flow profiles transverse to them (Figures 3-17*a*, 7-4, and 7-5). There have been several attempts to fit the observed heat flow to theoretical models of temperature distribution and convection.

The lithosphere acts as a thermal boundary layer supporting large temperature and density gradients because of its rigidity. There are two models for sea-floor spreading which relate to the ridge and its anomalies. Spreading is associated with (a) uprise of hot mantle material intruded in a narrow zone, or (b) uprise of hot mantle material in broad convective zone beneath the ridge. According to the first model the heat flow is not the surface expression of temperature anomalies in the upper mantle but a consequence of sea-floor spreading. The heat flow then cannot be used to infer mantle temperature distributions. This model was explored by D. P. McKenzie in 1967. According to the second model, as reviewed by E. R. Oxburgh and D. L. Turcotte in 1968, the geological features and geophysical anomalies associated with the ridges are surface expressions of the mantle conditions.

Thermal Structure

McKenzie presented a model for the temperature distribution in a lithosphere plate 50 km thick moving with constant velocity away from a vertical boundary at constant temperature. As the plates move apart hot material wells up from the mantle and forms a linear ridge along the boundary. He found good agreement with the observed width and shape of the heat flow anomalies and the free-air gravity anomalies transverse to the ridges. He concluded that both gravity and oceanic heat flow are probably controlled by the strength and thermal properties of the lithosphere and that it is not necessary for the upper mantle to be hotter beneath the ridges than it is elsewhere. In 1969 N. H. Sleep concluded from mathematical models that gravity, heat flow, and ridge topography cannot be expected to distinguish between McKenzie's preferred model and that of convective upwelling in the mantle.

A boundary-layer theory for the structure of two-dimensional convection cells was applied to mantle convection by Oxburgh and Turcotte, and the results beneath ocean ridges are shown in Figure 14-10. It is not

known whether the mantle behaves as a plastic or a viscous body. Oxburgh and Turcotte adopted a model with a constant, Newtonian viscosity independent of temperature, and they neglected the volume heat release due to radioactivity within the cell. K. E. Torrance and D. L. Turcotte have since obtained a numerical solution for mantle convection using a temperature and depth dependent viscosity appropriate for diffusion creep. They gave the resulting flow, temperature, and viscosity distributions. Oxburgh and Turcotte concluded that although their assumed viscosity relations were certainly wrong, the provisional agreement between the model and natural phenomena suggests that the simplifying assumptions had not too great an effect.

Figure 14-10 shows the steady-state distribution of isotherms beneath an ocean ridge predicted by the Oxburgh-Turcotte model. The temperature gradients are restricted to thin horizontal boundary layers and to thin vertical plumes on the boundaries between cells. The cores of cells are abiabatic. The mushrooming isotherm distribution for the upper ascending limb is not likely to be realized in detail because heat will be transferred rapidly upward by magmas from this region. The lines with arrows show the streamlines derived from velocity vectors for particle motion.

Petrological Structure

The general pattern of the petrology of the upper mantle beneath the ridges and oceanic lithosphere can be determined from this model for temperature distribution, and the schematic phase diagrams presented in Figures 6-18 and 6-19 for peridotite and gabbro, dry and in the presence of traces of water. Although modification of the temperature distribution and experimental determination of the phase diagrams will change the details, the general pattern may not be changed significantly. Each point in Figure 14-10 is fixed by a specific pressure (depth)

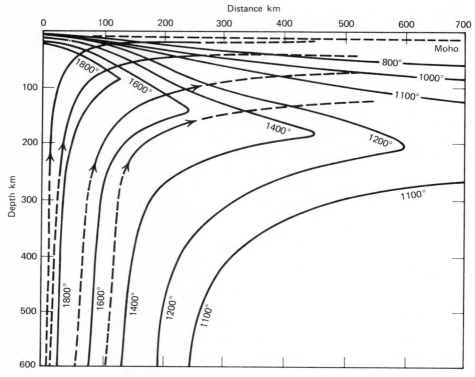

Figure 14-10.   Model for convection cells rising beneath a midoceanic ridge, with steady-state thermal structure computed for material with viscosity independent of temperature. Lines with arrows show the streamlines for particle motion (after E. R. Oxburgh and D. L. Turcotte, Jour. Geophys. Res., **73**, 2643, 1968, with permission).

and temperature, and the phase assemblage for a given material at each point can be determined from the appropriate figure, 6-18 or 6-19. Figure 14-11a shows the distribution of phase assemblages for a mantle composed of peridotite with a trace of water, and Figure 14-11b shows the phase assemblages produced in a hypothetical mantle of gabbroic composition with a trace of water. Figures 14-11a and 14-11b together provide a picture of the petrology of the suboceanic mantle and specifically the mantle beneath oceanic ridges assuming that the mantle is composed of peridotite together with material of gabbroic composition. The layered sequence for mantle some distance from the ridge is similar to that summarized in Figure 6-20. The

presence of a trace of water produces a zone of incipient melting which is equated with the low-velocity zone. Material of gabbroic composition exists as eclogite. The low-velocity zone increases in thickness considerably near the ridge crest extending downward to the level of the olivine-spinel transition (Chapter 6).

There is a large zone of partial melting where the temperature exceeds the dry solidus of peridotite, and gabbro in most of the corresponding zone in Figure 14-11b is completely melted. The zone is more than 400 km wide and 200 km thick. Oxburgh and Turcotte noted that only mantle material ascending along streamlines within 100 km of the plume center would pass through this fusion zone, and that this material then forms the upper 130 km of the horizontal limb. If all of the basaltic magma generated within the fusion zone escapes to the surface at or near the ridge, then this 130 km layer would be composed of residual peridotite. If some of the interstitial magma crystallizes as it is transported laterally through the upper boundary of the fusion zone then we have to consider phase assemblages in the upper mantle for material of basaltic composition.

Figure 14-11*b* shows that basaltic magma emerging from the fusion zone above 80 km depth crystallizes as gabbro, and that below 100 km it crystallizes as eclogite. Lateral transportation of the lithosphere causes cooling of the gabbro as shown by the streamlines in Figure 14-10, and if equilibrium is maintained the gabbro is transformed into eclogite through a wide zone of garnet granulite (Figure 5-15). This illustrates the dynamic model for the suboceanic mantle proposed by F. Press in 1969 (Figure 3-10). The low density layer of gabbro and garnet granulite exhibits a general pattern similar to that required by the models in Figure 7-5 for the structure of midoceanic ridges.

In 1967 F. Aumento presented the results of petrological study of a variety of extrusive rocks dredged from the first area of the mid-Atlantic ridge, at 45°N, to have been sampled and surveyed in detail. He reported a complete and continuous sequence from olivine tholeiites to alkali basalts together with high-alumina equivalents. Using available experimental data as a guide (Chapter 8) he developed a scheme of multiple cycles of partial melting beneath the ridge axis. A continuously variable decreasing degree of partial melting at 35–60 km depth could produce a continuous range in liquid compositions. With Figure 14-10 available Oxburgh and Turcotte considered similar models for magma generation, and P. W. Gast in 1968 placed the petrogenetic scheme on a more quantitative basis with special reference to the origin of tholeiitic and alkaline basalts and to trace element fractionation.

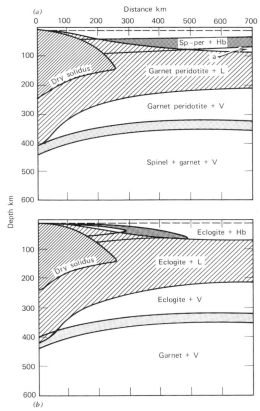

Figure 14-11. Schematic sections showing the petrology of peridotite (*a*) and eclogite (*b*) with traces of water in the suboceanic mantle, extending from the crest of a midoceanic ridge. These sections are derived from the temperature-depth (pressure) distribution in Figure 14-10, and the schematic isopleths for these materials shown in Figures 6-18(*b*) and 6-19(*b*). Compare Figure 6-20 for standard layered mantle sequences. Note the large zone in Figure 14-11(*b*) for partial melting above the dry solidus temperatures. The low-density layer of gabbro and garnet granulite in (*b*) exhibits a general pattern similar to that required by models in Figure 7-5 for the structure of midoceanic ridges. Compare Figure 5-18(*b*). (From P. J. Wyllie, Jour. Geophys. Res., **76**, 1328, 1971, with permission.)

LITHOSPHERE CONSUMPTION BENEATH ISLAND ARCS

The concept of plate tectonics requires that the lithosphere extends down into the mantle beneath island arcs (Figures 14-5 and 14-6). The intimate connection between surface structures and deep structures indicated by seismology is persuasive evidence. Whether the rigid slabs of lithosphere move down into the asthenosphere as a result of impelling forces or as a result of their own negative buoyancy (Figure 14-8), it is clear that the lithosphere constitutes and controls some kind of convective motion within the mantle. The descending lithosphere is a strong heat sink, yet the island arc areas above the cold slabs exhibit high heat flow anomalies and volcanic activity. The main features associated with sinking lithosphere which need explanation include:

1. Intermediate- and deep-focus earthquakes.
2. Ocean trenches 3 to 4 km deeper than the neighboring ocean basin floor.
3. Large negative gravity anomaly over the trench with smaller positive anomaly over the arc.
4. Abrupt increases in heat flow profiles along the line of the island arcs (Figures 3-17b and 3-17c).
5. Andesite volcanism in belts usually less than 300 km wide along island arcs, occupying the ocean side of the zone of high heat flow.

These topics were reviewed in 1969 by D. P. McKenzie and in 1970 by E. R. Oxburgh and D. L. Turcotte.

Thermal Structure

The factors influencing the temperature distribution in a downgoing slab of lithosphere include: (a) the descent rate, (b) adiabatic compression of the lithosphere as it moves into the mantle, (c) radioactive heating, (d) energy generated by phase changes in peridotite and gabbro of the lithosphere as the pressure increases, and (e) stress or frictional heating at the boundaries between lithosphere and asthenosphere.

McKenzie showed that there is a close relation between the temperature structure in a descending slab and that in the lithosphere near spreading ridges. He calculated the temperature distribution in a 50 km-thick slab with the geometry shown in Figure 14-12a, assuming that the spreading rate had been constant during the past 10 m.y. The thermal conductivity of the cold lithosphere is so small that the inner part of the slab is heated only very slowly (Figure 14-13). McKenzie gave temperatures on isotherms in dimensionless units, because of the uncertainty of the parameters used in the calculations, including the temperature of the adiabatic mantle. He concluded that earthquakes are restricted to those regions of the mantle which are colder than a certain, unspecified temperature. The earthquakes should thus occur in a thin layer within the lithospheric slab. The calculated temperature distribution could not explain the heat flow anomaly, and probably would not account for the volcanic activity either. McKenzie and other authors have appealed to frictional or stress heating at contacts between lithosphere and asthenosphere to explain these phenomena.

Movement of a large rigid slab into the asthenosphere would set up stresses and govern the flow of mantle material in the absence of other forces. Figure 14-12a shows McKenzie's calculated distribution of stream lines for flow in the mantle produced in this way. Shearing stresses are exerted on the lithosphere boundaries, and calculated contours are shown in Figure 14-12b. The stresses behind the arc are much greater than those in front. Viscous dissipation within the mantle produces heat from the mechanical energy which moves the plates against the viscous forces. The heat generated by friction between the mantle and the surface of the sinking slab would take 300 m.y. to be conducted to the surface, and this therefore cannot account for the surface heat flow anomaly. Others have appealed to diapiric uprise of hot material or magmas as factors contributing to the heat flow, but the magnitude of these effects appears to be small compared with the gross regional heat flux.

McKenzie cited three effects which could account for the heat flow anomaly:

1. The calculated contours for stress heating in Figure 14-12c show considerable heat generation at shallow depths, and this could diffuse to the surface in geologically reasonable times.
2. The calculated flow lines behind the island arc show hot mantle being carried closer to the base of the lithosphere.
3. Thinning of the lithosphere by this flowing mantle could possibly lead to even higher levels for the hotter material of the asthenosphere.

Oxburgh and Turcotte presented a possible distribution of isotherms under an island arc system, with the constraint that temperatures in the zone of movement between sinking lithosphere and mantle reached the melting temperatures of the oceanic sediments and crustal rocks, and were buffered by these temperatures. This assumption allowed the temperature along the fault zone to be prescribed. McKenzie concluded that the earthquake zone is not the physical boundary between lithosphere and asthenosphere, but a cooler layer within the slab, and this is consistent with the seismological data. The frictional zone of heating and magma generation then does not coincide with the earthquake zone, as Oxburgh and Turcotte and many others (Figure 8-1d) have assumed.

Minear and Toksöz used a quasi-dynamic scheme and a finite difference solution of the conservation of energy equation to determine the effects of the several factors listed above

on the temperatures in a downgoing slab. Thermal regimes were calculated for a 160 km-thick slab downwarping at 45° to the horizontal, with the complexity of the physical model being increased in a series of steps in order to demonstrate the effects of the various heat generating processes and the spreading rate. Figure 14-13a shows the distribution of isotherms taken from their Figure 9 which assumes a 1 cm/year spreading velocity, shear-strain heating of 1.6×10^{-4} ergs/cm^3 sec along the top edge of the slab and 1.0×10^{-5} ergs/cm^3 sec along the ends and bottom, and no contribution from phase changes or adiabatic compression. The dominant pattern is one of isotherms depressed deeply into the mantle. The zone of shear-strain heating with temperatures greater than those in normal mantle is less than 30 km wide above 200 km and the greatest temperature increases occur at depths less than 100 km (compare Figure 14-12c). Minear and Toksöz concluded that the temperatures were just high enough to produce melting along the upper edge of the slab, but this conclusion depends upon the results used for estimating the solidus temperature of the material.

Magma Generation

Using the phase diagram for peridotite in the presence of a trace of water given in Figure 6-18b, and assigning the appropriate phase assemblage to each pressure (depth)-temperature point in Figure 14-13a, we obtain Figure 14-13b. No melting occurs in peridotite (or in gabbro or eclogite) anywhere in this mantle section in the absence of water. Given a trace of water, incipient melting occurs through the shaded zone. If this layer of incipient melting does correspond to the low-velocity zone, as suggested in Figure 6-20, Figure 14-13b shows that the low-velocity zone is not continuous beneath island arcs. The upward migration of the olivine-spinel transition zone within the slab is shown.

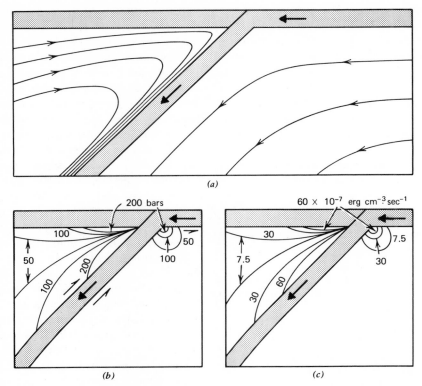

Figure 14-12. Stress heating caused by flow within the mantle associated with a downgoing slab of lithosphere (after McKenzie, 1969). (*a*) Stream lines for flow within the mantle. Motion is with respect to the 50 km thick plate behind the island arc, and is driven by the motion of the other plate and of the sinking slab. Thermal convection outside the slab is neglected. (*b*) Shear stresses in bars caused by the flow in Figure (*a*). The half arrows show the direction of the forces exerted by the fluid on the plates and slab. The stresses on the plate behind the island arc exceed those on the plate in front. (*c*) Stress heating caused by viscous dissipation within the flow in Figure (*b*), in units of 10^{-7} erg cm^{-3} sec^{-1}. The heating within the mantle is more intense behind than in front of the arc. (From Geophysical Journal, with permission.)

Magma may be generated beneath an island arc system in three ways:

1. The oceanic crust at the surface of the lithosphere, including siliceous and water-bearing sediments, could be partially fused by either thermal conduction from, the mantle or frictional heating. A small increment of temperature added to Figure 14-13*a* would cause partial melting of eclogite or peridotite, dry, in a zone near the surface of the slab at depths between about 150 and 300 km. Oxburgh and Turcotte discussed the mechanism for the formation of the main

members of the calc-alkaline suite by a process of continuous variation in temperature, pressure, composition of material being fused, and water content as melting occurs down the frictional heating zone and inward into the sinking lithosphere. T. Hatherton and W. R. Dickinson reviewed the relationship between andesitic volcanism and seismicity beneath island arcs in 1969 and concluded that the andesites are formed by fusion along the seismic zone, and that their potassium content increases as a function of depth of origin.

2. Dehydration of the lithosphere and upward migration of water into hotter mantle overlying the cold slab could produce partial melting of the mantle, as proposed, for example, by W. Hamilton and by A. R. McBirney in 1969. This process has been postulated as a source of intermediate and acid magmas for andesites and batholiths.

3. The hotter boundary zone comprising both mantle and upper lithosphere is potentially unstable, and diapiric uprise of solid material under adiabatic conditions may be initiated from this zone leading to magma generation according to the Green and Ringwood model in Figure 8-17. Figure 14-13*b* shows that if water is present along this high temperature boundary the peridotite of mantle and lithosphere will contain interstitial liquid, and influx of more water from the downgoing slab into any part of this layer would increase the percentage of liquid present, reduce the viscosity, and facilitate diapiric uprise according to the scheme shown in Figures 8-21 and 8-22. This process

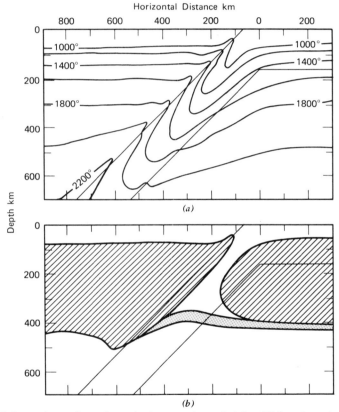

Figure 14-13. Schematic sections through the mantle and slab of lithosphere in mantle beneath an island arc. The lithosphere, 160 km thick, dips at 45° from the axis at 0 km. (*a*) Isotherms computed by Minear and Toksöz (1970), according to their Figure 9, which takes into account shear-strain heating along the edges of the lithosphere slab. (*b*) Petrology of the mantle section, assuming peridotite with traces of water (as in Figure 6-18*b*) and the temperature distribution in (*a*). The shaded zone shows incipient melting, interrupted by the slab. Uprise of the olivine-spinel transition (Figures 6-17 and 6-18*a*) within the cold slab is shown. The corresponding diagram for gabbroic material is similar. Note that normal fusion temperatures for the generation of basaltic magmas are not reached anywhere, according to this temperature distribution and the extrapolated dry peridotite solidus used (Figure 6-18(*a*)). (From P. J. Wyllie, Jour. Geophys. Res., **76**, 1328, 1971, with permission.)

would produce basaltic magmas from peridotite or andesites from eclogite at depths considerably above the Benioff zones (contrast Figure 8-1d). If water is abundant liquids of intermediate composition may be generated in similar fashion from peridotite. The compositions of the liquids produced depend upon a number of variables (Chapter 8).

Future correlation of the observed products of volcanoes with the compositions of liquids from various crystalline materials under known experimental conditions should permit us to locate the sources of various magmas beneath island arcs and beneath ocean ridges. For example it may prove possible to relate the variation in potash content across island arcs to the breakdown of hornblende, muscovite, and phlogopite in a partially hydrated downgoing lithosphere slab, as suggested by several investigators. If the sources of specific magmas can be located, this would provide invaluable fixed points in terms of depths (pressures) and temperatures for many geophysical calculations and models such as the temperature distribution shown in Figures 14-10 and 14-13a. This in turn would place limits on assumed physical properties of mantle materials. The way in which petrology, experimental petrology, and geophysics are coming to be mutually dependent is very heartening. Until more precise data are available, we will have to be satisfied with the general patterns such as Figures 14-11 and 14-13b based on extrapolated experimental data (Figures 6-18 and 6-19) and possible temperature distributions such as Figures 14-10 and 14-13a.

Structures Associated With Trenches

If slabs of lithosphere move down into the mantle beneath the island arcs we would expect to see evidence of compression. The seismic data show that the lithosphere beneath the trenches is in tension, so that gravity faults and vertical movements characterize this region, and that this is replaced beneath the inner margin of the trench by underthrusting (Figure 14-9). The details are not yet clear, and it is not known how or where the overthrust fault approaches the surface of the Earth.

Figure 14-14a is a summary by J. F. Dewey and J. M. Bird of the main geological structures associated with trenches. This shows the sites of blueschist metamorphism (glaucophane schist facies in Figure 9-10) and the injection of serpentinite from the descending slab. This is a mechanism suggested by R. S. Dietz in 1963 for the origin of ultramafic rocks of orogenic belts (Table 6-4). The sediments in the trench are undeformed, but they may show extreme deformation at the foot of the inner trench wall. It is generally assumed that some of the sediment is carried into the mantle with the oceanic plate, and that some of it is mechanically plastered by folding on to the continental margin or island arc.

There has been considerable discussion about the behavior of sediments in oceanic trenches, reviewed in 1970 by D. W. Scholl and associates. The problem is very complex, because it depends upon erosion rates from adjacent land masses, which depend in turn rather sensitively upon the climate and the paleogeography. Although some trenches exhibit clear evidence of compressional deformation as indicated in Figure 14-14a, the sediments at the base of the continental margin in the Peru-Chile trench and others around the Pacific basin appear to exhibit only extensional features. This observation is incompatible with the generally accepted model of spreading. C. K. Seyfert suggested in 1969 that the downward movement of sediments occurs landward of the axes of oceanic trenches and that the sediments are folded, faulted, and metamorphosed as they are carried beneath the continent or island arc. There is a sharp seismic contact between the undeformed sediments and the material underlying the lower part of the continental slope, which might be explained if the transition from undeformed to deformed and

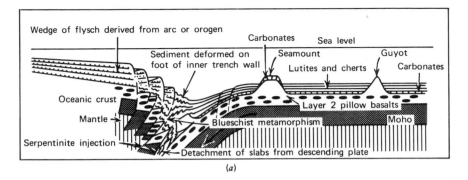

(a)

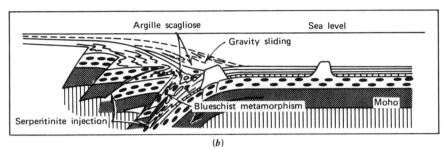

(b)

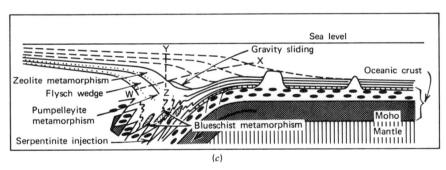

(c)

Figure 14-14. Structures associated with oceanic trenches (after Dewey and Bird, 1970). (*a*) Main geological structures. (*b*) and (*c*) Possible sequence of events in the growth of an island arc, as described in text. (From Jour. Geophys. Res., **75**, 2625, 1970, with permission.)

metamorphosed rocks is of tectonic form. This is a plausible explanation, but it is not altogether satisfactory to explain the apparently general lack of compressional features of sediments in trenches.

In their 1970 review Scholl and coauthors estimated the volume of undeformed trench sediments off Chile. If plate convergence had occurred at the 5 to 10 cm/year rates implied by the magnetic data during late Cenozoic time, and if those sediments deformed or consumed by the compression were undetect-

able because of the nature of their deformation as proposed by Seyfert, then the volume of undeformed sediments remaining should be considerably less than the volume of material supplied to the trenches during this period. Their estimates of the volume of terrigenous sediments contributed by continental erosion in late Cenozoic time are, in fact, quite close to the estimated volume in the trenches. They speculated that sinking of the oceanic lithosphere may take place seaward of the trench, with the trench fill and bedrock floor remaining partially or totally isolated from much of the differential movement between the plates.

During the Mesozoic and Cenozoic periods, according to J. Gilluly in 1969, the volumes of sediment derived from the North American continent should have been about the same in the Atlantic and Pacific Oceans. In fact it is estimated that the volume of terrigenous Atlantic sediment is at least six times as great as that of the Pacific sediment. The paucity of sediment off the Pacific coast is readily accounted for if North America began to drift westward in Triassic time overriding the Pacific and large volumes of sediment on a Benioff zone like that now presumed to underlie western South America. Gilluly suggested that the onset of the tremendous plutonism of the Mesozoic of western America might have been related to the beginning of continental drift, and that the missing sediments off the Pacific coast were dragged down into the mantle along a sinking plate of lithosphere, fused, and then intruded as batholiths. This theme was also presented by W. Hamilton in 1969 and W. G. Ernst in 1970.

Figures 14-14*b* and 14-14*c* illustrate the possible sequence of events in the growth of a single island arc according to Dewey and Bird. With the beginning of plate descent a trench forms by the intricate thrusting of wedges of oceanic crust and mantle to form a small ridge. Submarine gravity slides of sediment carry blocks of basic and ultrabasic rocks from the thrust wedges into the trench, and these are carried down where they undergo strong deformation and blueschist metamorphism. When the descending plate reaches a depth of over 100 km, the former oceanic crust fuses yielding calcalkaline magmas, with basaltic magmas being generated at higher levels. A volcanic island arc is thus developed and its erosion produces a wedge of flysch sediments as shown in Figure 14-14*c*.

During periods of fast plate consumption blueschist metamorphism is active producing chaotic mélanges, or "argille scagliose," which may rise isostatically during periods of slow movement. If the rate of plate consumption is slow the flysch may build across the trench onto oceanic sediments, producing enormous apparent thicknesses (measured along WX instead of YZ). Trenches and their associated sedimentary and metamorphic rocks are likely to be extremely complicated. The chaotic mélange terrains of many mountain belts are now being examined as possible positions of earlier trenches and Benioff zones.

GEOSYNCLINES, MOUNTAIN BUILDING, AND SEA-FLOOR SPREADING

The concept of plate tectonics is forcing geologists to reexamine nearly everything that they thought they knew about the origin of rocks and mountain ranges. The classical concepts of eugeosynclinal and miogeosynclinal sequences outlined in Chapter 9 appear to be too simple to derive maximum benefit from the plate tectonics model. A Penrose Conference of The Geological Society of America was held in December 1969 at Asilomar, Pacific Grove, California, to discuss "The meaning of the new global tectonics for magmatism, sedimentation, and metamorphism in orogenic belts." There is no printed record of the proceedings of these informal conferences, but reports of the meetings by W. R. Dickinson were published in *Geotimes* and *Science* in April and June respectively 1970. Dickinson noted the impact felt from the ideas of young research workers, citing specifically the rapt attention paid to the authoritative discussion of time and motion along the San Andreas Fault system by graduate student Tanya Atwater. More experienced workers were anxious to explore the implications of plate tectonic theory for their own interpretations. Marshall Kay, for example, inquired how his concepts of geosynclines (Chapter 9) should be translated to mesh with the fresh concepts of orogeny.

The term "subduction zone" was found useful by most of the 95 participants at the conference. This was applied to any elongate region along which crustal rocks are led to descend, by folding, faulting, or a combination of both, relative to masses in an adjacent block. The usage implied the presence of a plate juncture of convergence.

Eugeosynclines may include sedimentary sequences that have formed as (a) trench complexes, (b) island arc complexes, or (c) continental rise deposits. Miogeosynclines may include sedimentary sequences that have formed as (a) continental shelf and slope deposits, (b) deposits trapped in elongate traps between island arcs and trenches, or (c) clastic wedges in foredeep complexes (Kay's exogeosynclines, Table 9-1). If oceans can be shown to open and close, then in time any sediment wedge associated with a continental margin will eventually become involved with a subduction zone and orogeny. Dickinson concluded that, although the classic concept that geosynclinal downbowing exerts direct causative control on the position and extent of orogeny is no longer tenable, the geosynclinal theory remains valid if the causative function is replaced by the notion of coincidence, or consequence; the sediment piles relate to orogeny in the sense that "to exist is to be deformed." Each large orogenic belt should offer a different sequential history, although certain sequences are more apt to be duplicated.

The topics that received particular attention included: (a) andesite chains, (b) batholith belts, (c) ophiolite complexes, (d) glaucophane schist or blueschist metamorphism, (e) relationship of continental structures to subduction zones, (f) nomenclature of tectonic elements and stratigraphic facies in orogenic belts, and (g) the meaning of geosynclinal theory in the new conceptual framework.

Andesites and batholith belts were taken as extrusive and intrusive phases of magmatism related to inclined seismic zones in the mantle above intermediate- and deep-focus earthquakes (Figure 14-15). Ophiolite complexes and some other ultramafic rocks of the orogenic belts were interpreted as oceanic crust formed at midocean ridges at high temperatures, and incorporated tectonically at low temperatures into orogenic belts at subduction zones (Figures 14-16e and 14-16f). The formation of blueschists was attributed to rapid descent of cold lithosphere to great depths along subduction zones in trench complexes (Figures 14-14 and 14-15a).

In Chapter 9 we reviewed the classification of contemporary geosynclines proposed by Mitchell and Reading in 1969 as they are related to three types of continental margins. We can now relate them also to plate tectonics as in Figures 14-6, 14-15, 14-16, and 14-17. Atlantic-type geosynclines develop where there is no differential movement between continent and ocean; Andean-type geosynclines develop when there is differential movement; and island arc-type geosynclines develop where a lithosphere plate descends some distance from a continental margin, with Japan Sea-type geosynclines developing behind some island arcs. If spreading ceases or is renewed, or if the position of descending lithosphere changes, the type of geosyncline also will change and theoretically there are 12 possible changes. Mountain building associated with geosynclinal development may be Andean, island arc, or Himalayan. Himalayan orogeny is produced when two continents collide (Figure 14-17c). A comprehensive review was published by Dewey and Bird in 1970.

Dewey and Bird presented Figure 14-15 as a model for the evolution of a cordilleran-type mountain belt (Andean) formed where a trench originates near to a continental margin of Atlantic-type. This is a development and refinement of schemes presented early in the sea-floor spreading game by R. S. Dietz (1963). The general features of Figure 14-14 are incorporated in the trench-formation stage. The sequence is self-explanatory. It is

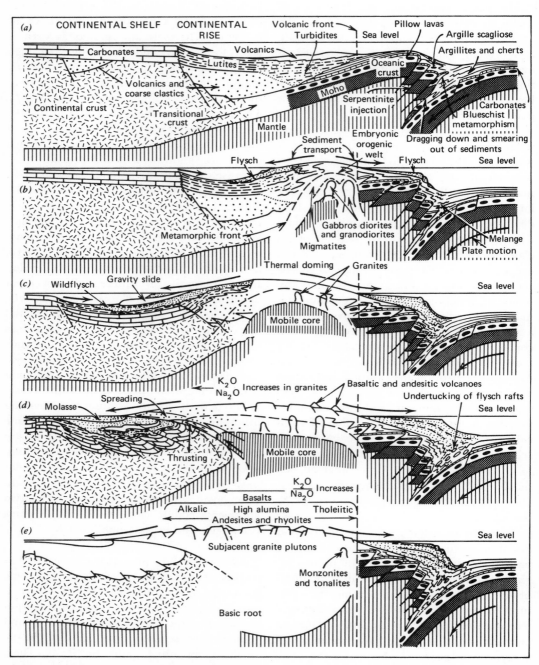

Figure 14-15. Schematic sequence of sections illustrating a model for the evolution of a cordilleran-type mountain belt developed by the underthrusting of a continent by an oceanic plate (after Dewey and Bird, 1970, Jour. Geophys. Res., **75**, 2625, with permission).

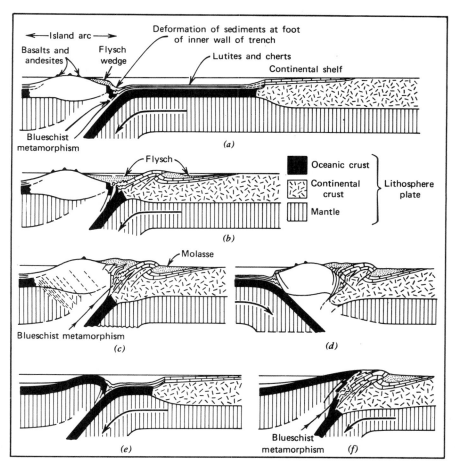

Figure 14-16. (a)–(d) Schematic sequence of sections illustrating the collision of a continental margin of Atlantic type with an island arc, followed by change in the direction of plate descent. (e)–(f) Proposed mechanism for thrusting oceanic crust and mantle onto continental crust (after Dewey and Bird, 1970, Jour. Geophys. Res., **75**, 2625, with permission).

based on an analysis of the north-western parts of the Appalachians for Ordovician time (Bird and Dewey, 1970) and on the general structure of the Mesozoic Cordilleran system of western North America. This figure should be compared with Figures 9-1, 9-3, and 9-4. The continental shelf edge in Figure 14-15a marks the boundary between the miogeosynclinal and the eugeosynclinal assemblages in Figure 9-1.

According to plate tectonics the continents are carried passively as part of the rigid lithosphere. Oceanic lithosphere moves down into the asthenosphere along lines marked by ocean trenches. Figures 14-1 and 14-6 show that continued spreading eventually brings continental masses to ocean trenches, and Figure 14-16 illustrates the consequence of the collision of a continental margin of Atlantic type with an island arc according to Dewey and Bird. The low density continental mass cannot be carried into the mantle because buoyancy forces will oppose the motion (Chapter 10). Continents once formed are therefore very difficult to destroy. McKenzie pointed out in 1969 that if an island arc attempts to consume a continent (Figures 14-16*b* and 14-16*c*) large stresses will develop and the island arc is likely to flip so that it consumes oceanic crust originally behind the arc, as in Figure 14-16*d*. Convection, as represented by the descending lithosphere plate, is now changed. McKenzie suggested that the whole pattern of mantle convection must depend on the motion of the continents.

Mitchell and Reading presented a simple model in which the development of continental margins and geosynclines was related to the oscillation of continents between ocean rises. This is illustrated in Figure 14-17. Each diagram shows two continents moving between ocean ridges considered as fixed in

position relative to one another, in both plan view and cross section. The geosynclines are distinguished as active if the lithosphere is descending, or passive if there is no differential movement between margin and ocean floor.

In Figure 14-17*a* continent I moves westward, approaches the ocean ridge, and in Figure 14-17*b* its motion has been reversed and it is moving east. The Andean-type geosyncline on the now passive western margin will be eroded and an Atlantic-type geosyncline will develop. An island arc-type geosyncline is formed near the eastern margin, and the former Atlantic-type geosyncline is now enclosed in a small ocean basin or geosyncline of Japan Sea-type.

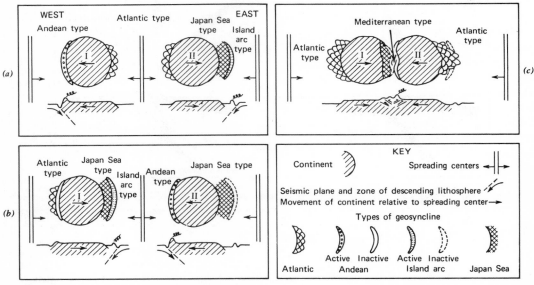

Figure 14-17. Simplified model to show how continents may oscillate between ocean rises as a result of changes in pattern of ocean-floor spreading (after Mitchell and Reading, 1969). Atlantic-type geosynclines develop on passive continental margins where the continent is moving with the lithospheric plate of the adjacent ocean floor. Andean-type geosynclines develop where spreading lithosphere descends beside a continent. Island arc-type geosynclines develop where lithosphere descends some distance from a continent. Japan Sea-type geosynclines occur between continents and islands arcs. The collision of two continents may produce Mediterranean-type geosynclines. Compare Figures 14-14, 14-15, and 14-16. See Figure 9-5 (*a*), (*b*), and (*c*). Each diagram shows both plan view and cross section of two continents moving between ocean ridges considered as fixed in position relative to one another. The geosynclines are distinguished as active if the lithosphere is descending, or passive if there is no differential movement between margin and sea floor. For discussion of spreading episodes and succession of one type of geosyncline by another, see text. (From Jour. Geology with permission. Copyright 1969, University of Chicago Press.)

Continent II in Figure 14-17*a* moves eastward with geosynclines equivalent to those near Continent I in Figure 14-17*b*. In Figure 14-17*b* it has reversed direction, and an active Andean-type geosyncline forms across the earlier Atlantic-type geosynclinal deposits. The now inactive island arc on the east is eroded to sea level and may subside. The Japan Sea-type geosynclinal basin persists.

Continued convergence of the two continents results in their collision in Figure 14-17*c* and the development of a Himalayan-type orogeny. The original arcuate shape of the Andean- and island arc-type geosynclines of Figure 14-17*b* may be destroyed. Small ocean basins of Mediterranean-type may remain where the two continents are not in contact.

Mitchell and Reading continued their examples one stage further, with the combined continent in Figure 14-17*c* separating again as a result of resumption of spreading along the line of the Himalayan-type orogenic belt. Figure 14-17 is sufficient to indicate the complexity of superimposed geosynclinal types that can be expected according to this model of oscillating continents, which is a necessary development of plate tectonics.

We have already referred to the application of plate tectonic theory to western North America by Hamilton, Gilluly, and Ernst. There are many other papers reevaluating geological interpretations. Two notable contributions of 1970 applied the theory to the geology on either side of the Atlantic. The paper by Bird and Dewey on the evolution of the Appalachian Caledonian orogenic belt is an outstanding demonstration of interpretation of complex geology in terms of plate tectonic theory. They examined in detail the sequences of sedimentation-deformation-metamorphism patterns of the New England and Newfoundland segments of the Appalachian orogen and its relationship to Atlantic Ocean plate tectonics. It is instructive to compare this paper with the 1963 contribution of Dietz on collapsing continental rises. The second paper is by A. M. Ziegler, who reviewed the geosynclinal development of the British Isles during the Silurian period. He presented a series of detailed paleogeographic maps, which gave insight into the positions of landmasses and shelf areas, and the areas of turbidite deposition and graptolitic shale deposition. He concluded that each side of the British geosynclinal complex might qualify as a geosyncline in its own right, and that the whole complex should be regarded as the marginal effects of two continents, unrelated until they collided. His conclusions are consistent with Wilson's 1966 suggestion that the Atlantic Ocean was open during the Lower Paleozoic, closed during the Upper Paleozoic and Lower Mesozoic, and opened again along a slightly different line during the present period of spreading from the mid-Atlantic ridge.

After examining the history of western America in terms of plate tectonics, Hamilton in 1970 turned his attention to Asia and the Ural Mountains. He reviewed the Soviet literature and attempted to explain the Uralides, the entire complex of late Precambrian and Paleozoic foldbelts between the Russian and Siberian platforms, as the result of collision of the two platforms. This was suggested as long ago as 1924 but most Soviet geologists have rejected long-distance horizontal transport; they postulate vertically moving elements, such as anticlinoria and synclinoria, assumed to have evolved through long periods. The paleomagnetic data for the Siberian and Russian platforms suggest that these land masses were widely separated in the Cambrian, converged during middle Paleozoic, and collided in the Permian or Triassic. Hamilton concluded that the geology of the Uralides accords with the collision concept with an intervening oceanic plate sliding down a subduction zone. From the igneous and metamorphic rocks of the Uralides, and the structure, he deduced a history of the continental margins before and during collision.

Dewey and Bird concluded that mountain building occurs in two ways:

1. The island arc/cordilleran (Figure 14-15) which is thermally driven and develops on the leading plate edges above a descending plate.

2. The result of collisions, continent/continent (Figure 14-17c) or continent/island arc (Figure 14-16) which are for the most part mechanically driven. They listed six main differences between these types of mountain belts, and emphasized the fundamental importance of the ophiolite suite (Table 6-4) in evaluating the stages of development of mountain belts. It is interesting to note that only a few years ago the ophiolite suite was largely ignored in petrogenetic discussion of ultramafic rocks by geologists not primarily concerned with Alpine geology [Wyllie, 1967 (p. 412), 1969, 1970]. Now these rocks are widely considered to be samples of the oceanic crust or upper mantle, and many occurrences of ultramafic rocks in orogenic belts are being closely re-examined to see if they can be identified as representatives from an ophiolite suite. These rocks are destined to play a key role in the interpretation of plate collisions. Some restraint is necessary, however, because not all ultramafic rocks represent oceanic crust or upper mantle (Chapter 6).

Ophiolites received particular attention from J. C. Maxwell in his 1966 review of Mediterranean geology (published in 1970). He concluded that ophiolite complexes represent ultramafic mantle material, partially melted and differentiated during diapiric uprise into and above lighter crustal rocks (Figure 6-6). He noted the similarity of larger ophiolite masses to the structure of the oceanic crust (Figure 7-8) and suggested that some parts of "oceanic areas" determined seismically may contain ophiolites overlying sedimentary rocks; false depths to the Moho may thus have been reported. Maxwell summarized his own work and that of European geologists active in the Mediterranean region, and concluded that there is convincing evidence for a continuous sialic basement between Europe and Africa in the later Carboniferous. The marine sediments of the Alpine system appear to have been deposited everywhere on continental crust. Maxwell interpreted the Mediterranean as a strip of continental crust in which massive injections of mantle material accompanied the generation of a dynamic ridge-and-basin movement. This process of oceanization by ophiolites is discussed further in the next chapter. The significance of ophiolites in this interpretation differs from that attributed to them in plate tectonics.

According to continental drift and plate tectonics, there was once a large oceanic area between Africa and Eurasia, which began to close in Permian or Cretaceous time. According to Maxwell the similarity and continuity of the geological histories of Africa and Europe is so striking that if the continents were not now close together, we would certainly postulate that they had once been together and had since drifted apart. The Mediterranean itself developed with widespread deposition of shallow-water Permo-Triassic sediments everywhere disconformable on older sialic rocks. The JOIDES drilling program in the Mediterranean should provide critical evidence for determination of the geological history of the region and for evaluation of Maxwell's conclusion that no lateral displacement of oceanic dimensions can reasonably be postulated between Africa and Eurasia, at least since Mid-Paleozoic.

15. *Global Geology in the 1970's*

INTRODUCTION

The revolution of the 1960's burst into the 1970's with a wave of enthusiasm for re-evaluation and reinterpretation of geology in terms of the new global tectonics. In Chapters 11, 12, 13, and 14 we reviewed the evidence that purports to prove the reality of continental drift in its new guise of plate tectonics. In October of 1970, using this evidence, R. S. Dietz and J. C. Holden presented a new version of Wegmann's original breakup of *Pangaea* (Figure 11-1), together with a prediction of what the world would look like 50 m.y. from now. Los Angeles is due to pass San Francisco in about 10 m.y., and to slide into the Aleutian trench in about 60 m.y.

What started as hypothesis is now hailed as theory. Articles in popular magazines state that the original geopoetry of H. H. Hess (Figure 12-22) is now geofact. Many items of evidence published since about 1966 were accompanied by confident claims that this was the virtual or definitive proof. Finally, in 1970, the results of the JOIDES drilling program brought evidence from the South Atlantic (Figure 14-7) that the overlapping of sediments and the ages of sediments were just as predicted (Figure 12-22). A single proof is sufficient for a theory to become fact. This model suffers from an overdose if one counts all the claims for virtual proof that have been made for it. The new global tectonics has enjoyed phenomenal success as a working hypothesis; so much success, in fact, that it is fast becoming a ruling theory.

There has been continuing opposition to the concept of continental drift. A. A. Meyerhoff has kept track of the pro-drift and anti-drift literature, and he informed me that of the papers published between about 1956 and 1970, antidrift papers comprise at least 21%. In the few years since the formulation of the plate tectonics model the papers with anti-drift arguments are rarely cited. Some of the problems were outlined in Chapter 14; these include the apparently undeformed sediments in many oceanic trenches and the properties of the deep seismic zones beneath the Andes. There are problems related to the geology of the Mediterranean. Other problems are presented by some paleontologic data and the paleoclimatic data reviewed in the following pages. Protagonists of plate tectonics tend to neglect the data and arguments not explained by the ruling theory, arguing that these can probably be ascribed to our lack of understanding rather than to inadequacies of the model, and that time and improved understanding will bring forth explanations.

Every revolution has its counterrevolution. The 1968 debate in *Geotimes* between J. T. Wilson, proclaiming the revolution, and V. V. Beloussov, presenting opposition, was mentioned in Chapter 11. Beloussov has consistently maintained that global tectonics are explained by vertical movements and oceanization rather than by horizontal movements and compression, and he appears to have a large following among Soviet geologists. A. A. Meyerhoff is a crusader, recently joined by C. Teichert and soon to be joined by 13 more coauthors, who is marshalling the rather scattered anti-drift arguments in counterrevolution. He maintains that unless the distribution of paleoclimate indicators and his interpretation of them can be ac-

counted for by the new global tectonics, the mobilist concepts must be relegated to the status of speculations.

This is yet another chapter that merits a whole book, but again I must satisfy myself with a brief outline.

PROGNOSIS FROM THE 1969 PENROSE CONFERENCE AND THE 1970 GEODYNAMICS COMMISSION

The topic of the Penrose Conference at Asilomar in December 1969 was "The meaning of the new global tectonics for magmatism, sedimentation, and metamorphism in orogenic belts." The reports of W. R. Dickinson were reviewed in Chapter 14. The classical views of orogenesis outlined in Chapter 9 appear very incomplete compared with the new concepts explored at the Penrose Conference. Stratigraphic approaches to the study of orogenic belts must involve evaluation of lateral movements and their effects, as well as the vertical movements. The topic that received particular attention at the conference was the relationship of continental structures to subduction zones, and the meaning of ophiolite complexes, stratigraphic facies, glaucophane schist or blueschist metamorphism, andesite chains, and batholith belts, with respect to this relationship. These are the rocks that will be studied diligently during the 1970's for clues in unravelling sequences of plate collisions.

The international Upper Mantle Project provided much of the stimulus for the development of plate tectonic theory. During the terminal phase of this Project, 1968-1970, an international Commission was appointed to plan a Geodynamics Project to facilitate extension of long-term research efforts initiated under the aegis of the Upper Mantle Project. Objectives were outlined at a meeting in Flagstaff, Arizona, in June 1970 and published in the November 1970 issue of the *Transactions of the American Geophysical Union*. The Geodynamics Project is an international, interdisciplinary program of research "on the dynamics and dynamic

history of the Earth with emphasis on deep-seated foundations of geological phenomena." The Commission of Geodynamics considered the program in four parts:

1. The movement of lithospheric blocks relative to each other, and the concentration of strong tectonic activity in a few relatively narrow mobile belts between blocks. Studies should be concentrated in active belts where movements and deformations can be measured.

2. Studies directed to characterization of the movements of the lithospheric blocks, and the driving forces within the Earth responsible for the movements.

3. Study of the primarily vertical movements that occur within lithosphere blocks, apparently independent of their horizontal movements. Sedimentary basins and elevated plateaus provide evidence for enormous vertical movements, often accompanied by block faulting and volcanism.

4. Systematic studies of past orogenic activity as reflected in the geological record, in order to determine past lateral movements of the lithosphere blocks. Igneous, metamorphic and tectonic relationships in ancient orogenic belts should be compared with similar relationships in recent active belts.

Initiation of the Geodynamics Project ensures that the international and interdisciplinary cooperation that flourished with the Upper Mantle Project will continue to grow. Whatever happens in the revolution and counterrevolution, there can be no doubt that our understanding of the Earth will advance rapidly.

V. V. BELOUSSOV AND R. W. VAN BEMMELEN PREFER "OCEANIZATION"

V. V. Beloussov struck a note of caution in his 1969 review of the results of the Upper Mantle Project, but this does not appear to have dampened the ardor of enthusiasts for the new global tectonics. Beloussov noted that, while much attention has been paid to data from the oceans in the formulation of the new global tectonics, our knowledge of the structure and development of the oceans is still much more sketchy than data from the continents. He considers that insufficient attention is being paid to continental geology, and historical perspective is thus being lost. The results of two centuries of data gathering from the continents should not be oversimplified in order to bring them down to the level of the schematic data available for the oceans. He suggested that the new oceanic results "have cast a hypnotic spell and thrown a shadow over much that is old and familiar."

Beloussov argued in favor of vertical movements initiated in the mantle as the major cause of folding, magmatism, and metamorphism. There are regions of slow uplift and subsidence on old continental platforms that have been maintained for hundreds of millions of years, and these are difficult to explain by theories requiring large horizontal displacements of the crust. He reviewed the contradictions involved in the idea of large transcurrent or transform fault displacements. The absence of a satisfactory scheme of worldwide convection to explain the apparent movements implied by paleomagnetic data leads him to question the assumption that the geomagnetic field has always had the same structure as it has now (Chapter 12).

The hypothesis of "oceanization" is an alternative to the hypotheses involving continental rift and ocean spreading, and Beloussov maintains that it meets most of the objections raised against the mobilist theories. The tectonosphere is divided into vertical blocks developing more or less independently. Continental crust is destroyed as sinking crustal blocks are replaced by basaltic lava, forming ocean floors. This explains the equality of continental and oceanic heat flow and avoids the conflict between oceanic and continental geology. The deep faults surrounding the Pacific separate areas of continental differentiation and of oceanic homogenization. If sediments older than the Cretaceous are distributed across the ocean floors, they must be contained in layer 2 or even layer 3. Beloussov suggested that intensified subsurface radioactive heating in oceanic regions at the end of the Paleozoic or beginning of the Mesozoic led to violent volcanic processes and oceanization which destroyed the older oceanic crust.

Beloussov emphasized the inadequacy of our knowledge of the Earth and advocated the balanced use of oceanic and continental data of geology, geophysics, and geochemistry. Apparent contradictions should not be artificially avoided by arbitrary assignment of more weight to one aspect or the other; all evidence must be studied and evaluated. No scientist would challenge this approach out of context, but in the context of plate tectonics there is at least a tendency among many geologists to neglect the method of multiple working hypotheses.

Beloussov and his followers are not alone in advocating a process of oceanization as a major factor in shaping the present Earth's surface. In 1968 R. W. van Bemmelen wrote a review paper on the origin and evolution of magmas and the Earth's crust. He argued against the hypothesis of continental growth by addition of magmas from the mantle (Chapters 7 and 8), and described a process of oceanization. The sialic crust was formed early in the Earth's history by the accretion of satellitic material. This crust has been transformed into oceanic crust by the emplacement of basic magmas, either locally within the crust or in regions of mega-shearing.

His views differ from those of Beloussov, because he believes that continental drift and sea-floor spreading do occur. According to his undation theory of the Earth's evolution endogenic energy flows outward from the core and is transformed into various types of energy according to the physicochemical conditions of the structural layers of the Earth through which the energy passes. Near surface phenomena such as volcanism, seismic activity, and heat flow are the effects of the deeper-seated processes caused by the outward flow of endogenic energy. Undations are accompanied by the flow of material within the Earth (mass-circuits) which deform the geoid: the greatest deformations are called "mega-undations." During the devel-

opment of a mega-undation the upper mantle is arched and the overlying layers spread under gravity. This can lead to removal of the continental shields from the crest of a mega-undation, with new ocean basin floored by upper mantle material being formed in the wake of the shields. Accumulating bodies of basaltic magma produce geo-undations such as the midoceanic ridges, the tops of which are subjected to gravitational spreading with tension and rifting occurring at its crest. According to van Bemmelen sea-floor spreading is caused by the spreading under gravity of the geo-undatory midocean rise (Figure 14-8c).

A. A. MEYERHOFF MAINTAINS THAT THE ATLANTIC OCEAN HAS BEEN OPEN FOR 800 MILLION YEARS

Not all geologists consider the new global tectonics to be a panacea. A. A. Meyerhoff published two major papers in 1970 marshalling geological and paleoclimatological evidence that appears to be inconsistent with the conclusion that the Atlantic Ocean has closed and opened again during the past 800 to 1000 m.y. In 1971 he was joined in a third paper by C. Teichert.

Meyerhoff reviewed the distribution of paleoclimate indicators. He plotted the distribution of marine evaporites, coal deposits, tillites, and proved desert eolian deposits on a series of world maps, one for each age from the Proterozoic through the Miocene, and interpreted them in terms of modern meteorology, physical oceanography, and climatology. This was emphasized because adherence to outdated climatological concepts has led to "myths widely believed by earth scientists," which Meyerhoff discussed and evaluated.

The maps show that:

1. Coal and evaporite belts are axisymmetric about the present rotation axis.

2. The Earth's two horse-latitude belts have remained coincident through time, as indicated by the desert eolian deposits and present desert belts.

3. Of the world's evaporite deposits, 95% by volume and area occur in regions now receiving less than 100 cm of annual rainfall; these regions reflect very closely the present planetary wind-circulation pattern, and the coincidence demonstrates that this pattern has not changed since middle Proterozoic time, except under local and temporary influences.

These and related facts show that the planetary wind-current and ocean-current pattern has been essentially the same for 800-1000 m.y. The only known explanation for this coincidence is that the rotational pole, the continents, and the ocean basins have been in the same positions since middle Proterozoic time.

Since the Devonian major coal deposits occupy two belts extending to high latitudes. Equatorward from the coal belts are the evaporite belts. Their widths fluctuate through time, indicating that there have been episodic changes in the world temperature. The Earth must have been very warm in "evaporite maxima," when the evaporite belts spread through up to 125° latitude. Glaciation is most common in periods of "glacial maxima," when the coal belts are broadest and the evaporites are restricted to about 40–60° latitude width. Meyerhoff presented a curve showing quantitatively the general distribution since the middle Proterozoic time of evaporite and glacial maxima, showing the fluctuation of world climate with corresponding fluctuations in the widths of climatic zones.

Ocean currents and wind currents may modify the positions of paleoclimate indicators anticipated from their latitude. Their main effect is a northward offset. In his second paper Meyerhoff described the northward deviation from axisymmetry of the Northern Hemisphere evaporite zone in the circum-Arctic and circum-North Atlantic Oceans which characterized several evaporite-maxima periods from later Proterozoic through Early Permian times. He explained these in terms of world-climate fluctuations and the development of major transoceanic sills within the framework of existing continents, ocean basins, and ocean-current systems.

Meyerhoff and Teichert reviewed the formation of coal and glacial deposits during the glacial maximum periods. They pointed out that coal is not a tropical deposit; coal requires cool winters in middle latitudes, abundant moisture, and large swampy areas. Every large coal basin is on one of two high-latitude coal zones or on the eastern sides of continents. Glaciation can occur only if there are elevated regions (regardless of latitude), and sufficient moisture for glaciers to build and ice caps to expand. This requires the presence of a nearby ocean and its associated moisture-laden air. During glacial maximum periods precipitation near ice-cap peripheries can be 45° or more from the poles,

and mountain glaciation can occur at any latitude.

Meyerhoff and Teichert pointed out that glaciers cannot form deep in the interiors of continents where there is no water supply. They gave reasons for believing that shallow epeirogenic seas cannot account for the ice-cap and coal distributions. A Gondwanaland reconstruction (Figures 11-1 and 11-5) requires that the Permo-Carboniferous glaciated areas be too far inland for moist ocean-air currents to reach them. "Thus the 'Gondwanaland' hypothesis, in this respect, defeats itself by making an impossibility of one phenomenon it purports to explain." This conclusion has been stated several times during the past 45 years, but the message has been lost.

The distribution of glacial and coal deposits during Carboniferous and Permian times requires certain moisture precipitation and ocean-current patterns, and these show that the Atlantic and Indian Oceans have existed since Late Carboniferous or earlier time.

These conclusions, based on "factual, observable data, in sharp contrast to speculations based on recent geophysical-oceanographic studies," convince Meyerhoff and Teichert that, until advocates of the new global tectonics find alternative explanations for the distribution of paleoclimate indicators, the mobilist concepts will have to be regarded as speculations supported by only a fraction of the known geological, paleontological, and paleoclimatological data. We have already noted Beloussov's concern that structural aspects of continental geology have been neglected for the sake of the ocean-based plate theory.

EPILOGUE

In the final sentence of their 1968 paper Isacks, Oliver, and Sykes wrote:

"Even if it is destined for discard at some time in the future, the new global tectonics

is certain to have a healthy, stimulating, and unifying effect on all the Earth sciences."

We have seen the stimulation and the unification of effort, and we have seen also the first major dissent. The new global tectonics has enjoyed such phenomenal success as a working hypothesis that it is becoming a ruling theory. The dangers of this situation are shown by a quotation from a distinguished paper by C. E. Wegmann in 1963 (p. 5):

"because commonly *the notions, concepts and hypotheses control the selection of facts recorded by the observers*. They are nets retaining some features as useful, letting pass others as of no immediate interest. The history of geology shows that a conceptual development in one sector is generally followed by a harvest of observations, since many geologists can only see what they are asked to record by their conceptual outfit."

This is not the first time in geological history that a theory has been acclaimed as virtually proven. It has been said that Abraham Werner's promulgation of the Neptunian theory elevated geology to the rank of a real science. For years his theory appeared to be unassailable. He maintained that all geological formations, and rocks of all types except for those actually observed to emerge from volcanoes as lavas, had originated as successive deposits or precipitates from a primeval ocean. Werner's thesis was challenged by the school of Plutonists and subsequently abandoned; this was one of the most celebrated and bitter controversies in science. James Hutton and the Plutonists presented evidence satisfying most geologists that many rocks were formed by the cooling and crystallization of hot material that had risen in a fused condition from subterranean regions. Neptunism, which dominated geological thought for many years, simply disappeared.

Another controversy of long duration concerned the question of whether or not fossils were of organic origin. When they were eventually recognized as the remains of living things, champions of the theological cause maintained that fossils had been carried to their present positions above sea level by the Noachian deluge. Indeed the existence of fossils in mountains was cited as scientific evidence proving that the biblical deluge had occurred. Baron Cuvier discovered that certain fossils were confined to specific rock formations, and he concluded that a series of widespread catastrophes had caused the disappearance of faunas characterizing certain formations. The doctrine of Catastrophism was dominant for nearly three centuries, with Cuvier's major contributions being published in 1811 and 1812. Toward the end of the eighteenth century, Hutton's theory of the Earth emphasized that in the geological record there is "no sign of a beginning—no prospect of an end." The idea that geological history should be explainable in terms of events occurring at present led to the controversy between Catastrophism and Uniformitarianism, as the competing principle was called by Sir Charles Lyell. In continental Europe the principle became known as Actualism. The publication in 1833 of Lyell's book *Principles of Geology* marked the end of Catastrophism, and Uniformitarianism became the new creed. It is now recognized, however, that the uniform flow of geological history has been interrupted by local intermittent catastrophes.

The conceptual development of plate tectonics is indeed gathering a harvest of data, observations, and interpretations. The pieces are fitting together very well, in a most persuasive fashion. But what about those paleoclimate indicators of Meyerhoff? Other dominant theories have been toppled: surely this could not happen to the new global tectonics —or could it?

References

Aubouin, J., 1965, *Geosynclines*, Elsevier, Amsterdam.

Ahrens, T. J., and Y. Syono, 1967, Calculated mineral reactions in the earth's mantle. *J. Geophys. Res.*, **72**, 4181–4188.

Akimoto, S., and H. Fujisawa, 1968, Olivine-spinel solid solution equilibria in the system Mg_2SiO_4-Fe_2SiO_4. *J. Geophys. Res.*, **73**, 1467–1479.

Akimoto, S., and E. Komada, 1967, Effect of pressure on the melting of olivine and spinel polymorph of Fe_2SiO_4. *J. Geophys. Res.*, **72**, 679–686.

Allard, G. O., and V. J. Hurst, 1969, Brazil-Gabon link supports continental drift. *Science*, **163**, 528–532.

Anders, E., 1968, "Chemical processes in the early solar system, as inferred from meteorites," in *Accounts of Chemical Research*, October 1968, American Chemical Society.

Anderson, D. L., 1967, "Latest information from seismic observations," in *The Earth's Mantle*, ed. T. F. Gaskell. Academic, New York.

———, 1967, Phase changes in the upper mantle. *Science*, **157**, 1165–1173.

———, 1968, Chemical inhomogeneity of the mantle. *Earth Planet. Sci. Letters*, **5**, 89–94.

———, 1970, "Petrology of the mantle," in *Fiftieth Anniversary Symposia*, ed. B. A. Morgan. Mineralogical Society of America, Special Paper No. 3.

———, and C. Sammis, 1969, The low velocity zone. *Geofisica Internacional*, **9**, 3–19.

Archambeau, C. B., E. A. Flinn, and D. G. Lambert, 1969, Fine structure of the upper mantle. *J. Geophys. Res.*, **74**, 5825–5865.

Aumento, F., 1967, Magmatic evolution on the mid-Atlantic ridge. *Earth Planet. Sci. Letters*, **2**, 225–230.

Ave'Lallemant, H. G., and N. L. Carter, 1970, Syntectonic recrystallization of olivine and modes of flow in the upper mantle. *Bull. Geol. Soc. Am.*, **81**, 2203–2220.

Badgley, P. C., 1965, *Structural and Tectonic Principles*, Harper and Row, New York.

Barazangi, M., and J. Dorman, 1969, World seismicity map of E.S.S.A., Coast and Geodetic Survey epicenter data for 1961-1967. *Bull. Seismol. Soc. Am.*, **59**, 369–380.

Barth, T. F. W., 1962, *Theoretical Petrology*, 2nd ed., Wiley, New York.

Bassinger, B. G., O. E. DeWald, and G. Peter, 1969, Interpretation of the magnetic anomalies off Central California. *J. Geophys. Res.*, **74**, 1484–1487.

Beloussov, V. V., 1968, An open letter to J. Tuzo Wilson. *Geotimes*, **13**(10), 17–19.

———, 1969, Earth's tectonosphere (results and further problems for investigation). *Intern. Geol. Rev.*, **11** (No. 12 December), 1368–1381.

Benioff, H., 1955, "Seismic evidence for crustal structure and tectonic activity," in *Crust of the Earth*, ed. A. Poldervaart. Geological Society of America Special Paper 62.

Bernal, J. D., 1936, Discussion. *Observatory*, **59**, 267–268.

———, 1961, Continental and oceanic differentiation. *Nature*, **192**, 123–125.

Birch, F., 1952, Elasticity and constitution of the earth's interior. *J. Geophys. Res.*, **57**, 227–286.

———, 1961, Composition of the earth's mantle. *Geophys. J.*, **4**, 295–311.

———, and P. LeComte, 1960, Temperature-pressure plane for albite composition, *Am. J. Sci.*, **258**, 209–217.

Bird, J. M., and J. F. Dewey, Lithosphere plate-continental margin tectonics and the evolution of the Appalachian orogen. *Geol. Soc. Am. Bull.*, **81**, 1031–1060.

Blackett, P. M. S., J. A. Clegg, and P. H. S. Stubbs, 1960, An analysis of rock magnetic data. *Proc. Roy. Soc. (London)*, **A-256**, 291–322.

Blackett, P. M. S., E. Bullard, and S. K. Runcorn, 1965, A symposium on continental drift. *Phil. Trans. Roy. Soc. London*, **1088**.

Boer, J. de, J. G. Schilling, and D. C. Krause, 1969, Magnetic polarity of pillow basalts from Reykjanes Ridge. *Science*, **166**, 996–998.

Boettcher, A. L., 1970, The system CaO-Al_2O_3-SiO_2-H_2O at high pressures and temperatures. *J. Petrol.*, **11**, 337–379.

———, and P. J. Wyllie, 1967, Hydrothermal melting curves in silicate-water systems at pressures greater than 10 kilobars. *Nature*, **216**, 572–573.

———, 1968, Melting of granite with excess water to 30 kilobars pressure. *J. Geol.*, **76**, 235–244.

——, 1968, The quartz-coesite transition measured in the presence of a silicate liquid and calibration of piston-cylinder apparatus. *Contr. Mineral. Petrol.*, **17**, 224–232.

——, 1968, Jadeite stability measured in the presence of silicate liquids in the system $NaAlSiO_4$-SiO_2-H_2O. *Geochim. Cosmochim. Acta.*, **32**, 999–1012.

——, 1969, Phase relationships in the system $NaAlSiO_4$-SiO_2-H_2O to 35 kilobars pressure. *Am. J. Sci.*, **267**, 875–909.

Bowen, N. L., 1928, *The Evolution of the Igneous Rocks*. Princeton University Press, Princeton.

——, 1940, Progressive metamorphism of siliceous limestone and dolomite. *J. Geol.*, **48**, 225–274.

——, and O. F. Tuttle, 1950, The system $NaAlSi_3O_8$-$KAlSi_3O_8$-H_2O. *J. Geol.*, **58**, 489–511.

Boyd, F. R., and J. L. England, 1960, Apparatus for phase-equilibrium measurements at pressures up to 50 kilobars and temperatures up to 1750°C. *J. Geophys. Res.*, **65**, 741–748.

——, 1962, Pyrope. *Carnegie Inst. Wash. Yearbook*, **61**, 109–112.

Brown, H., and C. Patterson, 1947, The composition of meteoritic matter II. The composition of iron meteorites and of the metal phase of stony meteorites. *J. Geol.*, **55**, 508–510.

Bruhnes, B., 1906, Recherches sur le direction d'aimantation des roches volcaniques. *J. Phys. Radium Paris*, **5**, 705–724.

Brune, J. N., 1969, "Surface waves and crustal structure," in *The Earth's Crust and Upper Mantle*, ed. P. J. Hart. Geophysical Monograph 13, American Geophysical Union, Washington, D.C.

Bull, C. E., E. Irving, and I. Willis, 1962, Further paleomagnetic results from South Victoria Land, Antarctica. *Geophys. J.*, **6**, 320–336.

Bullard, E. C., 1967, The removal of trend from magnetic surveys. *Earth Planet. Sci. Letters*, **2**, 293–300.

——, 1968, Reversals of the earth's magnetic field. *Phil. Trans. Roy. Soc. London*, **A-263**, 481–524.

——, J. E. Everett, and A. G. Smith, 1965. The fit of the continents around the Atlantic, *Phil. Trans. Roy. Soc. London*, **A-258**, 41–51.

Bullard, E. C., C. Freedman, H. Gellman, and J. Nixon, 1950, The westward drift of the Earth's magnetic field. *Phil. Trans. Roy. Soc. London*, **A-243**, 67–92.

Bullen, K. E., 1936, The variation of density and the ellipticities of strata of equal density within the earth. *Monthly Notices Roy. Astron. Soc. Geophys. Supplement*, **3**, 395–401.

——, 1967, "Basic evidence for earth divisions," in *The Earth's Mantle*, ed. T. F. Gaskell. Academic, New York.

Burk, C. A., 1968, Buried ridges within continental margins. *Trans. N.Y. Acad. Sci.*, **2-30**, 397–409.

Burnham, C. W., 1967, "Hydrothermal fluids at the magmatic stage," in *Geochemistry of Hydrothermal Ore Deposits*, ed. H. L. Barnes. Holt, Rinehart, and Winston, New York.

Cameron, A. G. W., 1968, A new table of abundances of the elements in the solar system, in *Origin and Distribution of the Elements*, ed. L. H. Ahrens. Pergamon, Oxford.

Carey, S. W., 1958, *Continental Drift, a Symposium*. Geology Department, University of Tasmania, Hobart.

Carmichael, I. S. E., 1963, The crystallization of feldspar in volcanic acid liquids. *Geol. Soc. London Quart. J.*, **119**, 85–131.

Carswell, D. A., 1968, Picritic magma—residual dunite relationships in garnet peridotite at Kalskaret near Tafjord, South Norway. *Contr. Mineral. Petrol.*, **19**, 97–124.

——, 1968, Possible primary upper mantle peridotite in Norwegian basal gneiss. *Lithos*, **1**, 322–355.

Carter, J. L., 1966, Chemical composition of primitive upper mantle. Geol. Soc. Amer. Annual Meeting, San Francisco, Program 35–36.

Carter, N. L., and H. G. Ave'Lallemant, 1970, High temperature flow of dunite and peridotite. *Bull. Geol. Soc. Am.*, **81**, 2181–2202.

Chayes, F., 1966, Alkaline and subalkaline basalts. *Amer. J. Sci.*, **264**, 128–145.

Cohen, L. H., and W. Klement, 1967, High-low quartz inversion: determination to 35 kbars. *J. Geophys. Res.*, **72**, 4245–4251.

Cohen, L. H., K. Ito, and G. C. Kennedy, 1967, Melting and phase relations in an anhydrous basalt to 40 kilobars. *Am. J. Sci.*, **265**, 475–518.

Cox, A., 1969, Geomagnetic reversals. *Science*, **163**, 237–245.

——, and R. R. Doell, 1960, Review of paleomagnetism. *Bull. Geol. Soc. Amer.*, **71**, 645–768.

——, 1961, Paleomagnetic evidence relevant

to a change in the Earth's radius. *Nature*, **189**, 45–47.

Clark, S. P., and A. E. Ringwood, 1964, Density distribution and constitution of the mantle. *Rev. Geophys.*, **2**, 35–88.

Clegg, J. A., M. Almond, and P. H. S. Stubbs, 1954, The remanent magnetism of sedimentary rocks in Britain. *Phil. Mag.*, **45**, 583–598.

Creer, K. M., 1967, "A synthesis of world-wide paleomagnetic data," in *Mantles of the Earth and Terrestrial Planets*, ed. S. K. Runcorn. Wiley-Interscience, New York.

Crook, K. A. W., 1969, Contrasts between Atlantic and Pacific geosynclines. *Earth Planet. Sci. Letters*, **5**, 429–438.

Dachille, F., and R. Roy, 1956, System Mg_2SiO_4-Mg_2GeO_4 at 10,000, 60,000, and about 300,000 psi. *Bull. Geol. Soc. Am.* **67**, 1682–1683.

———, 1960, High pressure studies of the system Mg_2GeO_4-Mg_2SiO_4 with special reference to the olivine-spinel transition. *Am. J. Sci.*, **258**, 225–246.

Dagley, P., R. L. Wilson, J. M. Ade-Hall, G. P. L. Walker, S. E. Haggerty, T. Sigurgeirsson, N. D. Watkins, P. J. Smith, J. Edwards, and R. L. Grasty, 1967, Geomagnetic polarity zones for Icelandic lavas. *Nature*. **216**, 25–29.

Dalrymple, G. B., A. Cox, R. R. Doell, and C. S. Grommé, 1967, Pliocene geomagnetic polarity epochs. *Earth Planet. Sci. Letters*, **2**, 163–173.

Dana, J. D., 1873, On some results of the earth's contraction from cooling, including a discussion of the origin of mountains and the nature of the earth's interior. *Am. J. Sci.*, **5**, 423–443; **6**, 6–14, 104–115, 161–171.

Dennis, J. G., and C. T. Walker, 1965, Earthquakes resulting from metastable phase transitions. *Tectonophysics*, **2**, 401–407.

Dewey, J. F., and J. M. Bird, 1970, Mountain belts and the new global tectonics. *J. Geophys. Res.*, **75**, 2625–2647.

Dickey, J. S., 1970, "Partial fusion products in alpine-type peridotites: Serrania de la Ronda and other examples," in *Fiftieth Anniversary Symposia*, ed. B. A. Morgan. Mineralogical Society of America, Special Paper No. 3.

Dickinson, W. R., 1970, Global tectonics. *Science*, **168**, 1250–1259.

Dickson, G. O., W. C. Pitman, and J. R. Heirtzler, 1968, Magnetic anomalies in the South Atlantic and ocean floor spreading. *J. Geophys. Res.*, **73**, 2087–2100.

Dietz, R. S., 1961, Continent and ocean basin evolution by spreading of the sea floor. *Nature*, **190**, 854–857.

———, 1963, Alpine serpentines as oceanic rind fragments. *Geol. Soc. Am. Bull.*, **74**, 947–952.

———, 1963, Collapsing continental rises: an actualistic concept of geosynclines and mountain building. *J. Geol.*, **71**, 314–333.

———, and J. C. Holden, 1970, Reconstruction of Pangaea: breakup and dispersion of continents, Permian to present. *J. Geophys. Res.*, **75**, 4939–4956.

———, 1970, The breakup of Pangaea. *Sci. Am.*, **223**, 30–41.

Dietz, R. S., and W. P. Sproll, 1970, Fit between Africa and Antarctica: a continental drift reconstruction. *Science*, **167**, 1612–1614.

Doig, R., 1969, An alkaline rock province linking Europe and North America. Programme and abstracts for joint Annual Meeting Geological Society of Canada and Mineralogical Association of Canada, Montreal, June 1969.

Dorman, J., 1969, "Seismic surface-wave data on the upper mantle", in *The Earth's Crust and Upper Mantle*, ed. P. J. Hart. Geophysical Monograph 13, American Geophysical Union, Washington, D.C.

Drake, C. L., and J. E. Nafe, 1968, "The transition from ocean to continent from seismic refraction data", in *The Crust and Upper Mantle of the Pacific Area*, eds. L. Knopoff, C. L. Drake, and P. J. Hart. Geophysical Monograph 12, American Geophysical Union, Washington, D.C.

Drake, C. L., J. I. Ewing, and H. P. Stockard, 1968, The continental margin of the eastern United States. *Can. J. Earth Sci.*, **5**, 993–1010.

Drever, H. I., and R. Johnston, 1967, Picritic minor intrusions, 71-82, in *Ultramafic and Related Rocks*, ed. P. J. Wyllie, John Wiley and Sons, New York.

DuToit, A., 1937, *Our Wandering Continents*, Oliver and Boyd, Edinburgh.

Egyed, L., 1957, A new dynamic conception of the internal constitution of the Earth. *Geol. Rundschau*, **46**, 101–121.

Emilia, D. A., and D. F. Heinrichs, 1969, Ocean floor spreading: Olduvai and Gilsa events in the Matuyama epoch. *Science*, **166**, 1267–1269.

Engel, A. E. J., 1963, Geologic evolution of North America. *Science*, **140**, 143–152.

———, C. G. Engel, and R. G. Havens, 1965, Chemical characteristics of oceanic basalts and the upper mantle. *Geol. Soc. Am. Bull.*, **76**, 719–734.

Ericson, D. B., and G. Wollin, 1968, Pleistocene climates and chronology in deep-sea sediments. *Science*, **162**, 1227–1234.

Ernst, W. G., 1970, Tectonic contact between the Franciscan mélange and the Great Valley sequence—crustal expression of a late Mesozoic Benioff zone. *J. Geophys. Res.*, **75**, 886–901.

Eskola, P., 1915, On the relations between the chemical and mineralogical composition in the metamorphic rocks of the Orijarvi region. *Bull. Comm. Geol. Finlande*, **44**, 109–145.

Evans, B. W., 1965, Application of a reaction rate method to the breakdown equilibria of muscovite and muscovite plus quartz. *Am. J. Sci.*, **263**, 647–667.

Ewing, J., and M. Ewing, 1967, Sediment distribution on the mid-ocean ridges with respect to spreading of the sea floor. *Science*, **156**, 1590–1592.

———, T. Aitken, and W. J. Ludwig, 1968, "North Pacific sediment layers measured by seismic profiling," in *The Crust and Upper Mantle of the Pacific Area*, eds. L. Knopoff, C. L. Drake, and P. J. Hart. Geophysical Monograph 12, American Geophysical Union, Washington, D.C.

Ewing, M., and F. Press, 1955, "Geophysical contrasts between continents and ocean basins," in *Crust of the Earth*, ed. A. Poldervaart. The Geological Society of America, Special Paper 62.

Fermor, L. L., 1913, Preliminary note on garnet as a geological barometer and on an infra-plutonic zone in the earth's crust. *Geol. Survey India*, **43**, Part 1.

Fischer, A. G., B. C. Heezen, R. E. Boyce, D. Bukry, R. G. Douglas, R. E. Garrison, S. A. Kling, V. Krasheninnikov, and A. C. Pimm, 1970, Geological history of the Western North Pacific. *Science*, **6**, 1210–1214.

Fisher, D. E., O. Joensuu, and K. Boström, 1969, Elemental abundances in ultramafic rocks and their relation to the upper mantle. *J. Geophys. Res.*, **74**, 3865–3873.

Fisher, O., 1889, *Physics of the Earth's Crust*, 2nd ed., Macmillan, London.

Fitch, T. J., and P. Molnar, 1970, Focal mechanisms along inclined earthquake zones in the Indonesia-Philippine region. *J. Geophys. Res.*, **75**, 1431–1444.

Foster, J. H., and N. D. Opdyke, 1970, Upper Miocene to Recent magnetic stratigraphy in deep-sea sediments. *J. Geophys. Res.*, **75**, 4465–4473.

Fujisawa, H., 1968, Temperature and discontinuities in the transition layer within the Earth's mantle: geophysical application of the olivine-spinel transition in the Mg_2SiO_4-Fe_2SiO_4 system. *J. Geophys. Res.*, **73**, 3281–3294.

Fyfe, W. S., 1960, Hydrothermal synthesis and determination of equilibrium between minerals in the subsolidus region. *J. Geol.*, **68**, 553–566.

Gast, P. W., 1968, Trace element fractionation and the origin of tholeiitic and alkaline magma types. *Geochim. Cosmochim. Acta*, **42**, 1057–1068.

Gilluly, J., 1969, Oceanic sediment volumes and continental drift. *Science*, **166**, 992–993.

Girdler, R. W., 1967, "A review of terrestrial heat flow," in *Mantles of the Earth and Terrestrial Planets*, ed. S. K. Runcorn. Wiley-Interscience, New York.

Goles, G. G., 1969, "Cosmic abundances," in *Handbook of Geochemistry*, Vol. 1., ed. K. H. Wedepohl. Springer-Verlag, Heidelberg.

Green, D. H., 1964, The petrogenesis of the high temperature peridotite intrusion in the Lizard area, Cornwall. *J. Petrol.*, **5**, 134–188.

———, 1967, "Effects of high pressure on basaltic rock," in *Basalts: The Poldervaart Treatise on Rocks of Basaltic Composition*, Vol. 1, eds. H. H. Hess and A. Poldervaart. Wiley-Interscience, New York.

———, 1969, The origin of basaltic and nephelinitic magmas in the earth's mantle. *Tectonophysics*, **7**, 409–422.

———, 1971, Compositions of basaltic magmas as indicators of conditions of origin: application to oceanic volcanism. *Phil. Trans. Roy. Soc. London*, Ser. A, **268**, 707–725.

———, and A. E. Ringwood, 1963, Mineral assemblages in a model mantle composition. *J. Geophys. Res.*, **68**, 937–945.

———, 1964, Fractionation of basalt magmas at high pressures. *Nature*, **201**, 1276–1279.

———, 1967, The stability fields of aluminous pyroxene peridotite and garnet peridotite and their relevance in upper mantle structure. *Earth Planet. Sci. Letters*, **3**, 151–160.

———, 1967, The genesis of basaltic magmas. *Contr. Mineral. Petrology*, **15**, 103–190.

————, 1967, An experimental investigation of the gabbro to eclogite transformation and its petrological applications. *Geochim. Cosmochim. Acta*, **31**, 767–833.

————, 1968, Genesis of the calc-alkaline igneous rock suite. *Contr. Mineral. Petrol.*, **18**, 105–162.

Grommé, C. S., R. T. Merrill, and J. Verhoogen, 1967, Paleomagnetism of the Sierra Nevada, California, and its significance for polar wandering and continental drift. *J. Geophys. Res.*, **72**, 5661–5684.

Guier, W. H., and R. R. Newton, 1965, The earth's gravity field as deduced from the Doppler tracking of five satellites. *J. Geophys. Res.*, **70**, 4613–4626.

Gutenberg, B., 1926, Untersuchungen zur Frage, bis zu welcher tiefe die erde Kristallin ist. *Zeit. Geophysik.*, **2**, 24–29.

Hales, A. L., 1969, Gravitational sliding and continental drift. *Earth Planet. Sci. Letters*, **6**, 31–34.

Hall, J., 1859, Description and figures of the organic remains of the lower Helderberg Group and the Oriskany Sandstone. Natural History of New York; paleontology. *Geol. Surv.*, **3**, Albany, New York.

Hamilton, W., 1969, Mesozoic California and the underflow of Pacific mantle. *Geol. Soc. Am. Bull.*, **80**, 2409–2430.

————, 1970, The Uralides and the motion of the Russian and Siberian platforms. *Bull. Geol. Soc. Am.*, **81**, 2553–2576.

Harris, P. G., and J. A. Rowell, 1960, Some geochemical aspects of the Mohorovicic Discontinuity. *J. Geophys. Res.*, **65**, 2443–2459.

Harris, P. G., A. Reay, and I. G. White, 1967, Chemical composition of the upper mantle. *J. Geophys. Res.*, **72**, 6359–6369.

Harrison, C. G. A., and B. M. Funnell, 1964, Relationship of paleomagnetic reversals and micropaleontology in two late Caenozoic cores from the Pacific Ocean. *Nature*, **204**, 566.

Hatherton, T., and W. R. Dickinson, 1969, The relationship between andesitic volcanism and seismicity in Indonesia, the Lesser Antilles, and other island arcs. *J. Geophys. Res.*, **74**, 5301–5310.

Haug, E., 1900, Les géosynclinause et les aires continentales. Contribution à l'étude des régressions et des trangressions marines. *Bull. Soc. Geol. France*, **28**, 617–711.

Heezen, B. C., M. Tharp, and M. Ewing, 1959, *The Floors of the Oceans. I. The North Atlantic.* The Geological Society of America, Special Paper 65.

Heirtzler, J. R., X. Le Pichon, and J. G. Baron, 1966, Magnetic anomalies over the Reykjanes Ridge. *Deep-Sea Res.*, **13**, 427–443.

Heirtzler, J. R., G. O. Dickson, E. J. Herron, W. C. Pitman, and X. Le Pichon, 1968, Marine magnetic anomalies, geomagnetic field reversals, and motions of the ocean floor and continents. *J. Geophys. Res.*, **73**, 2119–2136.

Herrin, E., 1969, "Regional variation of P-wave velocity in the upper mantle beneath North America," in *The Earth's Crust and Upper Mantle*, ed. P. J. Hart. Geophysical Monograph 13, American Geophysical Union, Washington, D.C.

Herz, N., 1969, Anorthosite belts, continental drift, and the anorthosite event. *Science*, **164**, 944–947.

Hess, H. H., 1955, "Serpentines, Orogeny, and Epeirogeny," in *Crust of the Earth*, Geological Society of America, Special Paper 62, 391–408.

————, 1962, "History of ocean basins," in *Petrologic Studies: A volume to Honor A. F. Buddington*, eds. A. E. J. Engel, H. L. James, and B. F. Leonard. Geological Society of America, New York.

————, 1964, "The oceanic crust, the upper mantle and the Mayaguez serpentinized peridotite," in *A Study of the Serpentinite*, ed. C. A. Burk. National Academy of Science-National Research Council Publication 1188.

————, 1964, The oceanic crust, the upper mantle and the Mayaguez serpentinized peridotite, in *A Study of the Serpentinite*, ed. C. A. Burk. National Academy of Science-National Research Council Publication 1188.

————, and G. Otalora, 1964, "Mineralogical and chemical composition of the Mayaguez serpentinite cores," in *A Study of Serpentinite*, ed. C. A. Burk. National Academy of Science-National Research Council Publication 1188.

Hibberd, F. H., 1962, An analysis of the positions of the Earth's magnetic pole in the geological past. *Geophys. J.*, **6**, 221–244.

Hilgenberg, O. C., 1962, Rock magnetism and the Earth's paleopoles. *Geofis. Pura. Appl.*, **53**, 52–54.

Hill, R. E. T., and A. L. Boettcher, 1970, Water in the earth's mantle: melting curves of basalt-water and basalt-water-carbon dioxide. *Science*, **167**, 980–981.

Holloway, J. R., and C. W. Burnham, 1969, Phase relations and compositions in basalt-H_2O-CO_2 under the Ni-NiO buffer at high temperatures and pressures. Abstracts with Programs for 1969, Part 7, The Geological Society of America, 104–105.

Holmes, A., 1931, Radioactivity and earth movements. *Trans. Geol. Soc. Glasgow*, **18**, 559–606.

———, 1965, *Principles of Physical Geology*, 2nd ed., Ronald, New York.

Hospers, J., 1967, "Review of paleomagnetic evidence for the displacement of continents, with particular reference to North America and Europe-northern Asia," in *Mantles of the Earth and Terrestrial Planets*, ed. S. K. Runcorn. Wiley-Interscience, New York.

Hsu, L. C., 1968, Selected phase relationships in the system Al-Mn-Fe-Si-O-H: a model for garnet equilibria. *J. Petrol.*, **9**, 40–83.

Hurley, P. M., and J. R. Rand, 1969, Pre-drift continental nuclei. *Science*, **164**, 1229–1242.

Hurley, P. M., F. F. M. de Almeida, G. C. Melcher, V. G. Cordani, J. R. Rand, K. Kawashita, P. Vandoros, W. H. Pinson, and H. W. Fairbairn, 1967, Test of continental drift by comparison of radiometric ages. *Science*, **157**, 495–500.

Hyndman, R. D., I. B. Lambert, K. S. Heier, J. C. Jaeger, and A. E. Ringwood, 1968, Heat flow and surface radioactivity measurements in the Precambrian shield of Western Australia. *Phys. Earth Planet. Interiors*, **1**, 129–135.

Irvine, T. N., 1966, ed. *The World Rift System*, Geological Survey Canada Paper 66-14, Queen's Printer, Ottawa.

Irving, E., 1964, *Paleomagnetism*, Wiley, New York.

Isacks, B., J. Oliver, and L. R. Sykes, 1968, Seismology and the new global tectonics. *J. Geophys. Res.*, **73**, 5855–5899.

———, 1969, *J. Geophys. Res.*, **74**, 2789–2790.

Ito, K., and G. C. Kennedy, 1967, Melting and phase relations in a natural peridotite to 40 kilobars. *Am. Jour. Sci.*, **265**, 519–538.

———, 1968, Melting and phase relations in the plane tholeiite-lherzolite-nepheline basanite to 40 kilobars with geological implications. *Contr. Mineral. Petrol.*, **19**, 177–211.

———, 1970, "The fine structure of the basalt-eclogite transition," in *Fiftieth Anniversary Symposia*, ed. B. A. Morgan. Mineralogical Society of America, Special Paper 3.

Jackson, E. D., 1969, Discussion on the paper "The origin of ultramafic and ultrabasic rocks" by P. J. Wyllie. *Tectonophysics*, **7**, 517–518.

Jacobs, J. A., R. D. Russell, and J. T. Wilson, 1959, *Physics and Geology*, McGraw-Hill, New York.

Jeffreys, H., 1936, The structure of the earth down to the 20° discontinuity. *Monthly Notices Roy. Astron. Soc. Geophys. Supplement*, **3**, 401–422.

Johannes, W., 1968, Experimental investigation of the reaction forsterite + H_2O = serpentine + brucite. *Contr. Mineral. Petrol.*, **19**, 309–315.

Joyner, W. B., 1967, Basalt-eclogite transition as a cause for subsidence and uplift. *J. Geophys. Res.*, **72**, 4977–4998.

Karig, D. E., 1970, Ridges and basins of the Tonga-Kermadec island arc system. *J. Geophys. Res.*, **75**, 239–254.

Kaula, W. M., 1968, *An Introduction to Planetary Physics*. Wiley, New York.

Kay, M., 1951, North American geosynclines. *Geol. Soc. Am. Memoir* **48**.

Kay, R., N. J. Hubbard, and P. W. Gast, 1970, Chemical characteristics and origin of oceanic ridge volcanic rocks. *J. Geophys. Res.*, **75**, 1585–1613.

Keays, R. R., R. Ganapathy, J. C. Laul, E. Anders, G. F. Herzog, and P. M. Jeffery, 1970, Trace elements and radioactivity in lunar rocks: implications for meteorite infall, solar-wind flux, and formation conditions of moon. *Science*, **167**, 490–493.

Kennedy, G. C., 1959, The origin of continents, mountain ranges, and ocean basins. *Am. Sci.*, **47**, 491–504.

———, G. J. Wasserburg, H. C. Heard, and R. C. Newton, 1962, The upper three-phase region in the system SiO_2-H_2O. *Am. J. Sci.*, **260**, 501–521.

Kennedy, W. Q., 1933, Trends of differentiation in basaltic magmas. *Am. J. Sci.*, **25**, 239–256.

———, and E. M. Anderson, 1938, Crustal layers and the origin of magmas. *Bull. Volcanol.*, **2-3**, 23–82.

Kitahara, S., S. Takenouchi, and G. C. Kennedy, 1966, Phase relations in the system MgO-SiO_2-H_2O at high temperatures and pressures. *Am. J. Sci.*, **264**, 223–233.

Knopoff, L., 1967, Thermal convection in the earth's mantle, in *The Earth's Mantle*, ed. T. F. Gaskell. Academic, New York.

Kornprobst, J., 1969, Le massif ultrabasique des Beni Bouchera (Rif Interne, Maroc): Etude des péridotites de haute température et de haute pression, et des pyroxénolites, à grenat ou sans grenat, qui leur sont associées. *Contr. Mineral. Petrol.*, **23**, 283–322.

Kosminskaya, I. P., and S. M. Zverev, 1968, "Deep seismic soundings in the transition zones from continents to oceans," in *The Crust and Upper Mantle of the Pacific Area*, eds. L. Knopoff, C. L. Drake, and P. J. Hart. Geophysical Monograph 12, American Geophysical Union, Washington, D.C.

Kosminskaya, I. P., N. A. Belyaevsky, and I. S. Volvovsky, 1969, "Explosion seismology in the USSR," in *The Earth's Crust and Upper Mantle*, ed. P. J. Hart. Geophysical Monograph 13, American Geophysical Union, Washington, D.C.

Krause, D. C., 1966, "Equatorial shear zone," in *The World Rift System*, ed. T. N. Irvine. Geology Survey of Canada, Paper 66-14, Queen's Printer, Ottawa.

Krumbein, W. C., and L. L. Sloss, 1963, *Stratigraphy and Sedimentation*, 2nd ed. Freeman, San Francisco.

Kuno, H., 1959, Origin of Cenozoic petrographic provinces of Japan and surrounding areas. *Bull. Volcanol.* **2-20**, 37–76.

Kushiro, I., 1968, Compositions of magmas formed by partial zone melting of the Earth's upper mantle. *J. Geophys. Res.*, **73**, 619–634.

———, 1969, "Clinopyroxene Solid Solutions Formed by Reactions Between Diopside and Plagioclase at High Pressures" in Mineralogical Society of America, Special Paper No. 2.

———, 1969, The system forsterite-diopside-silica with and without water at high pressures. *Amer. J. Sci.*, **267A**, 269–294.

———, 1969, Discussion of the paper "The origin of basaltic and nephelinitic magmas in the earth's mantle" by D. H. Green. *Tectonophysics*, **7**, 427–436.

———, 1970, Systems bearing on melting of the upper mantle under hydrous conditions. Carnegie Inst. Wash. Yearbook, **68**, 240–245.

———, 1970, Stability of amphibole and phlogopite in the upper mantle. Carnegie Inst. Wash. Yearbook, **68**, 245–247.

———, and H. S. Yoder, 1966, Anorthite-forsterite and anorthite-enstatite reactions and their bearing on the basalt-eclogite transformation. *J. Petrology*, **7**, 337–362.

Kushiro, I., Y. Syono, and S. Akimoto, 1968, Melting of a peridotite nodule at high pressures and high water pressures. *J. Geophys. Res.*, **73**, 6023–6029.

Lachenbruch, A. H., 1970, Crustal temperature and heat production: implications of the linear heat-flow relation. *J. Geophys. Res.*, **75**, 3291–3300.

Lambert, I. B., and K. S. Heier, 1967, The vertical distribution of uranium, thorium, and potassium in the continental crust. *Geochim. Cosmochim. Acta*, **31**, 377–390.

Lambert, I. B., and P. J. Wyllie, 1968, Stability of hornblende and a model for the low velocity zone. *Nature*, **219**, 1240–1241.

———, 1970, Melting in the deep crust and upper mantle and the nature of the low velocity layer. *Phys. Earth Planet. Interiors*, **3**, 316–322.

———, 1970, Low-velocity zone of the earth's mantle: incipient melting caused by water. *Science*, **169**, 764–766.

Lambert, I. B., J. K. Robertson, and P. J. Wyllie, 1969, Melting reactions in the system $KAlSi_3O_8$-SiO_2-H_2O to 18.5 kilobars. *Am. J. Sci.*, **267**, 609–626.

Larimer, J. W., and E. Anders, 1967, Chemical fractionations in meteorites—II. Abundance patterns and their interpretation. *Geochim. Cosmochim. Acta*, **31**, 1239-1270.

Larson, R. L., and F. N. Spiess, 1969, East Pacific Rise crest: a near-bottom geophysical profile. *Science*, **163**, 68–71.

Lee, W. H. K., and S. Uyeda, 1965, "Review of heat flow data," in *Terrestrial Heat Flow*, ed. W. H. K. Lee. Geophysical Monograph 8, American Geophysical Union, Washington D.C.

Le Pichon, X., 1968, Sea-floor spreading and continental drift. *J. Geophys. Res.*, **73**, 3611–3697.

———, and J. R. Heirtzler, 1968, Magnetic anomalies in the Indian Ocean and sea-floor spreading. *J. Geophys. Res.*, **73**, 2101–2117.

Lliboutry, L., 1969, Sea-floor spreading, continental drift and lithosphere sinking with an asthenosphere at melting point. *J. Geophys. Res.*, **74**, 6525–6540.

Lovering, J. F., 1958, The nature of the Mohorovicic Discontinuity. *Trans. Am. Geophys. Un.*, **39**, 947–955.

———, 1962, "The evolution of the meteorites—evidence for the coexistence of chondritic, achondritic, and iron meteorites in a typical parent meteorite body," in *Researches on Meteorites*, ed. C. B. Moore, 179–197.

Lubimova, E. A., 1967, "Theory of thermal state of the earth's mantle," in *The Earth's Mantle*, ed. T. F. Gaskell. Academic, New York.

Luyendyk, B. P., 1969, Origin of short-wavelength magnetic lineations observed near the ocean bottom. *J. Geophys. Res.*, **74**, 4869–4881.

———, 1970, Reply, *J. Geophys. Res.*, **75**, 6721–6722.

———, J. D. Mudie, and C. G. A. Harrison, 1968, Lineations of magnetic anomalies in the northeast Pacific observed near the ocean floor. *J. Geophys. Res.*, **73**, 5951–5957.

Lyell, C., 1833, *Principles of Geology*, Murray, London.

McBirney, A. R., 1969, "Compositional variations in Cenozoic calc-alkaline suites of Central America," in *Oregon Dep. Geol. Mineral Ind. Bull.*, **65**, 185–189.

McElhinny, M. W., and G. R. Luck, 1970, Paleomagnetism and Gondwanaland. *Science*, **168**, 830–832.

McKenzie, D. P., 1967, Some remarks on heat flow and gravity anomalies. *J. Geophys. Res.*, **72**, 6261–6273.

———, 1969, Speculations on the consequences and causes of plate motions. *Geophys. J.*, **18**, 1–32.

———, and W. J. Morgan, 1969, Evolution of triple junctions. *Nature*, **224**, 125–133.

McKenzie, D. P. and R. L. Parker, 1967, The North Pacific: an example of tectonics on a sphere. *Nature*, **216**, 1276–1280.

MacDonald, G. J. F., 1963, The deep structure of continents. *Rev. Geophys.*, **1**, 587–665.

———, and N. F. Ness, 1960, Stability of phase transitions within the earth. *J. Geophys. Res.*, **65**, 2173–2190.

MacGregor, I. D., 1964, The reaction 4 Enstatite + spinel = forsterite + pyrope. *Carnegie Inst. Wash. Yearbook*, **63**, 157.

Malahoff, A., 1970, Some possible mechanisms for gravity and thrust faults under oceanic trenches. *J. Geophys. Res.*, **75**, 1992–2001.

Markhinin, E. K., 1968, "Volcanism as an agent of formation of the Earth's crust," in *The Crust and Upper Mantle of the Pacific Area*, eds. L. Knopoff, C. L. Drake, and P. J. Hart. Geophysical Monograph 12, American Geophysical Union, Washington, D.C.

Mason, B., 1965, The chemical composition of olivine-bronzite and olivine-hypersthene chondrites. *Am. Museum Novitates*, 2223, 1–38.

———, 1966, Composition of the earth. *Nature*, **211**, 616–618.

Mason, R. G., 1958, A magnetic survey off the west coast of the United States between latitudes 32° and 36°N, longitudes 121° and 128°W. *Geophys. J.*, **1**, 320–329.

———, and A. D. Raff, 1961, Magnetic survey off the west coast of North America, 32°N latitude to 42°N latitude. *Bull. Geol. Soc. Am.*, **72**, 1259–1266.

Matsumoto, T., 1967, Fundamental problems in the circum-Pacific orogenesis. *Tectonophysics*, **4**, 595–613.

Maxwell, A. E., R. P. Von Herzen, K. J. Hsu, J. E. Andrews, T. Saito, S. F. Percival, E. D. Milow, and R. E. Boyce, 1970, Deep-sea drilling in the South Atlantic. *Science*, **168**, 1047–1059.

Maxwell, J. C., 1968, Continental drift and a dynamic earth. *Am. Sci.*, **56**, 35–51.

———, 1970, "The Mediterranean, ophiolites, and continental drift," in *What's New on Earth*. Rutgers University Press, New Jersey.

Maynard, G. L., 1970, Crustal layer of seismic velocity 6.9 to 7.6 kilometers per second under the deep oceans. *Science*, **168**, 120–121.

Melson, W. G., E. Jarosewich, V. T. Bowen, and G. Thompson, 1967, St. Peter and St. Paul Rocks: a high-temperature, mantle-derived intrusion. *Science*, **155**, 1532–1535.

Menard, H. W., 1964, *Marine Geology of the Pacific*, McGraw-Hill, New York.

———, 1967, Sea-floor spreading, topography, and the second layer. *Science*, **157**, 923–924.

———, 1967, Transitional types of crust under small ocean basins. *J. Geophys. Res.*, **72**, 3061–3073.

———, 1968, "Some remaining problems in sea-floor spreading," in *The History of the Earth's Crust*, ed. R. A. Phinney. Princeton University Press, New Jersey.

———, and T. Atwater, 1968, Changes in direction of sea-floor spreading. *Nature*, **219**, 463–467.

Menard, H. W., and S. M. Smith, 1966, Hypsometry of ocean basin provinces. *J. Geophys. Res.*, **71**, 4305–4325.

Merrill, R. B., J. K. Robertson, and P. J. Wyllie, 1970, Melting reactions in the system $NaAlSi_3O_8$-$KAlSi_3O_8$-SiO_2-H_2O to 20 kilobars compared with results for other feldspar-quartz-H_2O and rock-H_2O systems. *J. Geol.*, **78**, 558–569.

Meservey, R., 1969, Topological inconsistency of continental drift on the present-sized earth. *Science*, **166**, 609–611.

Meyerhoff, A. A., 1970, Continental drift: implications of paleomagnetic studies, meteorology, physical oceanography and climatology. *J. Geol.*, **78**, 1–51.

———, 1970, Continental drift, II: high-latitude evaporite deposits and geologic history of Arctic and North Atlantic Oceans. *J. Geol.*, **78**, 406–444.

———, and C. Teichert, 1971, Continental drift, III: late Paleozoic glacial centers, and Devonian-Eocene coal distribution. *J. Geol.*, **79**, in press.

Millhollen, G. L., 1971, Melting of nepheline syenite with H_2O and $H_2O + CO_2$, and the effect of dilution of the aqueous phase on the beginning of melting. *Am. J. Sci.*, **269**, in press.

Minear, J. W., and M. N. Toksöz, 1970, Thermal regime of a downgoing slab and new global tectonics. *J. Geophys. Res.*, **75**, 1397–1419.

Mitchell, A. H., and H. G. Reading, 1969, Continental margins, geosynclines, and ocean floor spreading. *J. Geol.*, **77**, 629–646.

Miyashiro, A., 1961, Evolution of metamorphic belts. *J. Petrol.*, **2**, 277–311.

———, F. Shido, and M. Ewing, 1969, Composition and origin of serpentinites from the mid-Atlantic ridge near 24° and 30° North latitude. *Contr. Mineral. Petrol.*, **23**, 117–127.

Molnar, P., and J. Oliver, 1969, Lateral variations of attenuation in the upper mantle and discontinuities in the lithosphere. *J. Geophys. Res.*, **74**, 2648–2682.

Molnar, P., and L. R. Sykes, 1969, Tectonics of the Caribbean and Middle America regions from focal mechanisms and seismicity. *Geol. Soc. Am. Bull.*, **80**, 1639–1684.

Morgan, W. J., 1968, Rises, trenches, great faults, and crustal blocks. *J. Geophys. Res.*, **73**, 1959–1982.

Morley, L. W., and A. Larochelle, 1964, "Paleomagnetism as a means of dating geological events," in *Geochronology in Canada*, ed. F. F. Osborne. The Royal Society of Canada Special Publications, 8. University of Toronto Press, Toronto.

Muehlberger, W. R., R. E. Denison, and E. G. Lidiak, 1967, Basement rocks in continental interior of United States. *Amer. Assoc. Petroleum Geol. Bull.*, **51**, 2351–2380.

Nagata, T., 1952, Reverse thermo remanent magnetism. *Nature*, **169**, 704–705.

Newton, R. C., 1966, Some calc-silicate equilibrium relations. *Am. J. Sci.*, **264**, 204–222.

Nicholls, G. D., 1967, "Geochemical studies in the ocean as evidence for the composition of the mantle," in *Mantles of the Earth and Terrestrial Planets*, ed. S. K. Runcorn. Wiley-Interscience, New York.

Ninkovich, D., N. Opdyke, B. C. Heezen, and J. H. Foster, 1966, Paleomagnetic stratigraphy, rates of deposition and tephrochronology in North Pacific deep-sea sediments. *Earth Planet. Sci. Letters*, **1**, 476–492.

O'Connell, R. J., 1968, Critique of a paper by W. J. van de Lindt, "Movement of the Mohorovicic discontinuity under isostatic conditions." *J. Geophys. Res.*, **73**, 6604–6066.

———, and G. J. Wasserburg, 1967, Dynamics of the motion of a phase change boundary to changes in pressure. *Rev. Geophys.*, **5**, 329–410.

O'Hara, M. J., 1965, Primary magmas and the origin of basalts. *Scottish J. Geol.*, **1**, 19–40.

———, 1968, The bearing of phase equilibria studies in synthetic and natural systems on the origin and evolution of basic and ultrabasic rocks. *Earth Sci. Rev.*, **4**, 69–133.

———, and H. S. Yoder, 1967, Formation and fractionation of basic magmas at high pressures. *Scottish J. Geol.*, **3**, 67–117.

Opdyke, N. D., B. Glass, J. D. Hayes, and J. Foster, 1966, Paleomagnetic study of Antarctic deep-sea cores. *Science*, **154**, 349–357.

Osborn, E. F., 1959, Role of oxygen pressure in the crystallization and differentiation of basaltic magma. *Am. J. Sci.*, **257**, 609–647.

———, 1969, The complementariness of orogenic andesite and alpine peridotite. *Geochim. Cosmochim. Acta*, **33**, 307–324.

Oxburgh, E. R., and D. L. Turcotte, 1968, Mid-ocean ridges and geotherm distribution during mantle convection. *J. Geophys. Res.*, **73**, 2643–2661.

———, 1970, Thermal structure of island arcs. *Geol. Soc. Am. Bull.*, **81**, 1665–1688.

Ozima, M., M. Ozima, and I. Kaneoka, 1968, Potassium-argon ages and magnetic properties of some dredged submarine basalts and their geophysical implications. *J. Geophys. Res.*, **73**, 711–723.

Pakiser, L. C., 1965, The basalt-eclogite transformation and crustal structure in the western United States. *U.S. Geol. Survey Prof. Paper*, **525-B**, B1–B8.

———, and R. Robinson, 1966, "Composition of the continental crust as estimated from seismic observations," in *The Earth Beneath the Continents*, eds. J. S. Steinhart and T. J. Smith. Geophysical Monograph 10, American Geophysical Union, Washington, D.C.

Pakiser, L. C., and I. Zietz, 1965, Transcontinental crustal and upper-mantle structure. *Rev. Geophys.*, **3**, 505–520.

Peter, G., 1970, Discussion of paper by B. P. Luyendyk, *J. Geophys. Res.*, **75**, 6717–6720.

———, and R. Lattimore, 1969, Magnetic structure of the Juan de Fuca-Gorda Ridge area. *J. Geophys. Res.*, **74**, 586–593.

Peter, G., B. H. Erickson, and P. J. Grim, 1970, "Magnetic structure of the Aleutian trench and northeast Pacific basin," in *The Sea*, Vol. 4, eds. A. E. Maxwell, E. C. Bullard, E. Goldberg, and J. L. Worzel, Wiley-Interscience, New York.

Phillips, J. D., and B. P. Luyendyk, 1970, Central North Atlantic plate motions over the last 40 million years. *Science*, **170**, 727–729.

Phinney, R. A., 1968, ed., *The History of the Earth's Crust*. Princeton University Press, New Jersey.

Pitman, W. C., and J. R. Heirtzler, 1966, Magnetic anomalies over the Pacific-Antarctic ridge. *Science*, **154**, 1164–1171.

Pitman, W. C., E. M. Herron, and J. R. Heirtzler, 1968, Magnetic anomalies and sea-floor spreading. *J. Geophys. Res.*, **73**, 2069–2085.

Piwinskii, A. J., 1968, Experimental studies of igneous rock series: central Sierra Nevada Batholith, California. *J. Geol.*, **76**, 548–570.

———, and P. J. Wyllie, 1968, Experimental studies of igneous rock series: a zoned pluton in the Wallowa Batholith, Oregon. *J. Geol.*, **76**, 205–234.

Poldervaart, A., 1955, "Chemistry of the Earth's Crust," in *Crust of the Earth*, ed. A. Poldervaart, Geological Society of America, Special Paper 62.

Press, F., 1968, Density distribution in the earth. *Science*, **160**, 1218–1221.

———, 1968, Earth models obtained by Monte Carlo inversion. *J. Geophys. Res.*, **73**, 5223–5234.

———, 1969, The suboceanic mantle. *Science*, **165**, 174–176.

Raff, A. D., 1968, Sea-floor spreading—another rift. *J. Geophys. Res.*, **73**, 3699–3705.

———, and R. G. Mason, 1961, Magnetic survey off the west coast of North America, 40°N latitude to 52°N latitude. *Bull. Geol. Soc. Am.*, **72**, 1267–1270.

Ramberg, H., 1967, *Gravity, Deformation and the Earth's Crust*. Academic, London.

Rezanov, I. A., 1968, Paleomagnetism and continental drift. *Intern. Geol. Rev.*, **10**, 765–776.

Richter, C. F., 1969, Comments on paper by B. Isacks, J. Oliver, and L. R. Sykes, "Seismology and the new global tectonics," *J. Geophys. Res.*, **74**, 2786–2788.

Ringwood, A. E., 1958, The constitution of the mantle—I. Thermodynamics of the olivine-spinel transition. *Geochim. Cosmochim. Acta*, **13**, 303–321.

———, 1958, The constitution of the mantle—II. Further data on the olivine-spinel transition. *Geochim. Cosmochim. Acta*, **15**, 19–28.

———, 1958, The constitution of the mantle—III. Consequences of the olivine-spinel transition. *Geochim. Cosmochim. Acta*, **15**, 195–212.

———, 1961, Chemical and genetic relationships among meteorites. *Geochim. Cosmochim. Acta*, **24**, 159–197.

———, 1962, A model for the upper mantle. *J. Geophys. Res.*, **67**, 857–868.

———, 1966, Genesis of chondritic meteorites. *Rev. Geophys.*, **4**, 113–175.

———, 1966, "The chemical composition and origin of the earth," in *Advances in Earth Sciences*, ed. P. M. Hurley. M.I.T. Press, Cambridge, Mass.

———, 1966, "Mineralogy of the mantle," in *Advances in Earth Sciences*, ed. P. M. Hurley. M.I.T. Press, Cambridge, Mass.

———, 1969, "Composition and evolution of the upper mantle," in *The Earth's Crust and Upper Mantle*, ed. P. J. Hart. Geophysics Monograph 13, American Geophysical Union, Washington, D.C.

———, 1969, Phase transformations in the mantle. *Earth Planet. Sci. Letters*, **5**, 401–412.

———, 1970, Phase transformations and the constitution of the mantle. *Phys. Earth Planet. Interiors*, **3**, 109–155.

———, and D. H. Green, 1964, Experimental investigations bearing on the nature of the

Mohorovicic discontinuity. *Nature*, **201**, 566–567.

———, 1966, An experimental investigation of the gabbro-eclogite transformation and some geophysical implications. *Tectonophysics*, **3**, 383–427.

———, 1966, "Petrological nature of the stable continental crust," in *The Earth Beneath the Continents*, eds. J. S. Steinhart and T. J. Smith. Geophysics Monograph 10, American Geophysical Union, Washington, D.C.

Ringwood, A. E., and A. Major, 1970, The system Mg_2SiO_4-Fe_2SiO_4 at high pressures and temperatures. *Phys. Earth Planet. Interiors*, **3**, 89–108.

Ringwood, A. E., and M. Seabrook, 1962, Olivine-spinel equilibria at high pressure in the system Ni_2GeO_4-Mg_2SiO_4. *J. Geophys. Res.*, **67**, 1975–1985.

Rittmann, A., 1962, *Volcanoes and Their Activity*. Wiley, New York.

Roberts, R. J., 1969, The cordilleran continental margin—continental collisions vs. geotectonic cycles. Abstracts with programs for 1969, Part 7, 286–288. The Geological Society of America.

Robertson, E. C., F. Birch, and G. J. F. MacDonald, 1957, Experimental determination of jadeite stability relations to 25,000 bars. *Am. J. Sci.*, **255**, 115–137.

Robertson, J. K., and P. J. Wyllie, 1971, Rock-water systems, with special reference to the water-deficient region. *Am. J. Sci.*, in press.

Ronov, A. B., and A. A. Yaroshevsky, 1969, "Chemical composition of the Earth's crust," in *The Earth's Crust and Upper Mantle*, ed. P. J. Hart, Geophysical Monograph 13, American Geophysical Union, Washington, D.C.

Rouse, G. E., and R. E. Bisque, 1968, Global tectonics and the Earth's core. *Mines Mag.*, 58, 2–8.

Roy, D. M. and R. Roy, 1954, An experimental study of the formation and properties of synthetic serpentines and related layer silicate minerals. *Am. Miner.*, **39**, 957–975.

Roy, R. F., D. D. Blackwell, and F. Birch, 1968, Heat generation of plutonic rocks and continental heat flow provinces. *Earth Planet. Sci. Letters*, **5**, 1–12.

Runcorn, S. K., 1962, editor, *Continental Drift*. Academic Press, New York.

Sacks, I. S., 1969, Distribution of absorption of shear waves in South America and its tectonic significance. *Carnegie Inst. Wash. Yearbook*, **67**, 339–344.

Scarfe, C. M., and P. J. Wyllie, 1967, Serpentine dehydration curves and their bearing on serpentinite deformation in orogenesis. *Nature*, **215**, 945–946.

Scheidegger, A. E., 1963, *Principles of Geodynamics*. Academic Press, New York.

Schmucker, V., 1969, "Geophysical aspects of structure and composition of the earth," in *Handbook of Geochemistry*, Vol. 1, ed. K. H. Wedepohl. Springer-Verlag, Heidelberg.

Scholl, D. W., M. N. Christensen, R. von Huene, and M. S. Marlow, 1970, Peru-Chile and trench sediments and sea-floor spreading. *Geol. Soc. Am. Bull.*, **81**, 1339–1360.

Schopf, J. M., 1970, Gondwana paleobotany. *Antarctic J. U.S.*, **5**, 62–66.

Seyfert, C. K., 1969, Undeformed sediments in oceanic trenches with sea-floor spreading. *Nature*, **222**, 70.

Shaw, H. R., 1970, Earth tides, global heat flow, and tectonics. *Science*, **168**, 1084–1087.

Sheridan, R. E., R. E. Houtz, C. L. Drake, and M. Ewing, 1969, Structure of continental margin off Sierra Leone, West Africa. *J. Geophys. Res.*, **74**, 2512–2530.

Sleep, N. H., 1969, Sensitivity of heat flow and gravity to the mechanism of sea-floor spreading. *J. Geophys. Res.*, **74**, 542–549.

Smith, A. G., and A. Hallam, 1970, The fit of the southern continents. *Nature*, **223**, 139–144.

Sollogub, V. B., 1969, "Seismic crustal studies in southeastern Europe," in *The Earth's Crust and Upper Mantle*, ed. P. J. Hart. Geophysical Monograph 13, American Geophysical Union, Washington D.C.

Sproll, W. P., and R. S. Dietz, 1969, Morphological continental drift fit of Australia and Antarctica. *Nature*, **222**, 345–348.

Stewart, A. D., 1968, Geology in British universities. *Nature*, **217**, 987–988.

Stewart, D. B., 1967, Four-phase curve in the system $CaAl_2Si_2O_8$-SiO_2-H_2O between 1 and 10 kilobars. *Schweiz. Miner. Petrogr. Mitt.*, **47**, 35–59.

Suess, H. E., and H. C. Urey, 1956, Abundances of the elements. *Rev. Mod. Phys.*, **28**, 53–74.

Sykes, L. R., 1966, The seismicity and deep structure of island arcs. *J. Geophys. Res.*, **71**, 2981–3006.

Takeuchi, H., S. Uyeda, and H. Kanamori, 1967, *Debate About the Earth*. Freeman, Cooper and Co. San Francisco.

Talwani, M., X. Le Pichon, and M. Ewing, 1965, Crustal structure of the midocean ridges, 2. *J. Geophys. Res.*, **70**, 341–352.

Tarakanov, R. Z., and N. V. Leviy, 1968, "A model for the upper mantle with several channels of low velocity and strength," in *The Crust and Upper Mantle of the Pacific Area*, eds. L. Knopoff, C. L. Drake, and P. J. Hart. Geophysical Monograph 12, American Geophysical Union, Washington, D.C.

Tatsch, J. H., 1964, Distribution of active volcanoes: summary of preliminary results of three-dimensional least-squares analysis. *Geol. Soc. Am. Bull.*, **75**, 751–752.

Taylor, F. B., 1910, Bearing of the Tertiary mountain belt on the origin of the earth's plan. *Geol. Soc. Amer. Bull.*, **21**, 179–226.

Taylor, H. P., 1968, The oxygen isotope geochemistry of igneous rocks. *Contr. Mineral. and Petrol*, **19**, 1–71.

Taylor, S. R., 1968, "Geochemistry of Andesites," in *Origin and Distribution of the Elements*, ed. L. H. Ahrens. Pergamon, Oxford.

———, 1969, "Trace element chemistry of andesites and associated calc-alkaline rocks," in *Proceedings of the Andesite Conference*, ed. A. R. McBirney. Oregon Dep. Geol. Mineral Ind. Bull., **65**.

Tilley, C. E., 1950, Some aspects of magmatic evolution. *Geol. Soc. London Quart. J.* **106**, 37–61.

———, and H. S. Yoder, 1964, Pyroxene fractionation in mafic magma at high pressures and its bearing on basalt genesis. Carnegie Inst. Wash. Yearbook, **63**, 114–121.

Tilling, R. I., D. Gottfried, and F. C. W. Dodge, 1970, Radiogenic heat production of contrasting magma series: bearing on interpretation of heat flow. *Geol. Soc. Am. Bull.*, **81**, 1447–1462.

Toksöz, M. N., J. Arkani-Hamed, and C. A. Knight, 1969, Geophysical data and long-wave heterogeneities of the earth's mantle. *J. Geophys. Res.*, **74**, 3751–3770.

Torrance, K. E., and D. L. Turcotte, 1971, Structure of convection cells in the mantle. *J. Geophys. Res.*, **76**, 1154–1161.

Tozer, D. C., 1967, "Towards a theory of thermal convection," in *The Earth's Mantle*, ed. T. F. Gaskell. Academic, New York.

Turcotte, D. L., and E. R. Oxburgh, 1969, Convection in a mantle with variable physical properties. *J. Geophys. Res.*, **74**, 1458–1474.

Turner, F. J., 1968, *Metamorphic Petrology*. McGraw-Hill, New York.

Tuthill, R. L., 1969, Effect of varying f_{O_2} on the hydrothermal melting and phase relations of basalt. *Trans. Am. Geophys. Un.*, **50**, 355.

Tuttle, O. F., and N. L. Bowen, 1958, Origin of Granite in the Light of Experimental Studies in the System $NaAlSi_3O_8$-$KAlSi_3O_8$-SiO_2-H_2O. *Geol. Soc. Am. Memoir*, **74**.

Umbgrove, J. H. F., 1947, *The Pulse of the Earth*. Martinus Nijhoff, The Hague.

Vacquier, V., S. Uyeda, M. Yasui, J. Sclater, C. Corry, and T. Watanabe, 1966, Heat flow measurements in the northwestern Pacific. *Bull. Earthquake Res. Inst. Tokyo Univ.*, **44**, 1519–1535.

Van Andel, Tj. H., and C. O. Bowin, 1968, Mid-Atlantic ridge between 22° and 23° North latitude and the tectonics of mid-ocean rises. *J. Geophys. Res.*, **73**, 1279–1298.

Van Bemmelen, R. W., 1968, On the origin and evolution of the earth's crust and magmas. *Sonderdruck Geolog. Rundschau*, **57**, 657–705.

Van de Lindt, W. J., 1967, Movement of the Mohorovicic discontinuity under isostatic conditions. *J. Geophys. Res.*, **72**, 1289–1297.

———, 1968, Reply. *J. Geophys. Res.*, **73**, 6607.

Van Hilten, D., 1968, Global expansion and paleomagnetic data. *Tectonophysics*, **5**, 191–210.

Vestine, E., I. Lange, L. Laporte, and W. Scott, 1947, The geomagnetic field, its description and analysis. *Carnegie Inst. Wash. Publ.*, **580**.

Verhoogen, J., 1965, "Phase changes and convection in the earth's mantle," in *A symposium on continental drift*, eds. P. M. S. Blackett, E. Bullard, and S. K. Runcorn. *Phil. Trans. Roy. Soc.* London, 1088.

Vine, F. J., 1966, Spreading of the ocean floor: new evidence. *Science*, **154**, 1405–1415.

———, 1968, "Magnetic anomalies associated with mid-ocean ridges," in *The History of the Earth's Crust*, ed. R. A. Phinney. Princeton University Press, New Jersey.

———, and H. H. Hess, 1970, "Sea-floor spreading", in *The Sea*, Vol. 4, eds. A. E. Maxwell,

E. C. Bullard, E. Goldberg, and J. L. Worzel. Wiley-Interscience, New York.

Vine, F. J., and D. H. Matthews, 1963, Magnetic anomalies over oceanic ridges. *Nature*, **199**, 947–949.

Vine, F. J., and J. T. Wilson, 1965, Magnetic anomalies over a young oceanic ridge off Vancouver Island. *Science*, **150**, 485–489.

Vogt, P. R., C. N. Anderson, D. R. Bracey, and E. D. Schneider, 1970, North Atlantic magnetic smooth zones. *J. Geophys. Res.*, **75**, 3955–3968.

Vogt, P. R., N. A. Ostenso, and G. L. Johnson, 1970, Magnetic and bathymetric data bearing on sea-floor spreading north of Iceland. *J. Geophys. Res.*, **75**, 903–920.

Von Herzen, R. P., and W. H. K. Lee, 1969, "Heat flow in oceanic regions," in *The Earth's Crust and Upper Mantle*, ed. P. J. Hart. Geophysical Monograph 13, American Geophysical Union, Washington, D.C.

Walker, C. T., and J. G. Dennis, 1966, Explosive phase transitions in the earth's mantle. *Nature*, **209**, 182–183.

Ward, M. A., 1963, On detecting changes in the Earth's radius. *Geophys. J.*, **8**, 217–225.

Washington, H. S., 1925, The chemical composition of the Earth. *Am. J. Sci.*, **9**, 351–378.

Watkins, N. D., 1968, Short period geomagnetic polarity events in deep-sea sedimentary cores. *Earth Planet. Sci. Letters*, **4**, 341–349.

————, 1968, Comments on the interpretation of linear magnetic anomalies. *Pure Appl. Geophys.*, **69**, 179–192.

————, and S. E. Haggerty, 1968, Oxidation and magnetic polarity in single Icelandic lavas and dikes. *Geophys. J. Roy. Astronom. Soc.*, **15**, 305–315.

Watkins, N. D., and A. Richardson, 1968, Comments on the relationship between magnetic anomalies, crustal spreading and continental drift. *Earth Planet. Sci. Letters*, **4**, 257–264.

Wegener, A., 1966, *The Origin of Continents and Oceans*. Translated by John Biram from the Fourth (1929) revised German edition. Dover, New York.

Wegmann, C. E., 1963, Tectonic patterns at different levels. *Geol. Soc. South Africa*, annexure to **66**, 1–78.

Wetherill, G. W., 1961, Steady-state calculations bearing on geological implications of a phase-transition Mohorovicic Discontinuity. *J. Geophys. Res.* **66**, 2983–2993.

White, I. G., 1967, Ultrabasic rocks and the composition of the upper mantle. *Earth Planet. Sci. Letters*, **3**, 11–18.

Wiik, H. B., 1956, The chemical composition of some stony meteorites. *Geochim. Cosmochim. Acta*, **9**, 279–289.

Wilson, J. T., 1959, Geophysics and continental growth. *Am. Sci.*, **47**, 1–24.

————, 1963, Hypothesis of Earth's behavior. *Nature*, **198**, 925–929.

————, 1965, Transform faults, oceanic ridges, and magnetic anomalies southwest of Vancouver Island. *Science*, **150**, 482–485.

————, 1966, Did the Atlantic close and then re-open? *Nature*, **211**, 676–681.

————, 1968, A revolution in earth science. *Geotimes*, **13**(10), 10–16.

————, 1968, A reply to V. V. Beloussov. *Geotimes*, **13**(10), 20–22.

Winkler, H. G. F., 1967, *Petrogenesis of Metamorphic Rocks*. 2nd ed. Springer-Verlag, New York.

Woollard, G. P., 1968, "The interrelationship of the crust, the upper mantle, and isostatic gravity anomalies in the United States," in *The Crust and Upper Mantle of the Pacific Area*, eds. L. Knopoff, C. L. Drake, and P. J. Hart. Geophysical Monograph 12, American Geophysical Union, Washington, D.C.

Wyllie, P. J., 1967, ed., *Ultramafic and Related Rocks*. Wiley, New York.

————, 1969, The origin of ultramafic and ultrabasic rocks. *Tectonophysics*, **7**, 437–455.

————, 1970, "Ultramafic rocks and the upper mantle," in *Fiftieth Anniversary Symposia*, ed. B. A. Morgan. Mineralogical Society of America Special Paper No. 3.

————, 1971, The role of water in magma generation and initiation of diapiric uprise in the mantle. *J. Geophys. Res.*, **76**, 1328–1338.

Wynne-Edwards, 1969, "Tectonic overprinting in the Grenville Province, southwestern Quebec." Geology Association of Canada Special Paper No. 5.

Yoder, H. S., 1955, "Role of water in metamorphism," in *The Crust of the Earth*, ed. A. Poldervaart, Geological Society of America Special Paper 62.

————, and G. A. Chinner, 1960, Almandite-pyrope-water system at 10,000 bars. *Carnegie Institut. Wash. Yearbook*, **59**, 81–84.

Yoder, H. S., and I. Kushiro, 1969, Melting of a hydrous phase: phlogopite. *Am. J. Sci.*, **267-A**, 558–582.

Yoder, H. S., and C. E. Tilley, 1962, Origin of basalt magmas: an experimental study of natural and synthetic rock systems. *J. Petrol.*, **3**, 342–532.

Ziegler, A. M., 1970, Geosynclinal development of the British Isles during the Silurian Period. *J. Geol.*, **78**, 445–479.

Author Index

Subject Index